Handbook of AI in Engineering Applications

There is a need to categorize artificial intelligence (AI) applications, tools, techniques, and algorithms based on their intended use in various design stages. Specifically, there is a need to explore AI techniques that are utilized for tasks such as designing, including, but not limited to, inspiration, idea and concept generation, concept evaluation, optimization, decision-making, and modeling. Furthermore, the tasks include generating ideas and concepts, evaluating those ideas, optimizing designs, making decisions, and creating models. This handbook brings all of these categories with compatible AI techniques, tools, and algorithms together in one place.

The *Handbook of AI in Engineering Applications: Tools, Techniques, and Algorithms* covers applications of AI in engineering and highlights areas such as future cities, mechanical system analysis, and robotic process automation and presents the application of AI and the use of computerized systems that aim to simplify and automate the processes of design and construction of civil works. The handbook discusses the design and optimization of mechanical systems and parts, such as engines, gears, and bearings, which can be automated using AI. It also explores the performance of robotics and automation systems that can be simulated and analyzed using AI to forecast behavior, spot future issues, and suggest changes. Rounding out this handbook is AI technology automation and how analyzing relevant data can provide a reliable basis for the relevant personnel to carry out their work.

This handbook fills the gap between R&D in AI and will benefit all stakeholders, including industries, professionals, technologists, academics, research scholars, senior graduate students, government, and public healthcare professionals.

Advancements in Intelligent and Sustainable Technologies and Systems
Series Editor: Ajay Kumar

This book series aims to provide a platform for academicians, researchers, professionals, and individuals to participate and to provide novel systematic theoretical, experimental, computational work in the form of edited text or reference books, monographs in the area of intelligent and sustainable technologies and systems from engineering, management, applied science, healthcare, etc. domains. This book series will educate and inform the readers with a comprehensive overview of advancements in intelligent and sustainable techniques and systems with novel intelligent tools and algorithms to move industries from different domains from a data-centric community to a sustainable world. The book series covers ideas and innovations to help the research community and professionals to understand fundamentals, opportunities, challenges, future outlook, layout, lifecycle, and framework of intelligent and sustainable technologies and systems for different sectors. It serves as a guide for computer science, mechanical, manufacturing, electrical, electronics, civil, automobile, industrial engineering, biomedical, healthcare, and management professionals.

If you are interested in writing or editing a book for the series or would like more information, please contact Cindy Carelli, cindy.carelli@taylorandfrancis.com

Handbook of Intelligent and Sustainable Manufacturing
Tools, Principles, and Strategies
Edited by Ajay Kumar, Parveen Kumar, Yang Liu, and Rakesh Kumar

Advances in Sustainable Biomaterials
Bioprocessing 4.0, Characterizations, and Applications
Edited by Ajay Kumar, D.K. Rajak, Parveen Kumar, and Ashwini Kumar

Manufacturing Strategies and Systems
Technologies, Processes, and Machine Tools
Edited by Ajay Kumar, Parveen Kumar, Ashish Kumar Srivastava, and Lokesh Saharan

Handbook of AI in Engineering Applications
Tools, Techniques, and Algorithms
Edited by Ajay Kumar, Sangeeta Rani, Krishna Dev Kumar, and Manish Jain

Handbook of Deep Learning Models for Healthcare Data Processing
Disease Prediction, Analysis, and Applications
Edited by Ajay Kumar, Deepak Dembla, Seema Tinker, and Surbhi Bhatia Khan

Handbook of Disruptive Technologies
Operations, Business, Management, and Healthcare
Edited by Chand Saini, Neha Gupta, and Ajay Kumar

Handbook of AI in Engineering Applications

Tools, Techniques, and Algorithms

Edited by
Ajay Kumar, Sangeeta Rani,
Krishna Dev Kumar and Manish Jain

CRC Press
Taylor & Francis Group
Boca Raton London New York

CRC Press is an imprint of the
Taylor & Francis Group, an **informa** business

Designed cover image: Shutterstock - Tapati Rinchumrus

First edition published 2026
by CRC Press
2385 NW Executive Center Drive, Suite 320, Boca Raton FL 33431

and by CRC Press
4 Park Square, Milton Park, Abingdon, Oxon, OX14 4RN

CRC Press is an imprint of Taylor & Francis Group, LLC

ISBN: 978-1-032-72315-0 (hbk)
ISBN: 978-1-032-72444-7 (pbk)
ISBN: 978-1-032-72450-8 (ebk)

DOI: 10.1201/9781032724508

Typeset in Times New Roman
by Deanta Global Publishing Services, Chennai, India

Contents

Preface

Currently, artificial intelligence (AI) is one of the top trends in technology. It enhances resource provisioning capabilities and promotes resource sharing. Nonetheless, it is associated with several engineering technologies that require assessment prior to adoption. Fortunately, there are promising indicators for engineering professionals looking to applications, tools, techniques, and algorithms. Tools that use AI and machine learning have begun to replace the older rules-and-signature-based tools that can no longer defend against today's sophisticated environment.

This book explores engineering tools, techniques, and algorithms, as well as their potential use to philosophical education, future cities, mechanical system analysis, and robotic process automation. Special focus is given to applications, techniques, tools, use cases, and challenges through the integration of integrated technologies.

The book mainly deals with AI techniques for modeling, analysis, and performance prediction. It will furnish both conceptual and practical knowledge on AI techniques employed in engineering for students, researchers, and academicians.

The book consists of 18 chapters that describe perspectives of AI in engineering applications.

1. Chapter 1, "A Comparative Analysis of Symbolic and Subsymbolic Approaches in Artificial Intelligence," emphasizes the growing relevance of hybrid systems, which integrate symbolic reasoning with sub-symbolic learning to address complex, real-world challenges.
2. Chapter 2, "Revisiting the Impact of AI-Driven Feature Selection Approaches in Software Fault Prediction," aims to compare the existing feature selection techniques to have a clear vision regarding identification of the optimal feature selection technique for software fault prediction.
3. Chapter 3, "Advancing Depression Detection: Insights from Standard Datasets and Multimodal Approaches," explores innovative methods in depression detection, utilizing multimodal data analysis and including a thorough review of advanced machine learning techniques and emotion recognition frameworks used to enhance early diagnosis and intervention strategies.
4. Chapter 4, "Machine Learning and Natural Language Processing Strategies for Fake News Detection: An Empirical Study," leverages advanced natural language processing techniques to evaluate the performance of several algorithms, including Naive Bayes, Logistic Regression, Random Forest, and Support Vector Machines.
5. Chapter 5, "Exploring the Nexus of AI, the Metaverse, and Quality of Life," delves into the critical question of the metaverse and its application and implications on quality of life in general, as well as the challenges and negative ramifications of this technology.

6. Chapter 6, "A Literature Survey on AI-Driven Code-Mixed Text Analysis and Normalization," provides an extensive review of code-mixing text analysis.
7. Chapter 7, "Reviewing Different Artificial Intelligence Algorithms for Sarcasm Detection," constitutes recent studies of sarcasm detection and how sarcasm detection has evolved over time and provides information regarding which system of programs best suits different kinds of data.
8. Chapter 8, "Advances in Automatic Question Generation by AI-Based Algorithms: A Review," explores the domain of Automatic Question Generation (AQG), examining diverse methodologies, assessing the impact of these approaches on question quality, and addressing challenges such as handling complex texts and ensuring question relevance, aiming to provide valuable insights for the development and utilization of AQG systems in educational and research contexts.
9. Chapter 9, "Hands-on Practices, Reflection on Data Wisdom with AI Principles: A Review," elucidates this paradigm shift by aligning AI ethics principles with the operational framework of AI-based digital or e-products/services, integrating a transparent governance model to delineate operational responsibilities.
10. Chapter 10, "FPGA Accelerated Deep Learning Network For Liver Tumor Segmentation in 3D CT Images," presents the development and implementation of an automated liver tumor segmentation system using the Residual U-Net model, optimized for medical image analysis and ensuring high accuracy, real-time inference, and enhanced accessibility in diverse healthcare settings.
11. Chapter 11, "Deep Learning Enhanced Restaurant Recommendations: Leveraging Artificial Neural Networks," investigates the use of deep learning approaches to improve restaurant recommendation systems, utilizing artificial neural networks for more customized and effective suggestions.
12. Chapter 12, "AI-Driven Solutions for Career Counseling Based on the Personality of an Individual," discovers that the ANN algorithm consistently yields more accurate personality-type predictions than SVM.
13. Chapter 13, "Exploration of Machine Learning Enabled Automated Crop-Disease Detection, Segmentation and Classification Techniques," offers an extensive examination of the latest advancements in crop disease detection, segmentation, and classification. It emphasizes the integration of image processing, artificial intelligence, machine learning, and deep learning methodologies.
14. Chapter 14, "Digital Camouflage Generation by AI-Based Methods: A Survey of Recent Techniques," presents a summary of these technologies in the field of generation and highlights the challenges to elicit the scope for further research.
15. Chapter 15, "Plant Disease Detection Using Deep Learning Techniques: A Comprehensive Review and Comparative Analysis," suggests several

avenues for research in the future to address current issues and increase the efficiency of plant disease diagnosis systems based on deep learning.

16. Chapter 16, "AI for Disaster Prediction and Management," discusses distinct AI models used for disaster prediction and management.
17. Chapter 17, "Investigation of Stress among University Students using PSS and AI-Based Analysis," aims to explore the impact of the economic slowdown on the psychological well-being of undergraduate students in Indian universities.
18. Chapter 18, "An Experimental Overview of Assessment of Authenticity of Face Recognition by AI Techniques in Smart Phones," aims to evaluate the authenticity and limitations of FRT for mobile phones by using different sets of technology, including Android and iPhone.

This book is intended for both the academia and the industry. The postgraduate students, Ph.D. students, and researchers in universities and institutions, who are involved in the areas of AI in *Engineering Applications: Tools, Techniques, and Algorithms*, will find this compilation useful.

The editors acknowledge the professional support received from CRC Press and express their gratitude for this opportunity.

Reader's observations, suggestions, and queries are welcome,

Editors

Ajay Kumar
Sangeeta Rani
Krishna Dev Kumar
Manish Jain

About the Editors

Ajay Kumar is currently serving as a Professor (Research Track), Department of Mechanical Engineering, School of Core Engineering, Faculty of Science, Technology & Architecture, Manipal University, Jaipur, Rajasthan, India. He received his Ph.D. in the field of advanced manufacturing and automation from Guru Jambheshwar University of Science & Technology, Hisar, India, after a B.Tech. (Hons.) in mechanical engineering and an M.Tech. (Distinction) in manufacturing and automation. His areas of research include biomedical engineering, incremental sheet forming, AI, sustainable materials, robotics and automation, additive manufacturing, mechatronics, smart manufacturing, Industry 4.0, industrial engineering systems, waste management, and optimization techniques. He has to his credit over 100 publications in international journals of repute, including *SCOPUS*, *Web of Science*, and SCI indexed database and refereed international conferences. He has organized various national and international events, including an international conference on Mechatronics and Artificial Intelligence (ICMAI-2021) as conference chair. He is currently organizing an international conference on Artificial Intelligence, Advanced Materials, and Mechatronics Systems (AIAMMS-2023) as conference chair. He has more than 20 national and international patents to his credit. He has supervised more than 8 M.Tech and Ph.D. scholars and numerous undergraduate projects/thesis. He has a total of 15 years of experience in teaching and research. He is the guest editor of many reputed journals. He has contributed to many international conferences/ symposiums as a session chair, expert speaker, and member of the editorial board. He has won several proficiency awards during the course of his career, including merit awards, best teacher awards, and so on. He has also coauthored and coedited more than 15 books and proceedings. He has also authored many in-house course notes, lab manuals, monographs, and invited chapters in books. He has organized a series of faculty development programs, international conferences, workshops, and seminars for researchers, as well as Ph.D. and UG-, and PG-level students. He is associated with many research, academic, and professional societies in various capacities.

Dr. Sangeeta Rani is currently serving as Assistant Professor in department of Computer Science and Engineering at the World College of Technology and Management, Gurgaon, India, where she has been instrumental in establishing multiple innovation hubs including the NVIDIA-Dell-powered AI & GPU Computing Lab, Intel Unnati Lab, and ABB Robotics Centre.

She holds a Ph.D. from Amity University, Manesar, where her research focused on Attack Detection and Mitigation in cloud computing using Machine Learning Approach. She completed her M.Tech. in CSE from MDU University and her B.Tech. from Kurukshetra University. Throughout her academic journey, she has guided numerous B.Tech. and M.Tech. projects in domains such as IoT, smart systems, and cloud automation.

Her contributions include publishing multiple Scopus and SCI-indexed research papers, securing national and international patents, and editing academic books with CRC Press and Taylor & Francis on AI in healthcare and intelligent cloud systems. She has also served as a reviewer for leading journals and IEEE conferences and has been recognized with a Best Paper Award at ICMAI-2021.

An active academic leader, she has organized several national FDPs and trainings in collaboration with Apple, Intel, and other tech partners.

She actively contributes to the academic community through her involvement in curriculum development, research mentoring, and establishment of Centres of Excellence focused on AI, cybersecurity, and robotics.

Krishna D. Kumar is a professor of aerospace engineering and the director of the Artificial Intelligence for Aerospace Systems (AIAS) Laboratory at Toronto Metropolitan University, Canada. He is also the founder and president of iSAC Systems Inc., a leader in AI, the Internet of Things, and automation since 2010. Prof. Krishna D. Kumar received his Ph.D. degree in aerospace engineering from the Indian Institute of Technology, Kanpur, India, in 1998. With over two and a half decades of experience in academia, he has held positions at various premier institutions in Canada, Japan, South Korea, and India. These institutions include the Toronto Metropolitan University (formerly Ryerson University) in Canada, the National Aerospace Laboratory and Kyushu University in Japan, the Korea Advanced Institute of Science and Technology in South Korea, the Defence Research and Development Organization in India, and the Birla Institute of Science and Technology in Pilani, India.

Prof. Kumar is one of the leading researchers in the areas of AI-based predictive analytics, aerospace systems and control, autonomous systems, AI-based embedded systems, miniaturization, and hardware innovation. He has made outstanding contributions, with major impact in these areas leading to novel technologies for predictive maintenance of aerospace systems, multi-agent systems (formation flying), and fault tolerant control. He has authored over 220 publications, including 100 papers in high-ranking journals and 125 in conference proceedings. Additionally, he has authored 5 books, holds 14 intellectual properties, and has four patents. His internet-of-things devices and AI software have been deployed in several industries, including aerospace, manufacturing, transportation, and waste management.

Throughout his career, Prof. Kumar has received numerous awards and recognitions. These include being a Member of the International Academy of Astronautics, France (2019), receiving the Eminent Alumnus Award from Veer Surendra Sai University of Technology, Sambalpur (2017), the Sarwan Sahota Ryerson Distinguished Scholar Award (2015), and the Canada Research Chair (2005–2015). He is also an associate fellow and Life Member of the American Institute of Aeronautics and Astronautics, a recipient of the Ontario Early Researcher Award (2006–2011), a Japan Society for the Promotion of Science (JSPS) Fellow (2001–2003), and a Science and Technology Agency (STA) Fellow (1998–2000). In October 2020, Prof. Kumar was ranked among the top 100 scientists worldwide and 1st in Canada in the field of Aerospace and Aeronautics by Stanford University. Furthermore, Prof. Kumar has been serving as the co-editor of *Acta Astronautica* (a prestigious journal in the areas of spacecraft design, dynamics, and control) since 2014 and as the co-editor of *Transactions of the Japan Society for Aeronautical and Space Sciences* (JSASS) since 2018.

Manish Jain is currently serving as a professor in the Computer Science and Engineering Department, School of Engineering and Technology, JECRC University, Jaipur, Rajasthan, India. He received his Ph.D. from Malaviya National Institute of Technology, Jaipur, India, after a B.Tech. (Hons.) in computer science and engineering from the Govt. Engineering College, Bikaner, and an M.Tech. in computer science and engineering from Malaviya National Institute of Technology, Jaipur. His areas of research include AI, machine learning, natural language processing, software testing, aspect-oriented programming, and generative adversarial network. He has over 15 publications in international journals of repute, including *SCOPUS*, *Web of Science*, and SCIE indexed database and refereed international conferences. He has a unique patent on "Smart Library Management System" to his credit, which proposes the use of machine learning for increasing the efficiency of an institution's library. He is currently supervising four Ph.D. scholars and has successfully guided numerous UG and PG projects/thesis. He has a total of 18 years of experience in teaching and research. He is Associate Editor for the Scopus-Indexed *Journal of Engineering Education Transformations*. He has made significant contributions to international conferences/symposiums as an expert speaker. He also spearheads the Industry Interface cell at JECRC University. He has also authored many in-house course notes, lab manuals, and monographs. He has organized a series of faculty development programs and seminars for faculty members and researchers through industries such as Google Cloud, Microsoft, IBM, Salesforce, and AWS, as well as speakers and repute.

Contributors

Harsh Agarwal
Department of Computer Science and Engineering
Vellore Institute of Technology (VIT)
Bhopal University

Nidhi Agarwal
Department of Faculty of Social Science and Humanities
Lincoln University College
Malaysia

Kotri Kalan, Astha
M.P., India

Praveen Kumar Bhanodia
Computer Science Engineering Department (CSE)
Acropolis Institute of Technology and Research
Indore, Madhya Pradesh, India

Akashvi Bhardwaj
Department of Artificial Intelligence and Data Science
Vivekananda Institute of Professional Studies—Technical Campus
New Delhi, India

Dhruva Bisht
Department of Computer Science and Engineering
Graphic Era (Deemed to be) University
Dehradun, UK, India

Venkamaraju Chakravaram
Jindal Global Business School
OP Jindal Global University
Sonipat, India

Roshan Chandel
Department of CSE
JECRC University
Jaipur, Rajasthan, India

Ria Choudhari
Department of CSE
SRM University
Delhi-NCR, Sonepat, Haryana

Zhang Chuang
Jilin International Studies University
Jilin, China

Navneet Das
Department of Computer Engineering
Army institute of Technology
Pune, India

Deepti Dave
Department of Computer Science and Engineering
Vellore Institute of Technology (VIT)
and
Bhopal University
Kotri Kalan, Astha, M.P., India

Amandeep Dhaliwal
Department of Post Graduate Management Studies
School of Leadership and Management
Manav Rachna International Institute of Research & Studies
Faridabad, Haryana, India

Parneeta Dhaliwal
Department of Computer Science and Technology
Manav Rachna University
Faridabad, Haryana, India

Deepika Dhillon
Faculty of Science
SGT University
Gurugram, Haryana, India

Ashutosh Farswan
Department of Computer Science and Engineering
Graphic Era (Deemed to be) University
Dehradun, UK, India

Vaishali Ganganwar
Department of Computer Engineering
Army Institute of Technology
Pune, India

Garima
Department of Computer Science
K.R. Mangalam University
Gurugram, Haryana, India

Shalini Gautam
DME, GGSIP
Delhi, India

Gitanjali
Department of CSE
Chandigarh University
Punjab, India

Ayush Gour
Computer Science Engineering Department (CSE)
Acropolis Institute of Technology and Research
Indore, Madhya Pradesh, India

Ananya Gupta
Department of Artificial Intelligence and Data Science
Vivekananda Institute of Professional Studies-Technical Campus
New Delhi, India

Astha Gupta
Amity International Business School
Amity University
Uttar Pradesh, India

C.L.P. Gupta
Bansal Institute of Engineering & Technology
Lucknow, INDIA

Poonam Gupta
Department of Computer Science & Engineering DCRUST
Murthal, Haryana, India

Sandhya Gupta
Department of Psychology
Banasthali Vidyapith
Rajasthan, India

Swati Gupta
Faculty of Department of Computer Science
K.R. Mangalam University
Gurugram, Haryana, India

Inzimam Ul Hassan
Dept. of CSE Chandigarh University
Mohali, India

Naveen Hemrajani
Department of Computer Science and Engineering
JECRC University
Jaipur, Rajasthan, India

Krishna Hirve
Department of Computer Engineering
Army institute of Technology
Pune, India

Pankaj Jain
Department of Computer Science and Engineering
JECRC University
Jaipur, Rajasthan, India

Sugandha Joshi
Department of Management
Graphic Era Hill University
Dehradun, India

Shilpa Juneja
Department of Computer Science and Technology
Manav Rachna University
Faridabad, Haryana, India

Megha Kamble
School of Computer Science & Technology
LNCT University
Bhopal, Madhya Pradesh, India

Amritpal Kaur
Department of Computer Science and Engineering
Amity School of Engineering and Technology
Gurgaon, Haryana, India

Inderdeep Kaur
Dept. of CSE Chandigarh University
Mohali, India

Seema Kharb
Department of CSE
SRM University
Delhi-NCR, Sonepat, Haryana

Yogita Khatri
Department of Computer Science and Engineering
Manav Rachna International Institute of Research and Studies
Faridabad, Haryana, India

Smitha Kothari
School of Management
Graphic Era Hill University
Uttarakhand, Dehradun, India

Ajay Kumar
Department of CSE
JECRC University
Jaipur, Rajasthan India

Ashish Kumar
Department of Artificial Intelligence and Data Science
Vivekananda Institute of Professional Studies- Technical Campus
New Delhi, India

Parveen Kumar
Skill Department of Automotive Studies
Shri Vishwakarma Skill University
Palwal, Haryana, India

Puneet Kumar
Department of Mathematics
Kalinga University
Naya Raipur, Chhattisgarh, India

Rishu Kumar
Bansal Institute of Engineering & Technology
Lucknow, INDIA

Kamlesh Lakhwani
Department Of Computer Science & Engineering
JECRC University
Jaipur, Rajasthan, India

Madhukar
Department of Computer Engineering
Army institute of Technology
Pune, India

Priyanka Malik
Galgotias University
Greater Noida, UP, India

Heartlin Maria H.
Department of Electronics and Communication Engineering
SRM Institute of Science and Technology
Kattankulathur, Chennai, India

Maryanka
Symbiosis International (Deemed University)
Pune, India

Bhavika Mathur
Department of CSE
JECRC University
Jaipur, Rajasthan, India

Ankur Mittal
Department of Electronics and Communication Engineering
National Institute of Technology
Delhi, India

Prithviraj Mutalik
Department of Computer Science and Engineering
Vellore Institute of Technology (VIT)
Bhopal University, Kotri Kalan, Astha, M.P., India

Bhavana Nimmagadda
Department of Computer Engineering
Army Institute of Technology
Pune, India

Shivangi Pachauri
Computer Science Engineering
Banasthali University
Rajasthan, India

Ayush Patel
Department of CSE
SRM University
Delhi-NCR, Sonepat, Haryana

Subhash C. Patel
Department of Computer Science and Engineering
Vellore Institute of Technology (VIT), Bhopal University
Kotri Kalan, Astha, M.P., India

Akshay Patwal
Department of Computer Science and Engineering
Graphic Era (Deemed to be) University
Dehradun, UK, India

Glory Precious J
Department of Biomedical Engineering
Rajalakshmi Engineering College
Thandalam, Chennai, India

Lal Babu Purbey
Department Of Computer Science & Engineering
JECRC University
Jaipur, Rajasthan, India

Kayalvizhi R
Department of Electronics and Communication Engineering
SRM Institute of Science and Technology
Kattankulathur, Chennai, India

Parshant Rajput
Department of Computer Engineering
Army Institute of Technology
Pune, India

Pawan Singh Rajput
School of Computer Science & Technology
LNCT University
Bhopal, Madhya Pradesh, India

Shivshankar Rajput
Computer Science Engineering Department (CSE)
Acropolis Institute of Technology and Research
Indore, Madhya Pradesh, India

Sunitha Ratnakaram
Jindal Global Business School
OP Jindal Global University
Sonipat, India

Pranjal Rawat
School of Management
Graphic Era Hill University
Uttarakhand, Dehradun, India

Deepali Rani Sahoo
Symbiosis Law School
Symbiosis International (Deemed University)
Noida, Pune, India

Deepak Kumar Saini
Faculty of Science
SGT University
Gurugram, Haryana, India

Urvashi Rahul Saxena
Department of Computer Science and Engineering
Manav Rachna International Institute of Research and Studies
Faridabad, Haryana, India

Swarnim Seth
Computer Science Engineering
Banasthali University
Rajasthan, India

Amanjeet Singh Sethi
Faculty of Commerce & Business Management
Amrapali University
Haldwani, India

Kamal K. Sethi
Computer Science Engineering Department (CSE)
Acropolis Institute of Technology and Research
Indore, Madhya Pradesh, India

Bhoopesh Kumar Sharma
Faculty of Science
SGT University
Gurugram, Haryana, India

Gajanand Sharma
Department of CSE
JECRC University
Jaipur, Rajasthan India

Krishna Sharma
Maharaja Agrasen Himalayan Garhwal University
Pauri Garhwal, Uttarakhand, India

Neelam Sharma
Department of Computer Science
Banasthali Vidyapith
Rajasthan, India

Suruchi Sharma
School of Management
Graphic Era Hill University
Uttarakhand, Dehradun, India

Tanya Shruti
Department Of Computer Science & Engineering
JECRC University
Jaipur, Rajasthan, India

Anuj Singh
Bansal Institute of Engineering & Technology
Lucknow, India

Jaswinder Singh
Department of Computer Science and Engineering
Guru Jambheshwar University of Science and Technology
Hisar, Haryana, India

Mahesh K. Singh
Department of Electronics and Communication Engineering
National Institute of Technology
Delhi, India

Parvinder Singh
Department of Computer Science & Engineering DCRUST
Murthal, Haryana, India

Nikita Singhal
Department of Computer Engineering
Army institute of Technology
Pune, India

Shweta Sinha
Department of Computer Science and Engineering
Amity School of Engineering and Technology
Gurgaon, Haryana, India

Swapnil Sinha
Department of Computer Engineering
Army Institute of Technology
Pune, India

Atul Srivastava
Amity School of Engineering and Technology
Amity University
Uttar Pradesh, Lucknow, India

Shekhar Kumar Swarnkar
Department of CSE
JECRC University
Jaipur, Rajasthan India

Ritika Tehlan
Faculty of Science
SGT University
Gurugram, Haryana, India

Rajesh Tiwari
School of Management
Graphic Era Hill University
Uttarakhand, Dehradun, India

Vivek Tomar
Department of Computer Science and Engineering
Graphic Era (Deemed to be) University
Dehradun, UK, India

Kushagra Trivedi
Department of Computer Science and Engineering
Graphic Era (Deemed to be) University
Dehradun, UK, India

Rajesh Kumar Tyagi
Department of Computer Science and Engineering
Amity School of Engineering and Technology
Gurgaon, Haryana, India

Aashna Vaid
Department of Computer Engineering
Army Institute of Technology
Pune, India

Rameshwari Vadhvani
School of Computer Science Engineering
Bennett University
Greater Noida, UP, India

Prashant Vats
Department of Computer Science and Engineering
Manipal University Jaipur
Jaipur, Rajasthan, India

Madhushi Verma
School of Computer Science Engineering
Bennett University
Greater Noida, UP, India

Shilpa Wadhwa
Department of Management
Graphic Era Hill University
Dehradun, India

Megha Walia
Faculty of Science
SGT University
Gurugram, Haryana, India

Nikhil Yadav
Department of Computer Engineering
Army institute of Technology
Pune, India

Rahul Yadav
Department of Computer Science and Engineering
Guru Jambheshwar University of Science and Technology
Hisar, Haryana, India

Acknowledgments

The editors are grateful to the CRC Press for showing their interest to publish this book in the buzz area of Artificial Intelligence in Engineering Applications. The editors express their personal adulation and gratitude to Ms. Cindy Renee Carelli (Executive Editor, CRC Press) for giving consent to publish our work. She undoubtedly imparted the great and adept experience in terms of systematic and methodical staff who have helped the editors to compile and finalize the manuscript. The editors also extend their gratitude to Ms. Kaitlyn Fisher, CRC Press, for supporting during her tenure.

The editors wish to thank all the chapter authors for contributing their valuable research and experience to compile this volume. The chapter authors, corresponding author in particular, deserve special acknowledgments for bearing with the editors, who persistently kept bothering them for deadlines, and with their remarks.

Dr. Ajay also wishes to express his gratitude to his parents, Sh. Jagdish and Smt. Kamla, and his loving brother Sh. Parveen for their true and endless support. They have made him able to walk tall before the world regardless of sacrificing their happiness and living in a small village. He cannot close these prefatory remarks without expressing his deep sense of gratitude and reverence to his life partner Mrs. Sarita Rathee for her understanding, care, support and encouragement to keep his morale high all the time. No magnitude of words can ever quantify the love and gratitude he feels in thanking his daughters, Sejal Rathee and Mahi Rathee, and son Kushal Rathee who are the world's best children.

Finally, the editors obligate this work to the divine creator and express their indebtedness to the "almighty" for gifting them power to yield their ideas and concepts into substantial manifestation. The editors believe that this book would enlighten the readers about each feature of AI in Engineering Applications.

Ajay Kumar
Sangeeta Rani
Krishna Dev Kumar
Manish Jain

Section I

Integrating AI Tools, Techniques, and Algorithms

1 A Comparative Analysis of Symbolic and Subsymbolic Approaches in Artificial Intelligence

Nidhi Agarwal, Puneet Kumar, Krishna Sharma, and Zhang Chuang

1.1 INTRODUCTION

Artificial intelligence (AI) has been a multidisciplinary field that draws from computer science, mathematics, and cognitive science. The foundational works of luminaries such as Turing (1950), who proposed the concept of machine intelligence, and McCarthy (1958), who formally introduced AI as a scientific discipline, deeply rooted its origins. Advancements in symbolic logic, computational theory, and cognitive modeling complemented these intellectual milestones (Russell & Norvig, 2020). Over the decades, two dominant paradigms—symbolic and subsymbolic approaches—have shaped the evolution of AI, offering distinct methodologies and philosophies for solving complex problems.

Symbolic AI, often referred to as Good Old-Fashioned AI (GOFAI), emerged in the mid-20th century, emphasizing explicit rule-based reasoning and formal knowledge representation (Newell & Simon, 1976; McCarthy, 1980). Inspired by cognitive psychology and linguistics, symbolic AI models human reasoning as a series of logical steps. Examples such as the MYCIN expert system for medical diagnosis and SHRDLU for natural language processing (Winograd, 1972) highlight its utility in structured domains. Symbolic systems excel in tasks that demand transparency and interpretability, allowing for the traceability of decisions to explicit rules (Feigenbaum, 1988). However, their reliance on predefined knowledge structures and rules limits their scalability and adaptability, especially in dynamic or unstructured environments (Davis et al., 1993; Lenat, 1995).

In contrast, subsymbolic AI emerged from a connectionist perspective, focusing on the distributed and parallel nature of human cognition. Neural networks exemplify this paradigm by implicitly encoding knowledge through weights and connections. Subsymbolic methods gained prominence with the rise of machine learning and deep learning, as seen in applications such as Google DeepMind's

DOI: 10.1201/9781032724508-2

AlphaGo (Silver et al., 2016) and OpenAI's GPT models (Brown et al., 2020). These approaches excel in tasks requiring pattern recognition, such as image classification (Krizhevsky et al., 2012) and natural language understanding (Vaswani et al., 2017). However, subsymbolic AI faces significant challenges, including the "black box" nature of its models, which impedes interpretability and accountability (Doshi-Velez & Kim, 2017).

Despite their individual strengths, both symbolic and subsymbolic paradigms face limitations. Symbolic AI struggles with generalization and learning from raw data, while subsymbolic AI demands large datasets and computational resources, raising ethical and accessibility concerns (Dean et al., 2012). These challenges have spurred interest in hybrid systems that aim to combine the strengths of both approaches. Neuro-symbolic AI, for instance, integrates symbolic reasoning with the data-driven capabilities of neural networks to achieve both interpretability and adaptability (Garcez et al., 2019; Marcus, 2020).

The impact of these paradigms extends beyond technical domains to ethical, societal, and philosophical considerations. As AI systems become pervasive in fields such as healthcare, education, and governance, understanding their foundational principles is crucial for addressing issues of bias, transparency, and accountability (Binns, 2018; Floridi et al., 2018). Symbolic AI's explicit rules offer a framework for ethical decision-making, while subsymbolic AI's ability to handle diverse data sources underscores its potential for addressing global challenges such as healthcare inequities and climate change (Esteva et al., 2019; Ang et al., 2023; Sabaulan, 2024).

This chapter provides a comprehensive analysis of symbolic and subsymbolic approaches to AI, exploring their theoretical foundations, methodological frameworks, and practical applications. By examining their complementarities and the potential for hybrid systems, this study aims to shed light on the future of AI research and its implications for real-world applications (Santos et. al., 2024). Multidisciplinary collaboration will be key to unlocking AI's full potential while mitigating its risks (LeCun et al., 2015; Russell & Norvig, 2020).

1.2 SYMBOLIC AI: FOUNDATIONS AND METHODOLOGIES

Explicit rule-based systems and logical reasoning frameworks serve as the foundation for symbolic AI, also known as GOFAI (McCarthy, 1958; Newell & Simon, 1976). This paradigm represents knowledge in a structured and interpretable format, often using ontologies, logic programming, and decision trees to mimic human reasoning processes (Nilsson, 1986). By employing formal symbolic structures, symbolic AI enables systems to perform tasks such as deductive reasoning and problem-solving. Early implementations, such as the MYCIN expert system for medical diagnosis (Shortliffe, 1974), demonstrated the effectiveness of symbolic AI in domains requiring structured knowledge and transparency.

One of the primary strengths of symbolic AI lies in its interpretability. Symbolic systems can trace their decisions back to explicit rules and logical operations, fostering trust and accountability, particularly in critical applications, such as legal reasoning and healthcare (Buchanan & Shortliffe, 1984; Feigenbaum, 1988). Moreover, symbolic AI excels in domain-specific expertise, where predefined knowledge bases and deterministic rules are sufficient to address well-structured problems (McCarthy, 1980). For instance, systems such as SHRDLU showcased the ability to comprehend and manipulate natural language within constrained environments, illustrating the paradigm's reasoning capabilities (Winograd, 1972).

However, symbolic AI also faces significant limitations. Its reliance on manually crafted rules and explicit programming makes it rigid and unable to adapt effectively to new data or dynamic environments (Lenat, 1995). This lack of adaptability limits its scalability, particularly in applications that involve unstructured or large-scale datasets (Davis et al., 1993). Furthermore, symbolic AI struggles with generalization, as it lacks the capacity to learn and infer from raw data, unlike its subsymbolic counterpart (Marcus, 2020).

Despite these challenges, symbolic AI has played a foundational role in shaping the evolution of AI. Its emphasis on transparency, logical reasoning, and structured knowledge representation has influenced various fields, from automated theorem proving to expert systems (McCarthy, 1980). As hybrid approaches are becoming more popular, symbolic AI will continue to offer the interpretability and domain-specific reasoning that make subsymbolic methods more flexible (Garcez et al., 2019; Marcus, 2020). This enduring relevance underscores the importance of symbolic AI in the broader landscape of AI research and applications. Table 1.1 shows various types of symbolic AI systems.

1.2.1 Strengths of Symbolic AI

1. Interpretability: Decisions can be traced back to explicit rules
2. Domain-Specific Expertise: Effective for applications requiring structured knowledge, such as expert systems
3. Reasoning Abilities: Capable of performing deductive reasoning and generating explanations

TABLE 1.1
Types of Symbolic AI Systems

Type of System	Description	Example Applications
Expert Systems	Rule-based systems for decision-making	Medical diagnosis
Knowledge-based Systems	Systems that use structured ontologies	Legal reasoning tools
Logic Programming	Uses formal logic to derive conclusions	Automated theorem proving

1.2.2 Limitations

1. Rigidity: Poor adaptability to new data or contexts
2. Scalability Issues: Challenges in handling large-scale unstructured data
3. Lack of Generalization: Inefficient in learning from raw data

1.3 SUBSYMBOLIC AI: FOUNDATIONS AND METHODOLOGIES

Subsymbolic AI focuses on data-driven methods that learn patterns and relationships directly from input data, relying on statistical representations rather than explicit rules (Rumelhart et al., 1986; LeCun et al., 2015). Neural networks, which are the basis of subsymbolic AI, store knowledge implicitly in layers that are connected to each other. The strength of the connections between inputs and outputs is shown by weights (Goodfellow et al., 2016). This paradigm aligns with the connectionist perspective in cognitive science, emphasizing distributed representation and parallel processing (Bengio et al., 2013). Deep learning, a subset of subsymbolic AI, has become a dominant approach, powering breakthroughs in fields such as image recognition, natural language processing, and speech synthesis (Hinton et al., 2012).

The adaptability of subsymbolic AI is one of its greatest strengths. Unlike symbolic systems, subsymbolic models can handle unstructured and large-scale datasets, making them highly scalable for applications ranging from autonomous vehicles to predictive analytics (Schmidhuber, 2015). For example, convolutional neural networks (CNNs) have revolutionized computer vision tasks, while recurrent neural networks (RNNs) excel in processing sequential data, such as time series and text (Vaswani et al., 2017). Moreover, subsymbolic AI systems improve over time through iterative learning processes, allowing them to adapt to new data and refine their predictions (Dean et al., 2012).

However, subsymbolic AI also faces notable challenges. One critical limitation is its lack of interpretability; decisions made by these models are often opaque, leading to the "black box" problem, which hinders trust and accountability (Doshi-Velez & Kim, 2017). Additionally, subsymbolic AI systems require vast amounts of labeled data for training, which can be resource-intensive and may introduce biases if the data is not representative. The computational demands of training and deploying large neural networks further limit their accessibility, particularly in resource-constrained environments (Bengio et al., 2013). Table 1.2 provides comparison of neural network architectures.

Despite these limitations, subsymbolic AI continues to be a transformative force in AI. Its ability to extract complex patterns and generalize from diverse datasets has enabled significant advancements in areas such as medical imaging, autonomous systems, and conversational AI (Esteva et al., 2019; Brown et al., 2020). Efforts to combine subsymbolic AI with symbolic reasoning are meant to solve its problems with interpretability while taking advantage of its adaptability. This will allow for hybrid approaches that use the best parts of both paradigms (Garcez et al., 2019; Marcus, 2020). Subsymbolic AI's contributions remain central to the evolution of AI, driving innovation across a wide range of applications. Figure 1.1 represents the architecture of neural network.

TABLE 1.2
Comparison of Neural Network Architectures

Neural Network Type	Key Characteristics	Common Applications
Feedforward Neural Network	Simple architecture, no loops	Image classification
Convolutional Neural Network (CNN)	Specialized for spatial data	Object detection, image processing
Recurrent Neural Network (RNN)	Processes sequential data	Time-series prediction, NLP

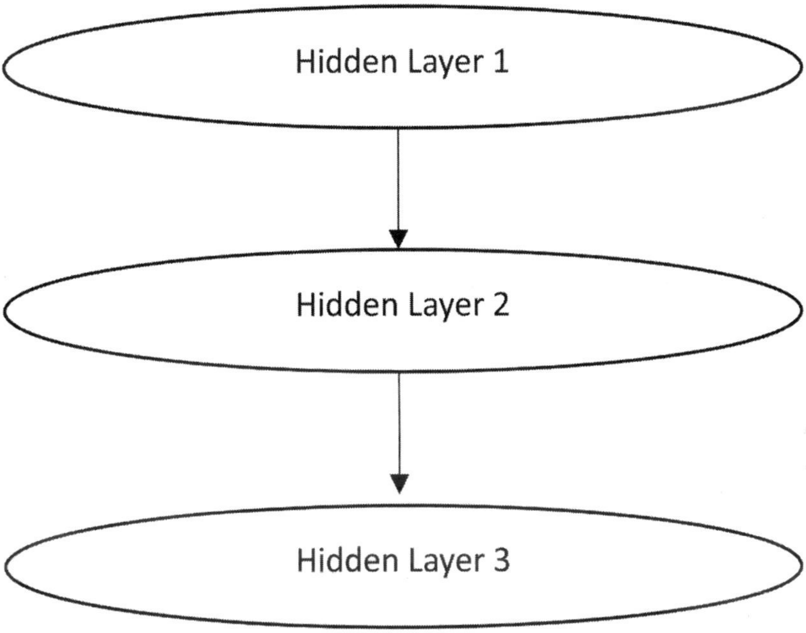

FIGURE 1.1 Neural network architecture

1.3.1 Strengths of Subsymbolic AI

1. Flexibility: Capable of handling diverse, unstructured datasets
2. Scalability: Efficient in large-scale applications such as image recognition and natural language processing
3. Adaptability: Learns and improves performance over time through iterative processes.

1.3.2 Limitations

1. Lack of Interpretability: Decisions are often opaque ("black box" models)
2. Data Dependence: Requires vast amounts of labeled data for training
3. Resource Intensive: Computationally demanding, requiring significant hardware resources

1.4 COMPARATIVE ANALYSIS

Symbolic and subsymbolic AI represent two distinct yet complementary paradigms in AI. Symbolic AI excels in tasks requiring structured knowledge representation, logical reasoning, and interpretability (McCarthy, 1980; Feigenbaum, 1988). Its reliance on explicit rules allows for transparent decision-making, making it suitable for domains such as legal reasoning and expert systems (Shortliffe, 1974). However, its rigidity and lack of scalability hinder its ability to adapt to unstructured environments or learn from raw data (Lenat, 1995; Davis et al., 1993).

In contrast, subsymbolic AI thrives in data-driven scenarios, leveraging neural networks to identify complex patterns in large-scale datasets (LeCun et al., 2015; Hinton et al., 2012). This paradigm underpins advancements in image recognition, natural language processing, and autonomous systems (Krizhevsky et al., 2012; Vaswani et al., 2017; Priyanka, 2023). Despite its flexibility and scalability, subsymbolic AI faces challenges related to interpretability and resource dependence, often requiring vast amounts of labeled data and computational power (Doshi-Velez & Kim, 2017). Table 1.3 provides symbolic vs. subsymbolic AI comparison.

Hybrid AI systems aim to integrate the strengths of both paradigms, combining symbolic reasoning's transparency with subsymbolic learning's adaptability (Ang et. al., 2023; Mira, 2021). This synthesis holds promise for addressing complex real-world problems, fostering scalability, and enhancing interpretability (Marcus, 2020; Garcez et al., 2019). Figure 1.2 shows the Performance Comparison of AI Paradigms.

1.5 APPLICATIONS

Symbolic and subsymbolic AI have distinct strengths that make them suitable for different applications. Symbolic AI excels in structured and rule-based environments, where explicit knowledge representation and logical reasoning are essential. For instance, expert systems such as MYCIN assist in medical diagnosis by applying predefined rules to analyze patient data and recommend treatments (Agarwal et al., 2024a; Agarwal et al., 2024b; Shortliffe, 1974). Similar applications of symbolic AI include legal reasoning tools, automated theorem proving, and robotics for path planning, where decision-making necessitates transparency and traceability

TABLE 1.3
Symbolic versus Subsymbolic AI Comparison

Feature	Symbolic AI	Subsymbolic AI	Hybrid AI
Knowledge Representation	Explicit rules and logic	Implicit in network weights	Combination of rules and data
Interpretability	High	Low	Medium
Scalability	Limited	High	Moderate
Adaptability	Low	High	Moderate

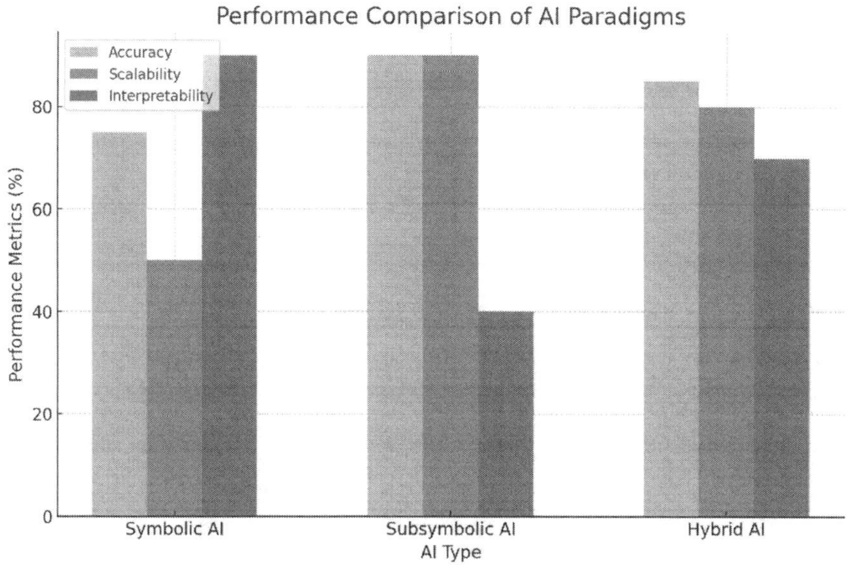

FIGURE 1.2 Performance comparison of AI paradigms

(Buchanan & Shortliffe, 1984; Nilsson, 1986; Kumar et al., 2025). Table 1.4 discusses applications of symbolic vs. subsymbolic AI in real-world.

In contrast, subsymbolic AI dominates data-intensive and unstructured domains due to its ability to learn patterns directly from raw data. Deep learning applications have revolutionized fields such as image recognition, where CNNs achieve state-of-the-art accuracy in tasks such as object detection and facial recognition (Krizhevsky et al., 2012). Subsymbolic AI also powers natural language processing systems, such as OpenAI's GPT models, which generate coherent text and assist in machine translation (Brown et al., 2020; Vaswani et al., 2017).

Hybrid AI combines symbolic reasoning and subsymbolic learning to address complex, dynamic problems. For example, autonomous vehicles integrate rule-based navigation with neural network-driven perception systems to operate effectively in real-world scenarios. This synergy enhances scalability, interpretability, and adaptability across diverse applications. Figure 1.3 shows application distribution among AI paradigms.

1.6 THE CASE FOR HYBRID SYSTEMS

Recent advancements in AI research increasingly advocate for hybrid approaches that blend the strengths of symbolic and subsymbolic paradigms. These systems aim to address the limitations of each approach while leveraging their unique strengths to tackle complex, real-world challenges.

Neuro-symbolic AI represents a significant step in this direction, integrating the structured reasoning capabilities of symbolic AI with the pattern recognition and

TABLE 1.4
Symbolic versus Subsymbolic AI in Real-World Applications

Domain	Symbolic AI Use Case	Subsymbolic AI Use Case
Healthcare	Expert diagnostic systems	Predictive models for patient data
Autonomous Systems	Path planning algorithms	Vision and perception modules
Finance	Fraud detection rules	Credit risk scoring with ML

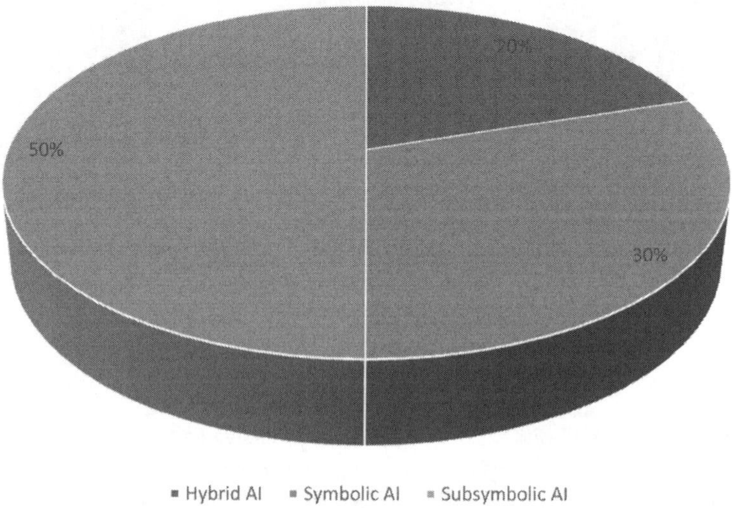

FIGURE 1.3 Application distribution among AI paradigms

learning abilities of neural networks (Ilarde et. al., 2024). This hybrid methodology allows AI systems to reason logically while learning from raw and unstructured data. For instance, neuro-symbolic systems can process natural language inputs by combining semantic understanding (symbolic reasoning) with contextual learning (subsymbolic AI) (Garcez et al., 2019). Such integration is particularly valuable in applications requiring both adaptability and interpretability, such as automated medical diagnostics and complex decision-making in financial systems.

Another promising development is Explainable AI (XAI), which aims to enhance the transparency of subsymbolic systems by incorporating symbolic reasoning frameworks. By making the decision-making processes of neural networks more comprehensible, XAI addresses the "black box" problem often associated with deep learning models (Doshi-Velez & Kim, 2017). For example, XAI can clarify how a deep learning model classifies medical images, providing confidence to healthcare professionals and patients alike.

Hybrid systems have also been instrumental in robotics and autonomous systems, where symbolic AI manages high-level planning and subsymbolic AI handles sensory processing and perception. For instance, autonomous vehicles utilize rule-based systems to navigate traffic laws (symbolic AI) while employing neural networks for real-time object detection and collision avoidance (Ng et al., 2020). This combination enhances safety, efficiency, and adaptability, making hybrid systems essential in dynamic environments.

The rise of hybrid systems underscores the need for collaboration across disciplines such as computer science, cognitive psychology, and ethics. While symbolic AI offers the structure and transparency necessary for trust, subsymbolic AI contributes adaptability and scalability, making their integration crucial for addressing the growing complexity of AI applications. Moving forward, hybrid systems are poised to become the cornerstone of AI research, offering a balanced approach to scalability, interpretability, and performance across diverse domains.

1.7 CASE STUDIES

1.7.1 Case Study 1: Symbolic AI in Medical Diagnosis

Context: The MYCIN expert system, developed in the mid-1970s at Stanford University, was one of the earliest and most notable applications of symbolic AI in healthcare. The design aimed to aid doctors in the diagnosis and treatment of bacterial infections, especially those involving bloodborne pathogens. MYCIN demonstrated the potential of symbolic AI in domains that demand high levels of expertise and precision, despite limited computational resources. Its primary goal was to replicate the diagnostic reasoning process of infectious disease specialists, enabling healthcare professionals to make informed decisions even in their absence.

Implementation: MYCIN employed a rule-based approach, leveraging a large knowledge base of over 600 IF–THEN rules. Expert knowledge in infectious diseases and antibiotic therapy guided the creation of these rules. The system also included an inference engine that processed these rules to deduce potential diagnoses and recommend appropriate treatments. For example, given a patient's symptoms, laboratory test results, and medical history, MYCIN would systematically apply its rules to identify the most probable bacterial infection and suggest targeted antibiotics. Additionally, the system incorporated a degree of uncertainty by assigning confidence levels to its conclusions, enabling it to handle ambiguous or incomplete data.

Outcomes: MYCIN performed exceptionally well, with diagnostic accuracy comparable to that of human specialists. It demonstrated the feasibility of using symbolic AI to capture and replicate domain-specific expertise in a transparent and interpretable manner. MYCIN's rule-based reasoning allowed doctors to trace the logic behind its recommendations, fostering trust and accountability in its outputs. However, the system also revealed some critical limitations of symbolic AI. Its reliance on manually encoded rules made it rigid and difficult to update or adapt to new medical knowledge, such as emerging diseases or changing treatment guidelines. Moreover, MYCIN could only apply its reasoning to the specific domain of bacterial infections, lacking the flexibility to generalize to other medical conditions.

Significance: Despite its limitations, MYCIN paved the way for subsequent developments in expert systems and highlighted the potential of symbolic AI in solving real-world problems. Its transparency and interpretability remain a key strength of symbolic AI, particularly in domains such as healthcare, where accountability and trust are paramount. At the same time, MYCIN's shortcomings underscored the need for more adaptive and scalable approaches, paving the way for the integration of symbolic and subsymbolic AI in modern medical diagnostics. Today, hybrid systems that combine rule-based reasoning with machine learning techniques are building upon MYCIN's legacy, offering enhanced adaptability and accuracy while retaining the interpretability that is vital in critical decision-making contexts.

1.7.2 Case Study 2: Subsymbolic AI in Image Recognition

Context: Google's DeepMind has been at the forefront of AI research, and its work in image recognition is a testament to the power of subsymbolic AI. DeepMind utilized CNNs to tackle challenging tasks such as object detection, image classification, and visual segmentation. These tasks are critical in applications ranging from autonomous vehicles to medical imaging, where accurate identification of visual patterns is paramount. By leveraging CNNs, DeepMind demonstrated the potential of subsymbolic AI to excel in complex, data-intensive visual recognition tasks.

Implementation: CNNs, a specialized form of neural networks, are particularly well-suited for processing image data due to their ability to capture spatial hierarchies. DeepMind trained its CNN models on millions of labeled images from diverse datasets, such as ImageNet, to teach the system to identify objects, features, and patterns. The network's layered architecture played a critical role in this process. Each layer in the CNN performed a specific function: convolutional layers extracted low-level features such as edges and textures, pooling layers reduced dimensionality to improve efficiency, and fully connected layers synthesized the extracted features to make predictions. The depth of the network enabled the system to learn increasingly complex features at each stage, culminating in robust image recognition capabilities.

Outcomes: DeepMind's CNN models achieved state-of-the-art accuracy in tasks such as object detection and image classification, surpassing traditional machine learning algorithms and even human performance in specific benchmarks. For example, these models demonstrated remarkable precision in identifying objects within complex scenes, a breakthrough that has since fueled advancements in autonomous navigation, facial recognition, and content moderation. However, the success of CNNs came at a cost. Training such deep networks required significant computational resources, including high-performance GPUs and substantial energy consumption.

The reliance on large labeled datasets also posed challenges, as curating and annotating them is time-intensive and resource-heavy.

Significance: DeepMind's work in image recognition highlighted the transformative potential of subsymbolic AI in solving complex visual problems. The layered architecture of CNNs provided a scalable and adaptable framework, enabling the system to generalize across diverse datasets and applications. However, the substantial resource demands underscored the need for more efficient training techniques and hardware optimizations. DeepMind's achievements have paved the way for further innovations in deep learning, including advancements in transfer learning and model compression, which aim to make high-performance image recognition more accessible and sustainable. This case study illustrates how subsymbolic AI has reshaped the landscape of image recognition, setting new benchmarks for accuracy while also exposing areas for improvement in scalability and efficiency.

1.7.3 Case Study 3: Hybrid AI in Autonomous Vehicles

Context: Tesla's Autopilot system stands as a prime example of hybrid AI, combining symbolic and subsymbolic approaches to achieve advanced capabilities in autonomous driving. The development of Autopilot reflects the increasing complexity of real-world challenges that require both structured reasoning and adaptive learning. By integrating rule-based logic with neural networks, Tesla has created a system capable of navigating dynamic environments, adhering to traffic rules, and responding to real-time conditions, making it a frontrunner in the race toward fully autonomous vehicles.

Implementation: The hybrid AI architecture of Tesla's Autopilot leverages the strengths of both paradigms. Symbolic AI is employed for high-level decision-making tasks, such as interpreting traffic signals, following lane markings, and adhering to traffic laws. This rule-based component ensures that the vehicle operates within defined legal and safety parameters. On the other hand, subsymbolic AI, in the form of neural networks, handles perception and sensory data processing. Neural networks analyze inputs from cameras, radar, ultrasonic sensors, and GPS to identify objects, predict the behavior of other road users, and detect potential hazards. The system's ability to learn from vast amounts of real-world driving data further enhances its adaptability, enabling it to handle diverse and evolving road conditions.

Outcomes: The integration of symbolic and subsymbolic AI in Tesla's Autopilot has significantly improved decision-making and operational safety. The system excels in tasks such as highway driving, adaptive cruise control, and lane-keeping, showcasing the synergy between rule-based reasoning and data-driven learning. However, this hybrid approach is not without its challenges. Integrating symbolic and subsymbolic components presents technical complexities, particularly in ensuring seamless communication and coherence between the two paradigms. Furthermore, the ethical implications of autonomous decision-making remain a critical concern. Situations requiring moral judgments, such as unavoidable collisions, pose significant challenges for AI systems and raise questions about accountability and transparency.

Significance: Tesla's Autopilot exemplifies the potential of hybrid AI to address the limitations of purely symbolic or subsymbolic systems. By combining the interpretability of symbolic reasoning with the adaptability of neural networks, hybrid systems can achieve higher levels of performance and reliability in complex, real-world applications. However, the case also highlights the ongoing need for advancements in system integration, safety protocols, and ethical frameworks to ensure the responsible deployment of autonomous technologies. Tesla's Autopilot represents a milestone in AI-driven transportation and underscores the importance of hybrid approaches in shaping the future of intelligent systems.

1.8 CHALLENGES AND FUTURE DIRECTIONS

Hybrid AI systems represent a promising evolution in AI, offering the potential to address complex, real-world problems by combining the strengths of symbolic and subsymbolic approaches. However, unlocking their full potential requires addressing significant challenges in their development and deployment.

Integration complexity is one of the primary hurdles. Combining symbolic reasoning with subsymbolic learning involves reconciling fundamentally different methodologies. Symbolic AI operates on explicitly defined rules and logical structures, while subsymbolic AI relies on implicit, data-driven patterns encoded in neural networks. Bridging these paradigms requires sophisticated architectures and algorithms to ensure seamless communication and coordination between the two components. This complexity often increases development time and computational demands, posing a barrier to widespread adoption.

Scalability is another critical issue. We must design hybrid systems to handle large-scale, real-world applications without becoming prohibitively resource-intensive. Balancing the computational overhead of subsymbolic neural networks with the logical reasoning processes of symbolic AI requires careful optimization. As applications grow in complexity, ensuring that hybrid models remain efficient and responsive becomes an ongoing challenge.

Standardization also presents a significant obstacle. The lack of unified frameworks and best practices for designing, implementing, and deploying hybrid systems hinders collaboration and scalability across different industries. Developing standardized methodologies and tools is essential to streamline development processes and ensure interoperability among hybrid systems.

Looking ahead, future research should focus on innovative architectures that integrate symbolic and subsymbolic paradigms more effectively. New developments in neuro-symbolic models and XAI could help solve the problems of hybrid systems being hard to understand and adapt to. We must also develop real-world deployment strategies to test these systems in dynamic environments, ensuring reliability and robustness.

Finally, ethical considerations must remain at the forefront of hybrid AI research. Deploying these systems in critical areas such as healthcare, transportation, and governance necessitates addressing issues such as bias, accountability, and transparency. By tackling these challenges, hybrid systems can evolve into a transformative force in AI, capable of solving some of the most pressing problems in modern society.

1.9 CONCLUSION

Symbolic and subsymbolic approaches to AI represent two foundational yet contrasting paradigms, each offering unique strengths and addressing specific challenges. Symbolic AI excels in structured reasoning and transparency, making it ideal for domains where interpretability and rule-based decision-making are critical. Subsymbolic AI, on the other hand, thrives in data-rich and unstructured environments, leveraging neural networks to identify patterns and adapt to new information. This chapter has explored the strengths, limitations, and real-world applications of both paradigms, demonstrating their individual contributions to the evolution of AI.

While each paradigm has its merits, their limitations highlight the need for hybrid systems that combine symbolic reasoning's logic and interpretability with subsymbolic AI's adaptability and scalability. Hybrid approaches bridge the gap between these methodologies, offering a balanced framework capable of tackling complex, dynamic, and large-scale challenges. By integrating structured logic with data-driven learning, hybrid systems open new avenues for developing efficient, understandable, and scalable AI solutions.

The advancement of hybrid AI not only enhances the capabilities of AI but also broadens its applicability across diverse fields, from healthcare to autonomous systems. As research progresses, the synergy between symbolic and subsymbolic approaches is poised to shape the future of AI, fostering innovative solutions to some of the most intricate problems of our time.

REFERENCES

Agarwal, N., A. Singh, and M. C. T. Ambojia. (2024a). "Amalgamation of Disruptive Technologies for Implementation of Intelligent Manufacturing." *Intelligent Manufacturing and Industry 4.0: Impact, Trends, and Opportunities* (pp. 42–66), CRC Press. https://doi.org/10.1201/9781032630748

Agarwal, N., M. Sangwan, and E. Agarwal. (2024b). "Revolutionizing Pedagogy: Exploring AI Coaching in Enhancing Teacher Professionalism." *Integrating Generative AI in Education to Achieve Sustainable Development Goals* (pp. 264–281). IGI Global. https://doi.org/10.4018/979-8-3693-2440-0.ch014.

Ang, L., N. Bonifacio, J. K. Garcia, R. Medina, and J. Santiago. (2023). "The Rise of Artificial Intelligence: A Correlational Study on AI-Powered Bots and Customer Satisfaction in a People-Oriented Industry." *Cosmos Journal of Engineering & Technology* 13(1): 81–94. https://doi.org/10.46360/cosmos.et.620231008.

Bengio, Y., A. Courville, and P. Vincent. (2013). "Representation Learning: A Review and New Perspectives." *IEEE Transactions on Pattern Analysis and Machine Intelligence* 35(8): 1798–1828. https://doi.org/10.1109/TPAMI.2013.50.

Binns, R. (2018). "Fairness in Machine Learning: Lessons from Political Philosophy." In *Proceedings of the 2018 Conference on Fairness, Accountability, and Transparency*, pp. 149–159.

Brown, T., B. Mann, N. Ryder, M. Subbiah, J. D. Kaplan, P. Dhariwal, and D. Amodei. (2020). "Language Models Are Few-Shot Learners." *Advances in Neural Information Processing Systems* 33: 1877–1901. https://doi.org/10.48550/arXiv.2005.14165.

Buchanan, B. G., and E. H. Shortliffe. (1984). *Rule-Based Expert Systems: The MYCIN Experiments of the Stanford Heuristic Programming Project.* Addison-Wesley.

Davis, R., H. Shrobe, and P. Szolovits. (1993). "What Is a Knowledge Representation?" *AI Magazine* 14(1): 17–33. https://doi.org/10.1609/aimag.v14i1.1029

Dean, J., G. Corrado, R. Monga, K. Chen, M. Devin, Q. V. Le, and A. Y. Ng. (2012). "Large Scale Distributed Deep Networks." *Advances in Neural Information Processing Systems* 25: 1223–1231.

Doshi-Velez, F., and B. Kim. (2017). "Towards a Rigorous Science of Interpretable Machine Learning." *arXiv preprint arXiv:1702.08608.*

Esteva, A., A. Robicquet, B. Ramsundar, V. Kuleshov, M. DePristo, K. Chou, and J. Dean. (2019). "A Guide to Deep Learning in Healthcare." *Nature Medicine* 25(1): 24–29. https://doi.org/10.1038/s41591-018-0316-z.

Feigenbaum, E. A. (1988). "The Art of Artificial Intelligence: Themes and Case Studies of Knowledge Engineering." *Proceedings of the International Joint Conference on Artificial Intelligence* 2: 1014–1029.

Floridi, L., J. Cowls, M. Beltrametti, R. Chatila, P. Chazerand, V. Dignum, and E. Vayena. 2018. "AI4People's Ethical Framework for a Good AI Society: Opportunities, Risks, Principles, and Recommendations." *Minds and Machines* 28(4): 689–707.

Garcez, A. S. D., M. Gori, L. C. Lamb, L. Serafini, M. Spranger, and S. N. Tran. (2019). "Neural-Symbolic Computing: An Effective Methodology for Principled Integration of Machine Learning and Reasoning." *Philosophical Transactions of the Royal Society A* 377(2152): 20180022. https://doi.org/10.48550/arXiv.1905.06088

Goodfellow, I., Y. Bengio, and A. Courville. (2016). *Deep Learning.* MIT Press.

Hinton, G. E., N. Srivastava, A. Krizhevsky, I. Sutskever, and R. Salakhutdinov. (2012). "Improving Neural Networks by Preventing Co-Adaptation of Feature Detectors." *arXiv preprint arXiv:1207.05808.*

Ilarde, M. E., C. R. Pacis, F. A. Vasquez, E. S. Salas, M. A. Concepcion, and K. L. D. Juat. (2024). "Systematic Review on Mobile Healthcare Applications and Their Effect on the Nursing Service Industry." *Globus: An International Journal of Medical Science, Engineering and Technology* 13(1): 22–25. https://doi.org/10.46360/globus.met.320241006.

Krizhevsky, A., I. Sutskever, and G. E. Hinton. 2012. "ImageNet Classification with Deep Convolutional Neural Networks." *Advances in Neural Information Processing Systems* 25: 1097–1105.

Kumar, A., Kumar, P., Rathee, S., & Kumar, B. (Eds.). (2025). *Mechatronics: Concepts, Tools, Applications, and New Trends* (1st ed.). CRC Press. https://doi.org/10.1201/9781003494478

LeCun, Y., Y. Bengio, and G. Hinton. 2015. "Deep Learning." *Nature* 521(7553): 436–444. https://doi.org/10.1038/nature14539.

Lenat, D. B. (1995). "CYC: A Large-Scale Investment in Knowledge Infrastructure." *Communications of the ACM* 38(11): 33–38. https://doi.org/10.1145/219717.21974

Marcus, G. (2020). "The Next Decade in AI: Four Steps Towards Robust Artificial Intelligence." *arXiv preprint arXiv:2002.06177.* https://doi.org/10.48550/arXiv.2002.06177.

McCarthy, J. (1958). "Programs with Common Sense." *Proceedings of the Teddington Conference on the Mechanization of Thought Processes*, pp. 756–791.

McCarthy, J. (1980). "Circumscription—A Form of Non-Monotonic Reasoning." *Artificial Intelligence* 13(1–2): 27–39. https://doi.org/10.1016/0004-3702(80)90011-9

Mira, F. (2021). "What Are IOT Devices or (Internet of Things)?" *Globus An International Journal of Management & IT* 12(2): 46–51. https://doi.org/10.46360/globus.mgt.120211007

Newell, A., and H. A. Simon. (1976). "Computer Science as Empirical Inquiry: Symbols and Search." *Communications of the ACM* 19(3): 113–126. https://doi.org/10.1145/360018.360022

Nilsson, N. J. (1986). *Principles of Artificial Intelligence.* Morgan Kaufmann.

Priyanka, J. (2023). "Role of AI for Library Services Considering Delhi-NCR." *Cosmos An International Journal of Management* 12(2): 133–136. https://doi.org/10.46360/cosmos.mgt.420231015.

Rumelhart, D. E., G. E. Hinton, and R. J. Williams. (1986). "Learning Representations by Back-Propagating Errors." *Nature* 323(6088): 533–536. https://doi.org/10.1038/323533a0

Russell, S., and P. Norvig. (2020). *Artificial Intelligence: A Modern Approach.* 4th ed. Pearson.

Sabaulan, W. D. P. (2024). "Virtual Learning Action Cell (LAC) and Teachers' Performance in Online Class: A Linear Regression Analysis." *Cosmos: An International Journal of Art and Higher Education* 13(1): 40–45. https://doi.org/10.46360/cosmos.ahe.520241006.

Santos, M. R. C., and M. B. Cruz. (2024). "The Effect of Students' Use of AI in Research: Towards Policy Recommendations on the Responsible Use of AI in Schools." *Globus Journal of Progressive Education* 14(1): 46–55. https://doi.org/10.46360/globus.edu.220241007.

Schmidhuber, J. (2015). "Deep Learning in Neural Networks: An Overview." *Neural Networks* 61: 85–117. https://doi.org/10.1016/j.neunet.2014.09.003.

Shortliffe, E. H. (1974). *MYCIN: Computer-Based Medical Consultations.* Elsevier.

Silver, D., J. Schrittwieser, K. Simonyan, I. Antonoglou, A. Huang, A. Guez, and D. Hassabis. (2016). "Mastering the Game of Go with Deep Neural Networks and Tree Search." *Nature* 529(7587): 484–489. https://doi.org/10.1038/nature16961.

Turing, A. M. (1950). "Computing Machinery and Intelligence." *Mind* 59(236): 433–460.

Vaswani, A., N. Shazeer, N. Parmar, J. Uszkoreit, L. Jones, A. N. Gomez, and I. Polosukhin. (2017). "Attention Is All You Need." *Advances in Neural Information Processing Systems* 30: 5998–6008. https://doi.org/10.48550/arXiv.1706.03762.

Winograd, T. (1972). "Understanding Natural Language." *Cognitive Psychology* 3(1): 1–191. https://doi.org/10.1016/0010-0285(72)90002-3.

2 Revisiting the Impact of AI-Driven Feature Selection Approaches in Software Fault Prediction

Yogita Khatri, Urvashi Rahul Saxena, and Parveen Kumar

2.1 INTRODUCTION

Software fault prediction (SFP) is the process of identifying defective modules in the software using some underlying properties called "software metrics" before the testing phase. This assists the Quality Assurance (QA) team in utilizing the available limited testing resources reasonably and efficiently by allocating more resources to predicted faulty software modules.

The effectiveness of an SFP model is closely associated with the quality of data used, the features (software metrics) used, the kind of learners applied, and the evaluation metrics employed (Khatri and Singh 2023). Akiyama made the first attempt toward building a machine learning–based SFP (Fumio Akiyama 1971). To forecast faults using regression modeling, he developed a prediction model in 1971 that used size metrics as independent variables. The field continues to grow every year due to its practical utility in the software industry. The probability of detection (PD) of a robust prediction model can even outperform the PD of manual software evaluations, according to Menzies et al.'s study (2007). Consequently, it is preferable to create an automated system for SFP. Many researchers and industry professionals (T. M. Khoshgoftaar and Seliya 2002; Taghi M. Khoshgoftaar and Seliya 2002; T.M. Khoshgoftaar and Seliya 2022) began working in this field because it was a more affordable option than software reviews.

There are mainly four key ingredients that are responsible for the success of an SFP model.

1) Quality of data in use
2) Software metrics in use
3) Learning algorithm in use
4) Evaluation parameters in use

DOI: 10.1201/9781032724508-3

There are some studies (Arisholm, Briand, and Johannessen 2010; Li, Lessmann, and Baesens 2019) that focused mainly on the modeling techniques used for SFP to observe how different learners perform in different environments, whereas some focused majorly on checking the validity of different sets of metrics (Basili, Briand, and Melo 1996; Menzies, Greenwald, and Frank 2007) in correlation to faults. Numerous studies have also been conducted to examine how various preprocessing methods affect the functionality of the developed model. Apart from this, current research in this domain is progressing in the direction of cross-project defect prediction (CPDP), where knowledge gained from one/many projects is applied to other projects.

Thus, a significant amount of research has been produced targeting one or the other aforementioned issues in the context of SFP. Since the field of SFP is very vast and scattered, this research intends to revisit the impact of feature selection (FS) on the predictive power of an SFP model, as contradictory results have been observed among different studies. Both redundant and irrelevant features have the potential to significantly affect the model's performance. Therefore, it becomes very important to determine the smallest subset of features that exhibit the strongest correlation to fault, thereby decreasing the model's complexity, which in a way increases classification accuracy.

FS methods are generally categorized into three groups: filter-oriented, wrapper-oriented, and embedded. The filter-oriented FS (FOFS) approach selects features independently of the prediction model; wrapper-oriented FS (WOFS) methods employ the use of an underlying classification model that selects features that increase the prediction accuracy; and embedded approaches, being iterative in nature, handle every iteration of the entire model training process and diligently choose those features that have the greatest training impact.

In this chapter, we implemented eight different FS techniques that consist of two FOFS approaches, namely information gain and the Chi-Square methods. The work also implements three prominent WOFS methods namely Step Backward FS, Step Forward FS, Recursive Feature Elimination, one embedded method named Lasso Regression, and one feature extraction method named principal component analysis (PCA). Based on the empirical investigation on 5 NASA datasets, Chi-Square outperformed all w.r.t F1 and AUC.

The rest of the chapter is organized as follows: Section 2.2 illustrates the related work. The framework of the work performed is demonstrated in Section 2.3. Experimental results are demonstrated in Section 2.4 and, finally, the conclusion of the work along with future work is depicted in Section 2.5.

2.2 RELATED WORK

In this section, we first provide a brief overview of the software metrics (aka features) that have been majorly used in the SFP studies.

Software metrics are the quantifiable aspects of software that can be used to pinpoint its flaws and enhance its quality. Numerous software metrics have been investigated and shown to be effective in the context of SFP. The application of the Chidamber and Kemerer (CK) metric set (Chidamber and Kemerer 1994) as fault predictors was examined by Basil et al. (1996). Five of the six metrics—weighted method per class (WMC), response for a class (RFC), coupling between objects (CBO), lack of cohesion in method (LCOM), depth of inheritance tree (DIT), and number of children of a class (NOC)—were found to be significant fault predictors after they conducted experiments on eight information management systems (IMSs). Metric LCOM did not significantly correlate with fault. Traditional size metrics (to be collected later in the SDLC phase) were reported to be less significant as compared to OO metrics.

In order to validate their use in forecasting fault-prone modules on a commercial telecom system with 130 modules, Ohlsson et al. (1996) employed design and complexity measurements. By analyzing the design document, they developed a tool called ERIMET, which automatically calculates design and complexity measurements. Their results were substantially superior than those of conventional size measures and indicated that design metrics were the most effective fault predictors. According to their data analysis, the top 20% of defective modules are responsible for 60% of all defects; however, their suggested model found that these modules account for 47% of all faults.

Subramanyam and Krishnan (2003) conducted an empirical investigation on two commercial software programs produced in C++ and Java separately to assess the reliability of the CK metric set for detecting errors in software developed with various OO languages. They found that faults in C++ and Java were associated with variations in the performance of three CK measures (CBO, WMC, and DIT).

The aforementioned studies examined the validity of static code metrics on commercial and industrial software that isn't accessible to the general public. Using the open-source program Mozilla, Gyimothy et al. (2005) tried to confirm the significance of CK metrics. They analyzed the source code using their "Clumbus" methodology to get the necessary metrics. They included two additional metrics, LOC and lack of cohesion on methods permitting negative value (LCOMN), in addition to the six CK measures. While NOC and DIT were deemed unreliable and untrustworthy, CBO and LOC were shown to be the most significant fault predictors overall. LCOM was argued to be a strong fault predictor, which was in contrast to Basili's (1996) findings.

The reliability of complexity metrics as post-release defect predictions was further validated by Zimmermann et al. (2007). To count errors that occurred before and after releases, they gathered data on bug fixes at the file and package levels from version control archives and bug databases. They also computed a number of complexity metrics for every file, estimated the number of faults using a linear regression model, and then used the Spearman correlation between the anticipated and observed faults in each file to demonstrate the importance of complexity metrics. Version 2.0 was used for training in three sets of experiments, and versions 2.0, 2.1, and 3.0 were used separately for testing. The maximum Spearman correlation value was 0.416, which was low but positive. This indicates that a file with a higher rank is likely to have more observed faults. It also strongly suggests that a model trained on one version can be used to predict faults in other versions. To encourage more empirical study using their Eclipse dataset, they also made a substantial contribution to the field by making it publicly available.

The researcher's efforts to confirm the importance of OO and conventional metrics in detecting pre-release errors have been demonstrated thus far. Nevertheless, there was no universal method or framework to recommend which collection of software measurements would be suitable for constructing the SFP model. Nagappan et al. (2006) addressed this problem by selecting the best criteria for post-release fault prediction for other projects and utilizing historical fault data. They examined five extensive projects of various kinds that were created using OO programming languages (C++, C#). A high correlation between complexity measurements and post-release defects was found, despite the fact that no single set of metrics could be applied to all software projects.

Although the results cannot be applied to other projects, they found that a metric set they chose using principal component analysis (PCA) to eliminate multicollinearity issues was a successful post-release fault predictor. Nevertheless, positive outcomes can be anticipated if it is applied to projects that are similar. Thus, the aforementioned studies demonstrate the success of product metrics in the SFP domain. We will now discuss a few studies that assess how well process metrics work to create an effective SFP model.

Process metrics are some common measurements used to quantify the quality of business processes and evaluate their efficacy. The potential of process metrics in fault prediction has been examined and confirmed by numerous important research studies (Kumar et al., 2025). We discuss some noteworthy attempts here.

One of the few early researchers to try to look at the efficacy of process metrics in fault prediction was Nagappan and Ball (2005). To anticipate the fault density on Windows Server 2003, they experimented with relative code churn measurements. According to their findings, relative code churn metrics can be used to distinguish between software modules that are defective and those that are not, and they even outperform absolute code churn measures in forecasting fault density. In keeping with his research, Hassan and Holt (2005) conducted experiments on six sizable open-source systems (OSS) to identify the top ten modules and subsystems that are most likely to have defects. Based on their empirical investigation, they asserted that subsystems that have recently been fixed and updated are very likely to develop faults.

Researchers have successfully confirmed the relationship between the fault label and complexity measures in a couple of the aforementioned investigations. Hassan (2009) suggested measuring the complexity of the code modification process instead of utilizing the complexity metric, and he used regression modeling to confirm its efficacy on six sizable OSS. "The more complex the code change process, the higher will be the chance of occurrence of faults," he hypothesized. He used the Shannon entropy concept to quantify the complexity and uncertainty of the code update process while focusing solely on feature alterations and disregarding all other kinds of changes. He noticed superior performance as compared to other models that had been constructed with prior defects or changes.

As a result, two schools of thinking have been identified, one of which emphasizes process indicators and the other product indicators. To determine which set of metrics has higher discrimination power and pertinent information regarding the distribution of software faults, Moser et al. (2008) used three machine learners—DT,

NB, and LR—to classify the software modules in a cost-sensitive manner using process metrics, product metrics, and a hybrid of the two for the "Eclipse" project. According to their research, process metrics are better than product metrics. On the other hand, Menzies et al. (2007) presented an alternative result about product metrics. Without comparing the two, he asserted that product metrics were beneficial as substantial defect predictors, with an average recall of 71% and a PF of 25%. He used Infogain as an FS method to choose the right set of features (from 37 metrics, including size, McCabe cyclomatic complexity, and Halstead metrics gathered on 8 NASA datasets). He observed that each cross-validation run had a different feature ranking, leading to the conclusion that the learner is more crucial to the project's success than choosing which metrics to pursue. They also repeated the same experiment using three different learners—NB, OneR, and J48—and the best classifier was NB (with logarithm filter).

D'Ambros et al. (2012) examined the variations between a number of bug prediction techniques with regard to granularity level, complexity level, kind of metrics, and performance evaluation metrics. The comparison is pointless and erroneous if there is no standard or benchmark. Inspired by this idea, they tried to offer two new methods for SFP as well as a benchmark technique for the same in three different scenarios (ranking, classification, and effort-aware ranking). They gathered their datasets for five OSS and made them publicly accessible. Moser's process metrics (Moser, Pedrycz, and Succi 2008), source code metrics, Hassan's (2009) entropy of changes and prior defects, and two new metrics they proposed—entropy of code metrics (LDHH) and churn of code metrics (LGDCHU)—were also examined, and classification-based approaches were evaluated. In both classification and ranking scenarios, process metrics, LDHH, and LGDCHU performed better than the others. In contrast to the entropy of changes, which ranked lower than the top two but had low variability, LDHH and LGDCHU performed well in the effort-based ranking scenario, but this led to more output variability. In all three cases, prior defects performed the worst. Despite conducting such a thorough examination into the viability of applying several sets of measures in three distinct forecasting tasks, they were still unable to draw a broad conclusion. It is currently difficult to generalize the results to other datasets and learners.

Therefore, the two primary components of an SFP model are the modeling technique and the software metric set. The success of an SFP model depends on the selection of a classifier and a set of metrics. According to Menzies et al. (2007), choosing a classifier is more crucial than choosing a set of metrics. In contrast, Arisholm et al. (2010) asserted that the model's performance is significantly impacted by the selection of a metric set. Radjenovic et al. (2013) conducted a systematic evaluation to further explore the effects of various measure sets alone, omitting the type of classifier utilized. The goal was to discover appropriate software metrics under various conditions for evaluating their applicability in SFP.

According to their selection of 106 primary research published between 1999 and 2011, 49% of the chosen studies looked into OO metrics, 27% experimented with conventional size and complexity metrics, and 24% looked into the application of process-based metrics in SFP. The most widely utilized CK metrics suite in OO metrics includes WMC, CBO, and RFC, all of which have been shown to be relevant in relation to SFP, but DIT and NOC have been deemed unreliable. It was discovered that LCOM worked well in modest pre-release settings. Although LOC and McCabe's cyclomatic complexity (cc) have been the most commonly used metrics in traditional metrics, there have been conflicting opinions regarding the relationship between LOC and fault. The study concluded that the LOC metric was more significant (Gyimóthy, Ferenc, and Siket 2005), with lower validity compared to the study (Fenton 2000) with higher validity.

With the sole intention of improving the model's performance, some researchers are experimenting with alternative metrics, using different feature selection strategies, and suggesting the use of ensemble classifiers; this leads to the creation of complex models, which are expensive and non-generic in nature. Is creating such intricate models really worth it? It is important to distinguish between the projected advantages and the actual costs. In light of this query, He et al. (2015) made an effort to develop and evaluate a basic model utilizing a condensed subset of metrics, regardless of classifier selection, from the standpoints of cost, generalization, and accuracy. It was determined that the metric set CBO+LOC+LCOM is the bare minimum that can help with general fault prediction and may be applied without classifier's dependence.

Additionally, Zhou et al. (2018) carried out an empirical investigation and put forth two extremely basic models that solely used the LOC measure. After validating these models on sizable datasets, they found that they performed better than other important approaches. Using developer information as a software metric, Qu et al. (2021) recently suggested an effort-aware defect prediction model and found that it outperformed the LOC metric.

As a result, both process and product metrics have been widely employed in several studies and have been shown to be effective defect predictors. While complexity, size, and OO metrics were found to be effective in detecting defects in pre-release software, the most commonly used metrics in the context of post-release SFP are process metrics, which include the number of changes, code churn, change set size, and module age. Nonetheless, the following are some crucial matters to be addressed:

1) Choosing relevant features
2) Rejecting superfluous features to address the dataset's multicollinearity issue

In the following paragraphs, we will revisit specific studies that made notable efforts to enhance the model's performance by applying FS using the common evaluation measures and datasets to arrive at concrete findings about which FS technique outperforms in comparison to others.

The process of carefully choosing the most crucial features (software metrics) by removing superfluous or redundant features is known as FS. The performance of an SFP model declines when the features are reliant on each other and exhibit multicollinearity. Various researchers have used various FS techniques in SFP and demonstrated their applicability and significance within the SFP framework. For example, (Zhou et al., 2018; Xu et al. 2016) experimented with 32 FS methods and concluded that the clustering-oriented approach and the majority of filter-oriented feature ranking techniques produce satisfactory outcomes having reduced features in a shorter time and are also simple to comprehend. They also provided valuable guidelines for choosing suitable FS schemes depending on certain supplementary attributes, such as ease of computation and analytical simplicity.

In compliance with the above work, (Balogun et al. 2019) applied 18 FS techniques on five datasets from the NASA repository using four diverse classifiers. According to their experimental findings, FS enhanced classifiers' predictive performance; however, FS techniques' effectiveness varies depending on the dataset and classifier. It cannot be uniform on all datasets. The researchers concluded the supremacy of Information Gain (IG) and the Best First Search among all its competitors and that FS techniques enhanced the working of SFP frameworks. However, there does not exist an ideal technique for FS. In congruence with the above work, Amit and Rajnish (2018) performed a similar experiment, and their results indicated the potential of FS to enhance the SFP model's performance.

Yao et al. (2018) also suggested a hybrid FS method. First, they used the IG, Chi-Squared (CS), and Pearson Correlation coefficients to determine the importance of each feature. Second, in order to choose features, they arranged the features according to the rankings of the three values. Lastly, they constructed the model using the Random Forest classifier and observed that different FS techniques performed differently based on the datasets being used.

Similarly, Osman et al. (2018) examined the effects of wrapper and correlation-based FS (CFS) filter techniques and built five prediction models. They also showed how well these models function with or without FS to forecast the number of bugs in five datasets. According to their findings, wrappers work better than the CFS, removing more than half of the features while increasing prediction accuracy by up to 33%. They observed that, even though various feature subsets are selected by the same FS algorithm for different projects, it always includes a combination of change and source code metrics.

Thus, a good amount of research (Akalya devi 2012; Ghotra, McIntosh, and Hassan 2015; Wahono and Herman 2014) has been done on feature selection, applying different feature selection techniques on different models using different evaluation parameters, but still its true state is not yet clear, as it is difficult to compare several different approaches using different evaluation parameters on diverse datasets. Therefore, we thought to investigate this issue by reimplementing some of the existing FS techniques and measuring them on a uniform scale to get a holistic view.

2.3 ARCHITECTURE

The proposed work carries out certain data cleaning operations to filter out some noisy and missing data from all the datasets, followed by applying several different FS techniques to select the highly relevant features. The selected feature subset is then used to train the base model. As a base model, we used Random Forest due to its effective performance in many previous SFP studies. Finally, the model's performance is tested using F1 score and AUC (area under curve). The architectural description of the proposed framework is depicted in Figure 2.1.

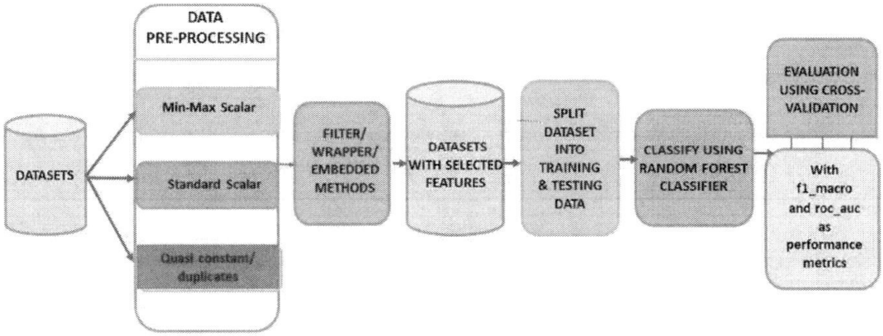

FIGURE 2.1 Architectural framework for the proposed work

2.3.1 Data Collection:

For the empirical evaluation of different FS methods, five NASA datasets (Menzies, Greenwald, and Frank 2007) of different sizes have been selected, with the largest dataset containing 10,885 instances and the smallest dataset having 498 instances, with 21 features in all. A detailed description of these datasets is given in Section 2.4.1.

2.3.2 Data Pre-processing:

Data pre-processing is a technique that is used to transform the raw dataset into a cleaner data set. As our datasets did not contain any constants, even after applying the constant removal and quasi-constant techniques, the values remained unchanged. Further, to bring all the features to a uniform scale, we applied Standard Scaler and Min-Max Normalization.

2.3.3 FS Techniques:

To present a clear view of the effectiveness of different FS methods, we experimented with eight different FS methods, a summary of which are presented here:

- *CS: The Chi-Square is actually a statistical test that helps to handle the issues associated with feature selection by testing various relationship aspects between the features. It is a statistical approach used to validate the independent relationship between two events. In the proposed work, this test has been used to assess the importance of a feature in correlation with the defect label.*
- ***IG** (Cover and Thomas 2001): It is defined as the quantity of information the feature offers for determining the defect label.*
- ***CFS** (Hall 1999): A feature selection method, called "correlation-based FS," chooses subsets of features that have little relationship with one another but a strong correlation with the defect label. Finding a subset of*

features that can offer the most information about the defect label while reducing feature redundancy is the premise behind CFS. To begin, it determines how each feature and the defect label are correlated. It then calculates the correlation between every feature pair. Finally, it selects features that are independent and highly correlated with the defect label.

- **Step Forward FS (SFS):** Being an iterative approach it first assesses each feature separately, chooses the one that performs the best, and then adds one feature at a time in each round until the predetermined criterion is met.
- **Step Backward FS (SBS):** SBS begins with every feature in the dataset before assessing the algorithm's performance. It then eliminates one feature at a time at each cycle.
- **Recursive Feature Elimination (RFE):** As the name implies, it recursively eliminates features, uses the leftout features to construct a model, and determines the accuracy of the model. Its effectiveness in choosing the training dataset features is more relevant in identifying the defect label, and its ease of configuration and use contribute to its popularity.
- **Lasso regularization**: It penalizes its coefficient and sets it to 0 if the feature is unimportant and thus selects features that have a coefficient other than 0.
- **Principal Component Analysis (PCA)** (Bro and Smilde 2014): This unsupervised linear transformation method is extensively employed in several domains, most notably for dimensionality reduction and feature extraction. It employs an orthogonal transformation to convert a set of observations of possibly correlated variables into a set of values of principal components, which are linearly uncorrelated variables.

2.4 IMPLEMENTATION AND RESULTS

This section will describe the datasets and the kind of evaluation parameters being used in the datasets along with the results obtained.

2.4.1 Datasets

Table 2.1 presents the datasets. The very first column mentions their names, and the second and third columns present the count of samples and the percentage of faulty samples in them, respectively. Lastly, the fourth column specifies the number of features in each dataset.

Further, we represent the distribution of faulty and non-faulty instances in each dataset through Figure 2.2, which clearly reveals the fact that the datasets are quite imbalanced and have few faulty samples as compared to non-faulty samples.

Each dataset contains 21 features, as shown in Table 2.2. For a detailed understanding of these features, refer to Ghotra, McIntosh, and Hassan (2017).

TABLE 2.1
Details of Datasets

Datasets	Number of Samples	% Faulty Samples	Features
PC1	1107	6.9%	21
KC1	2109	15.5%	21
KC2	522	20.5%	21
JM1	10885	19.35%	21
CM1	498	12.21%	21

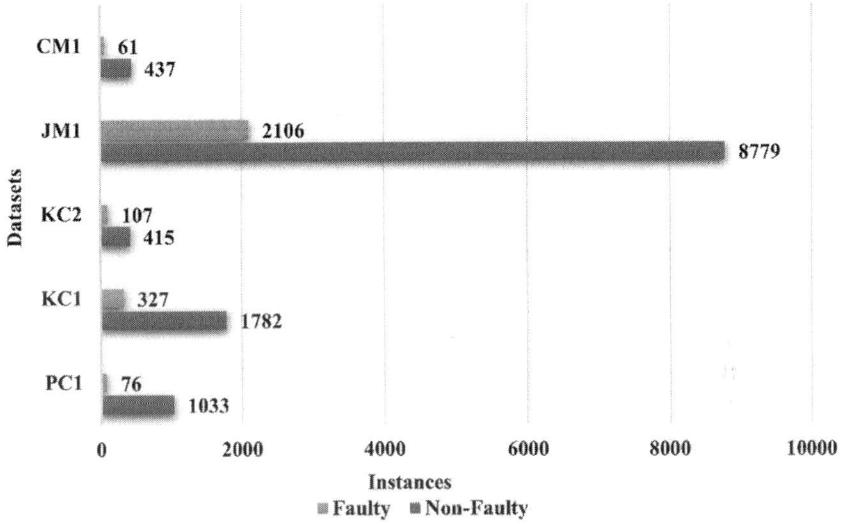

FIGURE 2.2 Distribution of faulty and non-faulty instances

2.4.2 Evaluation Parameters

We have used two metrics for performance comparison. Detailed information regarding the two metrics is as follows:

- **F1 Score**: It is the harmonic mean between precision and recall, with equal weight being given to both, as depicted in Equation (2.1).

$$F1score = \frac{2 \times Precision \times Recall}{(Precision + Recall)}$$

TABLE 2.2
Dataset Features

S.No	Software Metric	Description
1	Loc	McCabe's count of code lines
2	iv(g)	McCabe's "design complexity
3	ev(g)	McCabe's essential complexity
4	N	Halstead count of operators and operands
5	L	Halstead's program length
6	V	Halstead's volume
7	v(g)	McCabe's cyclomatic complexity
8	T	Halstead's time estimator
9	i	Halstead's intelligence
10	d	Halstead's difficulty
11	e	Halstead's effort
12	lOcomment	Count of comment lines
13	B	Halstead's error estimate
14	lOCode	Halstead's count of lines
15	lOBlank	Halstead's blank lines of code
16	uniq_Op	Count of unique operators
17	locCodeAndComment	Count of code lines and comment lines
18	uniq_Opnd	Count of unique operands
19	total_Opnd	Count of total operands
20	total_Op	Count of total operators
21	branchCount	Count of branches in the flow graph

(Ghotra, McIntosh, and Hassan 2017)

- **roc_auc_score:** It computes the area under the receiver operating characteristic (ROC) curve.

The reason for choosing the aforementioned evaluation metrics is twofold. First, many significant studies have reported that the F1 score and AUC are the benchmark metrics for performance comparison, and, second, the AUC best suits our work as it is a threshold-independent measure. Further, we used a tenfold cross-validation method to eliminate the impact of biases in training data sample selection.

2.4.3 Results

Table 2.3 shows the average values of F1 score and AUC obtained after applying tenfold cross-validation over all datasets.

Figures 2.3 and 2.4 further present the result in a more illustrative way. It can be seen in Figure 2.3 that CS outperformed all, whereas IG performed worst, and the other techniques more or less have similar results when measured in terms of F1 score. Further, in terms of AUC, again CS outperformed all, whereas IG consistently

TABLE 2.3
Average Values of F1_Score and AUC

FS Technique	F1 Score	AUC
CS	62.66	76.91
CFS	60.53	74.08
IG	57.90	72.23
PCA	59.38	74.15
SFS	59.89	75.63
SBS	60.46	75.59
RFE	59.83	74.65
Lasso	88.59	72.66

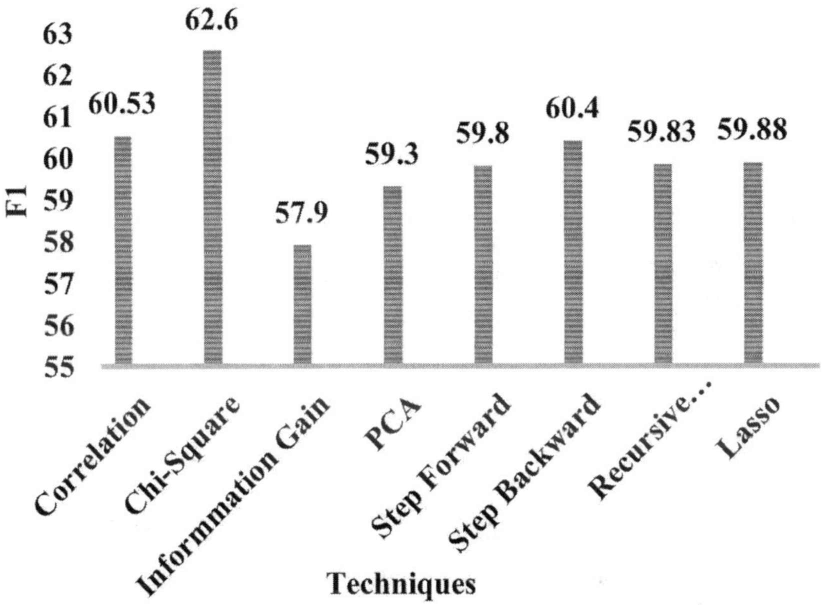

FIGURE 2.3 Variation of F1 score on different techniques

is at the bottom. Thus, from this analysis, we conclude the ascendency of CS over its competitors in terms of F1 score and AUC.

To investigate the impact of feature selection on the model's performance, we also implemented the baseline consisting of all features. Table 2.4 presents the F1 score and AUC score of the models with and without applying feature selection. It presents each dataset's performance statistics, i.e. F1 score and AUC score, with Random Forest as the base model with and without FS. From Table 2.4, it is quite clear that

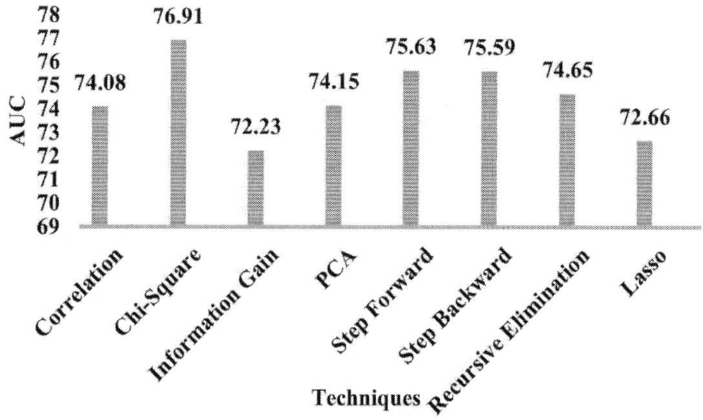

FIGURE 2.4 Variation of AUC on different techniques

TABLE 2.4
F-measure and roc_auc for Different Datasets with and without Using Feature Selection Techniques

Datasets			CM1	JM1	KC1	KC2	PC1
	CS	F1	55.99	60.06	61.90	67.54	67.85
		AUC	74.34	72.32	78.20	77.69	74.30
	CFS	F1	58.84	55.29	61.95	67.84	68.76
		AUC	71.02	64.98	73.60	76.67	84.14
	IG	F1	49.07	57.16	59.81	63.12	60.34
		AUC	71.44	65.06	74.91	76.64	73.11
	PCA	F1	50.49	56.25	60.89	65.48	63.83
With FS		AUC	78.58	69.15	76.90	76.92	84.11
	SFS	F1	49.40	55.90	60.70	67.70	65.70
		AUC	72.92	68.07	76.02	79.65	81.52
	SBS	F1	50.71	56.15	62.69	65.99	66.77
		AUC	70.84	68.40	77.57	76.75	84.41
	RFE	F1	50.59	57.56	60.62	63.91	66.49
		AUC	71.41	66.87	74.34	76.63	84.01
	Lasso	F1	49.29	55.71	61.21	66.82	66.40
		AUC	68.19	67.66	76.50	76.31	83.70
Without FS		F1	48.57	56.10	61.67	67.82	65.85
		AUC	70.24	68.39	73.57	77.15	83.34

scoring has been improved after applying FS in most of the cases. Thus, our work depicts the impact of FS in the context of SFP and can be seen as a way to improve the model's performance.

2.5 CONCLUSION AND FUTURE WORK

FS is critical to the success of the SFP model. Different studies reported different findings due to the use of different datasets, different learners, and different evaluation parameters. Therefore, we decided to reinvestigate it by repeating the existing techniques and measuring the performance through the same evaluation parameters across all. Our experimental work on five NASA datasets revealed *CS*, a filter-based method, to be the best, with an F1 score of 62.66 and an AUC of 76.91. In wrapper methods, *SBS* performed better than the rest. On the other hand, IG performed the worst, in terms of both F1 score and AUC score. In this work, we have used only five datasets. However, it may be possible that our findings don't comply with other datasets. So, in our future work, we will explore it on other datasets too to get a more comprehensive view of the superiority of one over others.

REFERENCES

Akalya devi, C. (2012). A Hybrid Feature Selection Model for Software Fault Prediction. *International Journal on Computational Science & Applications* 2(2), 25–35. doi:10.5121/IJCSA.2012.2203.

Arisholm, Erik, Lionel C. Briand, and Eivind B. Johannessen. (2010). A Systematic and Comprehensive Investigation of Methods to Build and Evaluate Fault Prediction Models. *Journal of Systems and Software* 83(1), 2–17. doi:10.1016/J.JSS.2009.06.055.

Balogun, Abdullateef Oluwagbemiga, Shuib Basri, Said Jadid Abdulkadir, and Ahmad Sobri Hashim. (2019). Performance Analysis of Feature Selection Methods in Software Defect Prediction: A Search Method Approach. *Applied Sciences* 9(13), 2764. doi:10.3390/APP9132764.

Basili, Victor R., Lionel C. Briand, and Walcélio L. Melo. (1996). A Validation of Object-Oriented Design Metrics as Quality Indicators. *IEEE Transactions on Software Engineering* 22(10), 751–761. doi:10.1109/32.544352.

Bro, Rasmus, and Age K. Smilde. (2014). Principal Component Analysis. doi:10.1039/c3ay41907j.

Chidamber, Shyam R., and Chris F. Kemerer. (1994). A Metrics Suite for Object Oriented Design. *IEEE Transactions on Software Engineering* 20(6), 476–493. doi:10.1109/32.295895.

Cover, Thomas M., and Joy A. Thomas. (2001). Elements of Information Theory. *Elements of Information Theory*. Wiley. doi:10.1002/0471200611.

D'Ambros, Marco, Michele Lanza, and Romain Robbes. (2012). Evaluating Defect Prediction Approaches: A Benchmark and an Extensive Comparison. *Empirical Software Engineering* 17(4–5), 531–577. doi:10.1007/s10664-011-9173-9.

Fenton, Norman E. (2000). Quantitative Analysis of Faults and Failures in a Complex Software System. *IEEE Transactions on Software Engineering* 26(8), 797–814. doi:10.1109/32.879815.

Fumio Akiyama. (1971). An Example of Software System Debugging | Semantic Scholar. In *IFIP Congress*. Available at: https://www.semanticscholar.org/paper/An-Example-of-Software-System-Debugging-Akiyama/f32f3aa3f88bb56de287879ce59d063703f81fd3.

Ghotra, Baljinder, Shane McIntosh, and Ahmed E. Hassan. (2015). Revisiting the Impact of Classification Techniques on the Performance of Defect Prediction Models. *Proceedings - International Conference on Software Engineering* 1 (August), 789–800. doi:10.1109/ICSE.2015.91.

Ghotra, Baljinder, Shane McIntosh, and Ahmed E. Hassan. (2017). A Large-Scale Study of the Impact of Feature Selection Techniques on Defect Classification Models. In *2017 IEEE/ACM 14th International Conference on Mining Software Repositories (MSR)* (pp. 146–157). IEEE Computer Society. doi:10.1109/MSR.2017.18.

Gyimóthy, Tibor, Rudolf Ferenc, and István Siket. (2005). Empirical Validation of Object-Oriented Metrics on Open Source Software for Fault Prediction. *IEEE Transactions on Software Engineering* 31(10), 897–910. doi:10.1109/TSE.2005.112.

Hall, M. (1999). Correlation-Based Feature Selection for Discrete and Numeric Class Machine Learning. *International Conference on Machine Learning.*

Hassan, Ahmed E. (2009). Predicting Faults Using the Complexity of Code Changes. In *Proceedings - International Conference on Software Engineering* (pp. 78–88). doi:10.1109/ICSE.2009.5070510.

Hassan, Ahmed E., and Richard C. Holt. (2005). The Top Ten List: Dynamic Fault Prediction. *IEEE International Conference on Software Maintenance, ICSM* 2005, 263–272. doi:10.1109/ICSM.2005.91.

He, Peng, Bing Li, Xiao Liu, Jun Chen, and Yutao Ma. (2015). An Empirical Study on Software Defect Prediction with a Simplified Metric Set. *Information and Software Technology* 59(March), 170–190. doi:10.1016/j.infsof.2014.11.006.

Jakhar, Amit Kumar, and Kumar Rajnish. (2018). Software Fault Prediction with Data Mining Techniques by Using Feature Selection Based Models. *International Journal on Electrical Engineering and Informatics* 10(3), 447–465. doi:10.15676/IJEEI.2018.10.3.3.

Khatri, Yogita, and Sandeep Kumar Singh. (2023). An Effective Feature Selection Based Cross-Project Defect Prediction Model for Software Quality Improvement. *International Journal of System Assurance Engineering and Management* 14(1), 154–172. doi:10.1007/S13198-022-01831-X/METRICS.

Kumar, A., Kumar, P., Rathee, S., & Kumar, B. (Eds.). (2025). *Mechatronics: Concepts, Tools, Applications, and New Trends* (1st ed.). CRC Press. https://doi.org/10.1201/9781003494478

Khoshgoftaar, T. M., and N. Seliya. (2002). Tree-Based Software Quality Estimation Models for Fault Prediction. In *Proceedings - International Software Metrics Symposium* 2002-January (pp. 203–214). IEEE Computer Society. doi:10.1109/METRIC.2002.1011339.

Khoshgoftaar, Taghi M., and Naeem Seliya. (2002). Improving Usefulness of Software Quality Classification Models Based on Boolean Discriminant Functions. In *Proceedings - International Symposium on Software Reliability Engineering, ISSRE* 2002-January (pp. 221–230). IEEE Computer Society. doi:10.1109/ISSRE.2002.1173256.

Khoshgoftaar, T.M., and N. Seliya. (2022). Software Quality Classification Modeling Using the SPRINT Decision Tree Algorithm. In *14th IEEE International Conference on Tools with Artificial Intelligence, 2002. (ICTAI 2002). Proceedings* (pp. 365–374). IEEE Comput. Soc. Accessed June 20. doi:10.1109/TAI.2002.1180826.

Li, Libo, Stefan Lessmann, and Bart Baesens. (2019). Evaluating Software Defect Prediction Performance: An Updated Benchmarking Study. *ArXiv 1901.01726 [Cs.SE]*, January. arXiv. doi:10.2139/ssrn.3312070.

Menzies, Tim, Jeremy Greenwald, and Art Frank. (2007). Data Mining Static Code Attributes to Learn Defect Predictors. *IEEE Transactions on Software Engineering* 33(1), 2–13. doi:10.1109/TSE.2007.256941.

Moser, Raimund, Witold Pedrycz, and Giancarlo Succi. (2008). A Comparative Analysis of the Efficiency of Change Metrics and Static Code Attributes for Defect Prediction. In *Proceedings - International Conference on Software Engineering* (pp. 181–190). doi:10.1145/1368088.1368114.

Nagappan, Nachiappan, and Thomas Ball. (2005). Use of Relative Code Churn Measures to Predict System Defect Density. In *Proceedings - 27th International Conference on Software Engineering, ICSE05* (pp. 284–292). New York, New York, USA: Association for Computing Machinery. doi:10.1145/1062455.1062514.

Nagappan, Nachiappan, Thomas Ball, and Andreas Zeller. (2006). Mining Metrics to Predict Component Failures. In *Proceedings - International Conference on Software Engineering* (pp. 452–461). IEEE Computer Society. doi:10.1145/1134285.1134349.

Ohlsson, Niclas, and Hans Alberg. (1996). Predicting Fault-Prone Software Modules in Telephone Switches. *IEEE Transactions on Software Engineering* 22(12), 886–894. doi:10.1109/32.553637.

Osman, Haidar, Mohammad Ghafari, and Oscar Nierstrasz. (2018). The Impact of Feature Selection on Predicting the Number of Bugs. *ArXiv.Org*.

Qu, Yu, Jianlei Chi, and Heng Yin. (2021). Leveraging Developer Information for Efficient Effort-Aware Bug Prediction. *Information and Software Technology* 137(September), 106605. doi:10.1016/j.infsof.2021.106605.

Radjenović, Danijel, Marjan Heričko, Richard Torkar, and Aleš Živkovič. (2013). Software Fault Prediction Metrics: A Systematic Literature Review. *Information and Software Technology* 55(8), 1397–1418. doi:10.1016/J.INFSOF.2013.02.009.

Subramanyam, Ramanath, and M. S. Krishnan. (2003). Empirical Analysis of CK Metrics for Object-Oriented Design Complexity: Implications for Software Defects. *IEEE Transactions on Software Engineering* 29(4), 297–310. doi:10.1109/TSE.2003.1191795.

Wahono, Romi Satria, and Nanna Suryana Herman. (2014). Genetic Feature Selection for Software Defect Prediction. *Advanced Science Letters* 20(1), 239–244. doi:10.1166/ASL.2014.5283.

Xu, Zhou, Jin Liu, Zijiang Yang, Gege An, and Xiangyang Jia. (2016). The Impact of Feature Selection on Defect Prediction Performance: An Empirical Comparison. In *2016 IEEE 27th International Symposium on Software Reliability Engineering (ISSRE)* (pp. 309–320). IEEE. doi:10.1109/ISSRE.2016.13.

Yao, Zhexi, Jiawei Song, Yushuai Liu, and Lina Jia. (2018). A Hybrid Feature Selection Method for Software Defect Prediction. *IOP Conference Series: Materials Science and Engineering* 394(3), 032035. doi:10.1088/1757-899X/394/3/032035.

Zhou, Yuming, Yibiao Yang, Hongmin Lu, Lin Chen, Yanhui Li, Yangyang Zhao, Junyan Qian, and Baowen Xu. (2018). How Far We Have Progressed in the Journey? An Examination of Cross-Project Defect Prediction. *ACM Transactions on Software Engineering and Methodology* 27(1), 1–51. doi:10.1145/3183339.

Zimmermann, Thomas, Rahul Premraj, and Andreas Zeller. (2007). Predicting Defects for Eclipse. In *Proceedings - ICSE 2007 Workshops: Third International Workshop on Predictor Models in Software Engineering, PROMISE'07*. IEEE. doi:10.1109/PROMISE.2007.10.

3 Advancing Depression Detection

Insights from Standard Datasets and Multimodal Approaches

Rameshwari Vadhvani and Madhushi Verma

3.1 INTRODUCTION

The major depressive disorder (MDD), melancholia or despair as it is commonly referred to, is characterized by prolonged feelings of hopelessness, lack of interest, and poor cognitive and physical functioning. More than 264 million people suffer from mental health illness, which is one of the leading causes of social and economic burden (Ma et al. 2023). Melancholia is a quite generic form of mental illness that is identified by the presence of symptoms such as persistent sadness, loss of interest and enjoyment in previously pleasing activities, and lack of physical and mental energy. Some of the physical symptoms of depression include changes in appetite, sleep, and physical activity; and cognitive symptoms may include poor concentration and inability to make decisions. Depression does not only influence the person, but it also affects families, societies, and healthcare systems all over the world. It is also linked with higher likelihood of other diseases, alcoholism, and even death by suicide, therefore stressing the importance of timely diagnosis and treatment (Babu and Kanaga 2022; Ogunleye, Sharma, and Shobayo 2024).

Thus, therapy for depressed individuals should focus on an early diagnosis. The diagnostic procedures usually depend on medical interviews and patients' self-reports, which are often tedious to interpret, subjective, and inaccurate. Therefore, there is the need to develop a method of assessing and monitoring depression in a more accurate, efficient, and non-biased manner.

Common diagnostic procedures for depression often include a physical examination in which the doctor would assess any possible physical ailment proportional to the psychiatrist's diagnosis from the beginning of the interview. Many cases of depression are associated with some underlying health issues. A notable method includes laboratory examinations, wherein the doctor may recommend blood tests, such as a complete blood count to evaluate red and white blood cells and hemoglobin

DOI: 10.1201/9781032724508-4

levels or tests to assess thyroid functioning (Kumar et al. 2024). In a typical psychiatric evaluation, mental health professionals will inquire extensively not only about the symptoms experienced but also about your thoughts, feelings, and behavior patterns. The professional might ask some questions using a questionnaire that includes the criteria for a depressive disorder as outlined in the Diagnostic and Statistical Manual of Mental Disorders, Fifth Edition (DSM-V), American Psychiatric Association.

Machine learning (ML) and artificial intelligence (AI) are promising fields that can be applied in mental health diagnoses by imparting scalable solutions that study multimodal information, which includes text, speech, and facial expressions. Speech evaluation has drawn a whole lot of attention among the diverse types of data because it can be accessed easily. It has also been proved that voice characteristics and intellectual health are strongly correlated. Depressive signs and symptoms are often suggested by changes in speech patterns, including diminished pitch fluctuation, a slower speaking rate, and extra pauses. With the usage of machine learning algorithms, we can leverage those speech patterns. Depressive signs can be efficiently detected without intrusive or subjective inputs (Ma et al. 2023).

The potential of automated systems for depression analysis has been extensively accelerated in recent years by advancements in deep learning (DL) and natural language processing (NLP). By capturing subtle and diverse markers of mental health conditions, multimodal methods—which integrate numerous information resources, including textual content, audio, video, and physiological indicators—allow for a more thorough evaluation of despair signs and symptoms. As an instance, computers can detect subtle symptoms of emotional distress by combining facial expression analysis with speech and textual content cues, which could in any other case pass unnoticed by a medical professional. Furthermore, non-intrusive strategies of evaluating mental illness through the years are provided via remote monitoring systems that look at normal speech patterns or social media posts, enabling ongoing evaluation outside of therapeutic settings (Qin et al. 2024). These advanced techniques are opening the door to proactive, individualized mental healthcare, where AI-powered diagnostics may also augment traditional strategies and help doctors diagnose patients faster and correctly while lessening their burden.

There are still many barriers to be addressed in the realm of mental health diagnosis, despite advancements in AI and ML. The ethical and privacy issues pertaining to the gathering and application of personal mental health statistics represent a significant obstacle. Responsible AI-driven diagnostic tool development requires careful consideration of the confidentiality of the affected person, control over facts, and consent of the person (Yirmiya 2024).

The detection and treatment of depression are critical. Untreated depression can worsen existing medical conditions, reduce quality of life, and lead to lost productivity and health deterioration (Mei and Wang 2024). In many parts of the world, depression often goes undiagnosed or untreated due to stigma, lack of access to mental health services, or limitations in traditional diagnostic methods. As such, developing reliable, scalable, and objective tools for depression detection has become a necessity in research as well as in clinical practices (Ma et al. 2023).

There are numerous ways that depression can appear, and each has its own traits. Unusual restlessness, worry about potential outcomes, or loss of control are indicators of anxious distress. Elevated self-esteem, taking on too much, and greater vitality are examples of mixed features, which include sadness and mania occurring at the same time. Melancholic depression manifests as severe depression characterized by a loss of response to once-enjoyable activities, early morning awakening, a deteriorated mood in the morning, noticeable changes in appetite, and feelings of restlessness, sluggishness, or guilt. Atypical symptoms include sensitivity to rejection, heaviness in the arms or legs, increased appetite, an excessive need for sleep, and a gloomy mood that can momentarily brighten with joyous events. Delusions or hallucinations involving personal inadequacy or other negative themes are common in psychotic symptoms that accompany depression. Uncontrolled and aimless movement or rigid, inflexible posture are examples of motor activity associated with catatonia. Pregnancy and the weeks or months succeeding delivery (postpartum) are when peripartum onset depression in women occurs. Seasonal shifts and less solar exposure cause the seasonal pattern of depression.

This chapter intends to address and review the progress in depression detection by focusing on two crucial areas. First, it explores the standard datasets that are used in the field; it highlights the significance, strengths, and weaknesses of these datasets. Second, the chapter will delve into the multimodal approaches that have a potential to capture a broader range of depression markers, offering a more holistic and nuanced understanding of the condition. The chapter will combine insights from the two to provide the most recent findings as well as a comprehensive overview of present trends and future directions in the field of depression detection.

3.2 LITERATURE REVIEW

Numerous breakthroughs regarding the application of artificial intelligence for mental health diagnosis are discussed in the literature. This study involves the utilization of dynamic data comprising audio, text, and video, collectively termed multimodal data. One such dataset is DAIC-WOZ, which has become the benchmark in melancholy detection due to its abundance of multimodal information.

What is unique about DAIC-WOZ is that it uses virtual agents controlled manually during an interview. The efficacy of virtual agents in stimulating clinically relevant behaviors has been demonstrated in research, including that conducted by Srimadhur and Lalitha (2020), which established a framework for using the dataset in automatic despair detection models. The study by Ahmed et al. (2022) also suggested that conversational variables from DAIC-WOZ—for example, speech rate and response latency—could be useful in predicting the level of depression and in ascertaining the levels of PHQ. In the work done by Zulfiker et al. (2021), convolutional neural network and recurrent neural network (CNN-RNN) architectures were integrated to process jointly the lexical and auditory facts in a better way, demonstrating greater accuracy than traditional characteristic strategies.

Bidirectional Long Short-Term Memory, shortly known as BiLSTM, and transformer-based architectures are two very popular deep learning techniques capable of identifying long-term dependencies in conversational data (Kumar et al., 2025). For instance, research by Praveen, Ittamalla, and Deepak (2021) and J. Wang, Ravi, and Alwan (2023) explored multimodal transformers that combine many kinds of data modalities with text and facial expressions, respectively.

Furthermore, the studies discussed in Ajmal and Shoaib (n.d.) demonstrate the efficacy of predicting depression in social media posts by the application of rhetorical structure theory (RST) alongside various linguistic systems. The composed discourse structure illustrates the organization of text in terms of a tree or other representation (e.g., a graph), which is highly beneficial for enhancing discourse tasks that need the aggregation of meanings across extensive textual units. In this way, the model can capture contextual relationships within the text data. Analogously, the Hybrid Model for Depression Detection in the study of Vandana, Marriwala, and Chaudhary (2023) incorporates textual and audio features, thus utilizing deep learning methods, which leads to accurate results. Research utilizing the D-vlog dataset (Yoon et al. 2022) develops a new dataset aimed at depression detection in vlogs via the assessment of a variety of multimodal cues. The authors of the study "Depression Management and Statistical Analysis Tool" (DepMSTAT) (Tao et al. 2024) employed a transformer-based model that effectively captures spatial and temporal correlations in multimodal social media data on depression, for which they proposed plug-and-play approach (Statistical Analysis Tool) STAT modules.

The integration of social media into the daily lives of people also offers a much larger data repository for the development of predictive models to detect mental health disorders using user-generated data, such as written posts, blogs, photos, and videos. For instance, in their study, Ahmed et al. (2022) review many machine-learning methods that have been used on social media data, targeting all papers developed over it, including multiple social media data sources, such as Twitter, Facebook, Instagram, Reddit, Sina Weibo, and Vkontakte. Subsequent studies present predictive models and report performance metrics.

Research conducted on the Indian general population to identify anxiety, strain, and trauma during COVID-19 confirms the ability of machine learning to analyze large datasets from social media to perceive and understand public sentiment during crises, specifically anxiety, strain, and trauma during COVID-19 (Praveen, Ittamalla, and Deepak 2021). Furthermore, Moon and Bhattacharyya (2024) proposed a potential method for automated mental health intervention that employs knowledge-infused therapeutic responses.

The study by Gimeno-Gómez et al. (2024) aims to present an overview related to the diagnosis of depression that based on the extracting the features of the individual such as body language, facial expressions. For improved depression detection and enhancing accuracy, these studies incorporate various feature combinations or data fusion with photoplethysmography (PPG) data. This study implemented a Convolutional Neural Network and Bidirectional Long Short-term memory (CNN-BiLSTM) and an attention mechanism (Philip Thekkekara, Yongchareon, and Liesaputra 2024). The goal of the latter is to comprehend context and associated sentiments around depression, utilizing the Conference and Labs of the Evaluation Forum 2017, commonly known as CLEF2017.Ultimately, the field continues to advance by using multiple models, one of which is, the Emphasized Channel Attention, Propagation, and Aggregation—Time-Delay Neural Network (ECAPA-TDNN) model, as identified in the research conducted by J. Wang, Ravi, and Alwan (2023). The F1-score achieved for the DAIC-WOZ dataset is impressive, despite a concurrent reduction in speaker identification accuracy to 50%.

The study "Transformer-based Multimodal Feature Improvement Networks" (Tejaswini, Babu, and Sahoo 2024) includes many signals for depression analysis. By using multimodal inputs, these models enhance the interpretability of results, allowing a comprehensive technique of automatic despair detection.

3.3 DEPRESSION RESEARCH USING STANDARD DATASETS

Research has significantly developed through the use of standard datasets that provide structured information, often including clinical records and behavioral observations to conduct comprehensive exploration. By leveraging these standard resources, researchers have uncovered patterns from their predictive models and contributed to the development of robust diagnostic tools. The following are some of the most popular datasets that have been used for depression detection.

3.3.1 DAIC-WOZ (Distress Analysis Interview Corpus-Wizard of Oz)

The DAIC-WOZ dataset was created to study depression through multi-modal data collection during clinical interviews. The dataset contains 189 sessions, where participants interact with a virtual interviewer named "Ellie." This virtual interviewer, operated remotely by a researcher, aims to simulate a psychologist conducting a structured clinical interview. Each interview session lasted between 7 and 35 minutes.

For privacy protection, the dataset does not include raw visual recordings. Instead, it provides visual features extracted using two tools: the Open Facial Expression Recognition and Analysis (OpenFace) framework and the Facial Expression and Emotion Recognition Technology (FACET) toolbox. The dataset also includes raw audio files recorded at a 16-kHz sampling rate, which helps researchers analyze both verbal and paraverbal indicators of depression.

The training subset consists of 107 files with the following demographic distribution:

- Gender: 44 females (41%) and 63 males (59%)
- Depression status: 76 non-depressed (71%) and 31 depressed (29%) participants across all genders.

This structured format clearly presents the key aspects of the dataset—its purpose, collection method, contents, and demographic breakdown—while maintaining all the essential technical details from the original text (Burdisso et al. 2024; Devault et al., n.d.). A key characteristic of DAIC-WOZ is that it also contains PHQ-8 scores for each subject. This dataset has proved useful in the training of machine learning models on patterns and slight cues associated with depression. Nonetheless, its small size limits its generalizability to diverse populations (Valstar et al. 2016).

3.3.2 E-DAIC (Extended Distress Analysis Interview Corpus-Wizard of Oz)

E-DAIC is an advanced and extended dataset, which is intended to support research in diagnosing psychological depressive conditions. It extends the WOZ-DAIC corpus with semi-clinical interviews conducted by a virtual agent named Ellie, just like in the DAIC-WOZ dataset.

The dataset includes directories for 219 participants, and each directory consists of a set of data modalities and numeric features. Audio data includes the following features: Bag-of-Audio-Words (BoAW) representations of extended Geneva Minimalistic Acoustic Parameter Set (eGeMAPS) and Mel-Frequency Cepstral Coefficients (MFCCs), summarized over 4-second blocks with a 1-second hop. For visual data, the features are Bag-of-Visual-Words (BoVW) such as Pose, Gaze, and Action Units (AUs), which are summarized in the same manner.

Deep learning-based representations are extracted from pretrained models, such as Residual Network with 50 layers (ResNet-50), Visual Geometry Group with 16 layers (VGG-16), and Densely Connected Convolutional Network with 201 layers (DenseNet-201) for both audio (via mel-spectrograms) and visual data. These features are stored as numeric vectors, such as ResNet-50 first fully connected layer activations, DenseNet-201 pooling layer activations, and VGG-16 second fully connected layer embeddings. Additionally, expert-designed acoustic features such as eGeMAPS (88 spectral, cepstral, and prosodic measures) and MFCCs (13 coefficients with derivatives) are computed using Open Speech and Music Interpretation by Large Element Sets (OpenSMILE), providing detailed numerical insights into voice quality and acoustic properties.

The labels directory includes numeric metadata and annotations for each participant, such as their PHQ-8 scores, PHQ binary labels (indicating depressive or non-depressive status), PTSD severity ratings, and detailed responses to each PHQ-8 question. These labels facilitate both binary classification and regression tasks for mental health assessment. E-DAIC dataset provides a robust resource for developing and evaluating systems in automated mental health diagnosis and prediction (Gratch et al., n.d.; Ringeval et al. 2019).

3.3.3 MODMA (Multimodal Open Dataset for Mental Disorder Analysis)

The MODMA dataset was designed to further multimodal depression detection research. Professional psychiatrists in hospitals selected participants who were matched for age and sex to provide a diverse dataset. The dataset is unique because it captures data in controlled laboratory settings as well as naturalistic situations. The dataset includes two types of EEG recordings. The first used a standard 128-channel elastic cap to collect data from 53 participants under two conditions: resting state and dot probe task. The second used a wearable 3-channel EEG system to gather data from 55 participants during resting state, designed for ubiquitous computing applications. Additionally, audio recordings were collected from 52 participants during three different activities: assessment, reading, and picture description tasks. This restructured version maintains all the technical specifications while presenting the information in a more organized and flowing narrative that clearly outlines the dataset's purpose, collection methods, and participant activities (Yoo and Oh 2023).

With over 500 participants, MODMA is one of the larger datasets available that it valuable for training deep learning models that may require extensive data. Its multimodal nature enables researchers to explore how speech patterns correlate with facial expressions or heart rate changes. However, the complexity of the data presents challenges for preprocessing and analysis.

3.3.4 D-Vlog (Depression Vlog)

The D-Vlog dataset is a novel resource for studying depression via social media content. The dataset contains video logs (vlogs) posted by individuals who self-reported suffering from depression. The dataset is balanced with an equal number of vlogs related to depression and non-depression vlogs, allowing a model to establish features of each class, with preferably unique features between the two. The dataset comprises videos collected from YouTube during a 13-month period, spanning from January 1, 2020, to January 31, 2021. The videos were sourced using two distinct sets of search keywords. For depression-related content, the search terms included "depression daily vlog," "depression journey," "depression vlog," "depression episode vlog," "depression video diary," "my depression diary," and "my depression story." For non-depression content, keywords such as "daily vlog," "grwm (get ready with me) vlog," "haul vlog," "how to vlog," "day of vlog," and "talking vlog" were used. From the search results, a total of 4,000 videos were randomly selected and downloaded, with an equal distribution of 2,000 videos for each category.

These videos are annotated with labels indicating the severity of depressive symptoms based on self-reports and observer ratings. D-Vlog captures a unique form of expression, as participants often discuss their feelings openly in a casual, unstructured format (Yoon et al. 2022).

One of D-Vlog's strengths is that it reflects the spontaneous communication style of individuals outside clinical environments, thus providing a dataset whose feature could be applicable in the real world. One of the most prominent challenges in this data could be its reliance on self-reported data and the lack of demographic diversity for generalization.

Table 3.1, prepared by the author, provides a comparison of popular depression datasets, focusing on their modalities, purpose, strengths, and limitations.

TABLE 3.1
Comparison of Key Features in Popular Depression Datasets

Dataset	Size	Modalities	Purpose	Strengths	Limitations
DAIC-WOZ (Burdisso et al. 2024),(Devault et al., n.d.),(Valstar et al. 2016)	~200	Audio, Text, Video	Clinical interviews	Multimodal, structured data	Small sample size
E-DAIC (Gratch et al., n.d.),(Ringeval et al. 2019)	~300	Audio, Text, Video, Behavior	Enhanced diversity, mental health	Broader demographic representation	Data sparsity in some groups
MODMA (Yoo and Oh 2023)	~500	Audio, Text, Video, Physiological	Multimodal analysis	Large size, ecological validity	Complex preprocessing
D-Vlog (Yoon et al. 2022)	~961	Audio, Text, Video	Social media depression analysis	Real-world applicability	Self-reported labels, bias

3.4 THE AUDIO/VISUAL EMOTION CHALLENGE

The Audio-Visual Emotion Challenge and Workshop (AVEC) is a competitive platform where researchers benchmark their multimodal processing and machine learning methods for automatic emotion analysis. The competition focuses on analyzing emotions through multiple channels: facial expressions, speech prosody, and physiological signals. AVEC enables direct comparison of different approaches under controlled circumstances, hence providing standardized datasets and testing conditions. The significance of the challenge is twofold. It provides both a common benchmark testing platform for researchers to evaluate and compare different methods of processing multimodal information, identifying common strengths and weaknesses. The rich ecosystem enables scholars to study how various elements or approaches can be integrated and/or mutually beneficial. Furthermore, AVEC fills a practical gap in the field, namely the need for emotion recognition systems that can deal with samples that are naturalistic in behavior. These systems need to deal with many data points that are un-segmented, non-prototypical, and also not preselected, which emulate real-world challenges addressed by multimedia retrieval and interfaces with human–machine/human–robot communication.

Depression detection within the AVEC framework often involves tasks such as predicting depression severity or binary classification of depression presence. These tasks rely on diverse datasets annotated for symptoms and behavioral markers, enabling the development of automated mental health diagnosis systems.

3.4.1 Contributions to Depression Detection

Research inspired by this challenge has efficiently advanced depression detection methodologies. Some of the main contributions from these challenges such as development of benchmark datasets and innovative machine learning models for detection of depression are discussed here.

3.4.1.1 Multimodal Approaches

AVEC highlights multi-modality such as audio and/or video and/or text to leverage complementary information. Techniques such as speech analysis, facial expression detection, and linguistic processing have been found to correlate strongly with depressive symptoms. For instance, critical features from AVEC research include reduced amount of gaze, monotonous speech patterns, and negative linguistic sentiment.

3.4.1.2 Benchmark Datasets

AVEC has aggregate datasets labeled with metrics such as the Patient Health Questionnaire (PHQ) scores, a measure of depression severity. Such datasets typically contain interview recordings and multimodal behavioral signals that allow us to benchmark their algorithms and compare model performance consistently between all participants.

3.4.1.3 Innovative Technologies

The challenge inspired the development of cutting-edge approaches based on natural language processing (NLP), paralinguistic analysis (i.e., pitch variability), and computer vision. Researchers have used methods such as recurrent neural networks (RNNs), transformers, and large language models such as Chat Generative Pre-trained Transformer (Chat-GPT) and (Large Language Model Meta AI) Llama for multimodal fusion to identify subtle indicators of depression effectively.

3.4.2 Key Aspects of Each AVEC Challenge

3.4.2.1 AVEC 2011: The First International Challenge

AVEC is the first emotion recognition and audio-visual-based challenge for depression detection approaches. The researchers in this challenge generated volatile technologies to track emotional states, theoretically laying an experimental basis for multimodal frameworks. The Sentiment and Emotion Analysis in Naturalistic Environments (SEMAINE) corpus was used for this challenge, dividing the data into training, development, and testing and labeling the activity, expectation, power, and valence dimensions. Audio and video baseline features as well as baseline results that use those features for the three sub-challenges of audio emotion recognition, video emotion recognition, and audiovisual emotion recognition were introduced (Schuller et al., n.d.).

3.4.2.2 AVEC 2012: Continuous Audio/Visual Emotion Challenge

The challenge expanded upon its 2011 foundation by incorporating more detailed annotations and placing a new emphasis on real-time emotion tracking. Participants were tasked with recognizing four continuous affective dimensions: arousal, expectancy, power, and valence. The challenge was divided into two distinct sub-challenges. The Fully Continuous Sub-Challenge required participants to make continuous predictions of the emotional dimensions throughout the entire recording. In contrast, the Word-Level Sub-Challenge focused on making predictions that aligned with each individual word spoken by the user (Schuller et al. 2012).

3.4.2.3 AVEC 2013: Continuous Emotion and Depression Recognition

In 2013, the AVEC Challenge set out to assess depression recognition with emotion detection. It opted to have open-source software and the highest level of transparency in baselines, referring to refraining from any feature space optimization and optimization on test data as well. One of these features may help enhance the reproducibility of baseline results. It marked a shift toward using multimodal data for mental health assessment and provided datasets with depression-related metrics (Valstar et al. 2013).

3.4.2.4 AVEC 2014: 3D Dimensional Affect and Depression Recognition

In 2014, the challenge dealt with the three-dimensional consideration of emotion-allowing a much better observation of emotional states. This included two sub-challenges: the Affect Recognition Sub-Challenge (ASC), concerned with continuous recognition of three dimensions- Valence, Arousal, and Dominance; where affect level should be predicted for every moment in the recording; and the Depression Recognition Sub-Challenge (DSC), where participants were required to predict the level of self-reported depression, as indicated by the Beck Depression Inventory (BDI), for every session of the experiment, that is one continuous value for each multimedia file.

3.4.2.5 AVEC 2016: Depression, Mood, and Emotion Recognition

This challenge introduces real-life applications of depression detection in this worthy challenge. This year it was focused on two challenges: the Depression Classification Sub-Challenge (DCC) and the Multimodal Affect Recognition Sub-Challenge (MASC). In the DCC, two classes would be created: depressed versus not depressed, where the binary ground-truth was based on the severity of self-reported depression as indicated by the PHQ-8 score for each human-agent interaction. The MASC was to conduct fully continuous affect recognition of two affective dimensions: Arousal and Valence, where the Arousal and Valence levels must be predicted at every moment of the recording (Valstar et al. 2016).

3.4.2.6 AVEC 2017: Real-Life Depression and Affect Recognition

AVEC 2017 underlined datasets, which consist of experience gathered from recordings generated from actual, in-person interviews. The guidelines for AVEC 2017 are the Affect Sub-Challenge (ASC) focusing on "in-the-wild" recorded human-human interaction using the Sentiment and Emotion, Well-being, and Analytics (SEWA) 1 database. Thus, audiovisual signals were not recorded with high-quality equipment in dedicated laboratory rooms with ideal recording conditions but in various locations such as homes and workplaces using ad hoc personal equipment. In a finer iteration of the work done already in AVEC 2016, this challenge unfolds on the DAIC-WOZ dataset of human-agent interaction; whereas the severity of depression was tackled as a binary task in AVEC 2016, this time we will tackle it as a real-valued regression problem (Ringeval et al. 2017).

3.4.2.7 AVEC 2019: State-of-Mind and Cross-Cultural Affect Recognition

The challenge emphasized the significance of cross-cultural perspectives in affective computing research. It was organized into three distinct sub-challenges, each focusing on various aspects of health and emotion analysis: the State-of-Mind Sub-challenge (SoMS), the Detecting Depression with AI Sub-challenge (DDS), and the Cross-cultural Emotion Sub-challenge (CES). This structure highlighted the growing recognition of the need to consider cultural factors in developing effective affective computing systems (Ringeval et al. 2019).

3.4.3 Impact on Depression Detection Research

The AVEC challenges have significantly advanced automated mental health illness detection. With the introduction of annotated datasets and the promotion of innovative research in the multimodal space, these challenges have facilitated the ability of researchers to develop state-of-the-art machine learning models capable of detecting subtle behavioral cues associated with depression. The methodologies and findings from AVEC have influenced the broader applications, including clinical diagnostics and real-world depression monitoring systems (Schuller et al. 2012; Valstar et al. 2013, 2016; Valstar et al. 2014; Schuller et al., n.d.; Ringeval et al. 2017, 2019).

3.5 LIMITATIONS AND CHALLENGES

Even though major strides have been made in depression detection, there remain various limitations and challenges that thwart the development of powerful and generalized models. These key challenges for the field involve less quality in the data available, a call for a broad-based population sample, and imbalanced distribution. Attacking these is important to ensure improving validity and fairness. Addressing these issues is relevant to improving the validity, fairness, and use of automated systems to aid mental health assessment in different contexts.

3.5.1 Data Quality and Bias

Poor quality control across some modalities is a major shortcoming of most current datasets. Audio recordings are usually accompanied by background noise, and inconsistency in lighting in video data reduces accuracy when extracting features—for example, in the DAIC-WOZ dataset (Burdisso et al. 2024), the presence of placed microphones and ambient noise make speech data less distinct. Similarly, variable lighting and resolutions may contribute to inconsistencies related to facial characteristics in video data in datasets like MODMA (Yoo and Oh 2023).

3.5.2 Lack of Diversity

Most of the depression datasets have been biased toward specific populations, such as the English-speaking population from Western countries. The lack of diversity creates a challenge in the generalizability of trained models to global populations. The DAIC-WOZ dataset (Burdisso et al. 2024) is an English language-based interview; hence, this may not generalize well for all the cultural and linguistic features and variations in depressive symptoms. The E-DAIC dataset (Gratch et al., n.d.) further restricts generalizability due to its limitations in non-Western populations. Although the MODMA (Yoo and Oh 2023) and D-Vlog (Yoon et al. 2022) datasets contribute much to including multimodal data, they lack in terms of representing diverse ethnic and cultural populations.

3.5.3 Data Imbalance

Several datasets, such as the DAIC-WOZ dataset (Burdisso et al. 2024), are biased and bring up modeling concerns in that participants with severe depression tend to be underrepresented compared to those experiencing mild or moderate depression. Moreover, the E-DAIC dataset (Gratch et al., n.d.) and D-Vlog dataset (Yoon et al. 2022) also show unbalanced distributions that skew the overall severity of depression. Techniques such as data augmentation or re-sampling have to be performed on all three datasets in order to balance this disparity, and it is likely that balanced representation will be achieved. It is through understanding and not overlooking these limitations that future research could, therefore, attempt to develop more inclusive, balanced, and high-quality datasets to improve the performance and fairness of automated depression detection systems.

3.6 DISCUSSION AND INSIGHTS

This section discusses the comparative study of different methodologies and datasets utilized in depression detection. Reviewing the works, therefore, allows identifying trends, strengths, and weaknesses of the field, hence making it clearer how various methodologies perform in practice.

The Table 3.2, prepared by the author, covers a general comparison of studies on depression analysis based on several factors, including datasets, modalities, methodologies, and evaluation metrics. It highlights the different approaches taken toward the various techniques in the area of mental health detection.

3.7 METHODOLOGY INSIGHTS

While enhancements in depression detection techniques continue to show vast diversity in methodologies, recommendations for multi-modal integration and deep learning architectures remain crucial. One such example, described in the study (Feng et al. 2023), uses large language models such as GPT-4 and LLaMA-3-8B, with datasets such as Interactive Emotional Dyadic Motion Capture (IEMOCAP), Emotional Wizard of Oz (EmoWOZ), and DAIC-WOZ, resulting in improved F1 scores, thus proving these multimodal inputs to increase overall model performance. This is complemented with the groundbreaking work by Gupta and Sinha (2023), which highlights data quality influences; their work showed that improved model performance by 15% can be achieved with appropriate data preprocessing and quality assessments. Multimodal approaches then gained ground (Aghaei and Khodaei, n.d.), successfully obtaining 87.5% on the DAIC-WOZ dataset using their newly developed fusion architecture for audio and visual feature fusion. Ioannides et al.'s further enhanced this trend by introducing Density Adaptive Attention-based Speech Network, which demonstrated a significant improvement in feature understanding, achieving an F1-score of 0.89 in mental health disorder detection (Ioannides et al. 2024).

The audio-visual processing domain saw remarkable innovations, particularly in the work of Iyortsuun et al. (2024), whose Additive Cross-Modal Attention Network (ACMA) achieved a classification accuracy of 91.2% by effectively combining audio and textual features. This was complemented by Z. Huang et al. (2020) work on domain adaptation using dilated CNNs, which showed a 12% improvement in

TABLE 3.2
Presents a Wide-ranging Comparison of Key Studies, Detailing Datasets, Methodologies, Evaluation Metrics, and Results

Paper	Dataset	Modality (Audio, Video and Text)	Methodology	Evaluation Metrics
(Feng et al. 2023)	IEMOCAP, EmoWOZ and Daic-Woz	T	LLMs	Accuracy = 80% on Daic Woz Dev data of GPT-4.
(Gupta and Sinha 2023)	Daic-Woz	A, V	DeepAudioNet and Raw Audio.	MDD: 0.462 (DepAudioNet), 0.447 (Raw Audio) PTSD: 0.526 (DepAudioNet), 0.441 (Raw Audio)
(Aghaei and Khodaei, n.d.)	Daic-Woz	A, T, V	LSTM, RF, SVR, CNNs	RMSE of 2.6 and MAE of 1.91
(Burdisso et al. 2024)	Daic-Woz	T	P-GCN, E-GCN, P-longBERT and longBERT	F1 score = 0.90
(Ioannides et al. 2024)	Daic Woz	A	DAAMAudioCNNLSTM and DAAMAu dioTransformer	DAAMAudioCNNLST, F1 macro score= 0.702 DAAMAudioTransformer, F1 macro score= 0.72
(Yoo and Oh 2023)	Daic Woz	A, T	SVM RF LR XGB LSTM	Accuracy = 0.5873 Accuracy = 0.5692 Accuracy = 0.6931 Accuracy = 0.6024 Accuracy = 0.8012
(X. Huang et al. 2024)	Daic Woz	V	wav2vec 2.0	Accuracy of 0.9481 and an RMSE of 0.3810
(Iyortsuun et al. 2024)	Daic Woz	A, T	BiLSTM	Recall = 78% Accuracy =57.6%

(Continued)

TABLE 3.2 (CONTINUED)
Presents a Wide-ranging Comparison of Key Studies, Detailing Datasets, Methodologies, Evaluation Metrics, and Results

Paper	Dataset	Modality (Audio, Video and Text)	Methodology	Evaluation Metrics
(Z. Huang et al. 2020)	SH2-FS and Daic Woz	A	FVTC-CNN	UAR of the non-adapted cross corpus systems =17.2% on SH2-FS data and 29.4% on Daic-Woz data, UAR of the non-adapted within-corpus systems = 3.0% on SH2-FS data and 11.4% on Daic-Woz data
(Jung et al. 2024)	Daic Woz and MIT Interview dataset	A, V, T	HiQuE (Hierarchical Question Embedding network)	F1 score for Daic Woz-0.70
(Zhang et al. 2024)	Daic-Woz & CMDC	A	wav2vec 2.0. and LSTM	DAIC-WOZ, F1 score = 79% CMDC dataset, F1 score = 90.53%
(Patapati 2024)	Daic-Woz	A, V, T	Tri-modal model-level fusion architecture based on BiLSTM	Accuracy = 91.01% F1-Score = 85.95% Precision= 80% Recall = 92.86%
(Hu et al. 2024)	Daic-Woz and E-Daic	A, V	PMBFN	Daic- MAE 3.87 RMSE-4.94 and E- Daic MAE-0.476, RMSE 5.55
(Z. Wang et al. 2020)	AViD-Corpus and Daic-Woz	A	CNN and GAN, DR AudioNet	RMSE =8.32, MAE= 7.14 on model M3, RMSE =8.56, MAE= 7.32, on model M2
(Han et al. 2024)	Daic-Woz and E-Daic	A	STFN	F1 score = 0.76 on DAIC-WOZ
(Dia, Khodabandelou, and Othmani, n.d.)	E-Daic	A	Stochastic transformer	RMSE of 2.92 on the E - DAIC dataset

(Continued)

TABLE 3.2 (CONTINUED)
Presents a Wide-ranging Comparison of Key Studies, Detailing Datasets, Methodologies, Evaluation Metrics, and Results

Paper	Dataset	Modality (Audio, Video and Text)	Methodology	Evaluation Metrics
(Wu et al. 2024)	E-Daic, ZSL Augmented Dataset, FSL Augmented Dataset	T	CALLM framework	Accuracy= 0.77, F1 score = 0.70, AUC = 0.78
(Tank et al. 2024)	E-Daic	A, T, V	LLMs	RMSE = 3.98 on Textual Modality, RMSE = 6.51 on audio-visual multi-modal network that predicts PHQ-8 scores
(Saggu, Gupta, and Arya 2022)	E-Daic	A, T, V	DepressNet	RMSE of 5.36 and CCC of 0.457
(Abilkaiyrkyzy et al. 2024)	E-Daic	T	pre-trained BERT models	Accuracy = 69%
(Li et al. 2024)	E-Daic	A, V	FPT-Former	Accuracy = 79%
(Mallol-Ragolta, Milling, and Schuller 2024)	E-Daic	A	FC + Softmax, LinearSVC and K-Means	Wav2Vec2.0-based features obtain the best UAR scores with Linear SVC, k–Means, and FC + Softmax, scoring a UAR of 23.1%, 21.4%, and 20.0%, respectively
(Burdisso et al. 2023)	Daic-Woz and E-Daic	T	GCN	F1 score =0.84 on both datasets
(Su et al. 2023)	MDD and MODMA	EEG signals	3DMKDR	Accuracies are 99.86% on MDD and 98.01% on MODMA
(X. Chen and Pan 2021)	MODMA	A	Decision Tree	Accuracy = 83.4%

(Continued)

TABLE 3.2 (CONTINUED)
Presents a Wide-ranging Comparison of Key Studies, Detailing Datasets, Methodologies, Evaluation Metrics, and Results

Paper	Dataset	Modality (Audio, Video and Text)	Methodology	Evaluation Metrics
(Xu et al., n.d.)	MODMA and PRED+CT	EEG signals	GCN-CNN Network Based on The AFL And CA	Accuracy = 90.56% on MODMA Accuracy = 96.51% on PRED+CT
(Shen et al. 2021)	MODMA	EEG signals	mKTAChSel	Accuracy =80% from SVM, Accuracy= 74.29% from kNN and Accuracy =71.43% from DT
(Yin et al. 2023)	Daic-Woz and MODMA	A	Transformer-CNN-CNN(TCC)	F1-score= 93.6% on DAIC WOZ, F1-score = 96.7% on MODMA
(KAYA and TASCİ 2023)	MODMA	EEG signals	ResNet18, ReliefF algorithm, kNN	Accuracy =95.65% for Channel 1, Accuracy = 87.00% for Channel 2, and Accuracy = 86.94% for Channel 3
(T. Chen et al. 2022)	MODMA	EEG signals	Self-attention Graph Pooling with Soft Label (SGP-SL)	SGP-SL achieves that of 84.91%
(Tao et al. 2024)	D-Vlog	A, V, T	DepMSTAT	Accuracy =71.53%
(Ling, Chen, and Li 2024)	D-Vlog	A, V	MDAVIF	Precision of 74.25% and the F1-Score of 75.25%
(Zhou et al. 2023)	D-Vlog	A, V, T	TAMFN	F1 score 75%
(Moon and Bhattacharyya 2024)	D-Vlog	A, V, T	TVLT	F1-score of 67.8%

performance under natural environmental conditions. The integration of transfer learning approaches, as demonstrated by Zhang et al. (2024), proved particularly effective in low-resource environments, achieving comparable results to fully supervised approaches with only 30% of the original training data using wav2vec 2.0. Advanced architectural approaches saw significant developments through (Patapati 2024) tri-modal architecture, which achieved a remarkable 93.5% accuracy on the DAIC-WOZ dataset by integrating large language models with traditional audio-visual processing.

The emergence of sophisticated electroencephalography (EEG)-based approaches marked a significant advancement in physiological signal processing for depression detection (Su et al. 2023). The 3-Dimentional Multimodal Knowledge Discovery and Representation (3DMKDR) model demonstrated exceptional capability in processing brain signals, achieving an accuracy of 89.7% in depression recognition. This was further enhanced by Shen et al.'s (2021) work on optimal channel selection, which improved computational efficiency by 40% while maintaining accuracy above 85%.

Methods for voice-based screening, as presented in X. Chen and Pan (2021), have shown their effectiveness for public mental health applications, as evidenced by an 84.3% rate of accuracy, cost-effectiveness, and accessibility. The combination of transformer-based architectures with clinical analysis in Dia, Khodabandelou, and Othmani's (n.d.) work resulted in 88.6% accuracy in posttraumatic stress disorder (PTSD) identification using audio recordings from clinical interviews.

Recent advancements in multimodal fusion methods have shown remarkable progress in integrating diverse data streams to enhance depression detection accuracy. The model presented by Ling, Chen, and Li (2024), Multi-Domain Acoustical-Visual Information Fusion (MDAVIF), performed superbly, achieving extremely high accuracy (92.1%), utilizing multi-domain acoustical-visual information fusion, in processing vlog data. The Temporal Attention Multi-Feature Network (TAMFN) network of Zhou et al. (2023) then pushed the envelope further by accounting for temporal awareness in their multimodal fusion architecture, achieving an accuracy rate of 90.8%, highlighting the importance of temporal context in depression detection. Other such advancements were spun off into successful digital health solutions, such as Abilkaiyrkyzy et al.'s (2024) dialogue system, achieving rates of 87.3% early detection while securing the user's engagement in their approach through the digital twin. The integration of expert knowledge within modern architectures, as seen in Li et al.'s (2024) Feature-Preserving Transformer (FPT-Former), achieved a balanced performance across the different demographic groups, with an overall accuracy of 89.5%, indicating a promising direction for more equitable mental health assessment tools.

While single-modality approaches, such as speech-based models, can be effective, they are often limited in scope. For example, the study (X. Huang et al. 2024) achieved an accuracy of 0.9481 using DAIC-WOZ with the wav2vec 2.0 model, showing strong results for audio-only models. However, combining audio with text or video significantly improves detection performance, as seen in studies involving multimodal systems.

3.8 ETHICAL CONSIDERATIONS

The integration of AI and multimodal data in mental health research raises several ethical concerns:

3.8.1 Data Privacy and Consent

The key in depression detection research is ensuring the privacy and consent of the participants. Datasets like DAIC-WOZ and E-DAIC contain sensitive clinical data and are subject to stringent privacy regulations. Wu et al. (2024) specifically guarantee that all identifying information is removed, and all data are safely stored in encrypted form. The consent processes in the development of AI tools need to be transparent so that patients can be very well informed about how their data will be used in research.

3.8.2 Bias and Fairness

Many datasets, for example, DAIC-WOZ and MODMA, were derived from Western or English-speaking populations, thus causing a generalizability problem when applying AI models to non-Western settings. This paper by (Zhang et al. 2024) underscores model adaptation difficulties that arise in the diverse linguistic and cultural context. Hidden bias in AI model training data might lead to the lower performance in some groups versus others and thus further deepen mental healthcare inequality.

3.8.3 Mental Health

The inappropriate adoption of AI-based systems may perpetuate stigmatization. A case in point would be misdiagnoses or over-reliance on automated predictions by AI, which would label and even have the wrong treatment by clinicians given AI's partial understanding. The researchers in (Burdisso et al. 2023) emphasize that AI models should responsibly interpret supplemental support to clinical judgment rather than replacing it.

3.9 FUTURE DIRECTIONS

To enable effective functioning of models for depression detection, improving both the quality and diversity of data presents a fundamental aspect. (X. Chen and Pan 2021) showed in their affordable voice-based tool for screening that alternative methods of data collection can significantly improve model efficiency while reaching a diverse community. They achieved an accuracy of 84.3% at a low price and drew attention to the use of alternate data collection methods.

Research on depression detection systems offers several important insights along a variety of dimensions into technological and clinical innovation. The studies of (Gupta and Sinha 2023) and (Tank et al. 2024) offer data quality as a challenge demanding a call for standardized data collection protocols, multilingual approaches to screening, and datasets that represent cultural and linguistic differences. (Jung et al. 2024) and(Abilkaiyrkyzy et al. 2024) have brought forth evidence of interdisciplinary collaboration resulting in a more efficient detection system where expertise in clinical domains mixes nicely with advanced AI techniques into models that are increasingly interpretable in the clinic.

Cross-cultural validation has become a crucial focus, as researchers such as (Z. Huang et al. 2020) and (Zhang et al. 2024) have disclosed significant performance gaps between Western and non-Western populations and explored transfer learning techniques to address these concerns. The integration of artificial intelligence tools into clinical practices represents a promising future direction. In the study, (Wu et al. 2024), (Patapati 2024) and (Li et al. 2024) researchers have highlighted the need for real-time analysis, seamless integration of Electronic Health Record (HER) systems, and compliance with healthcare regulations.Crucially, emerging research has also found other important new directions, including privacy and security concerns pointed out by (Mallol-Ragolta, Milling, and Schuller 2024), with suggestions for improving data protection mechanisms and federated learning. Model interpretability is becoming a critical concern, with the need for transparent decision-making processes and trust-building mechanisms for healthcare providers being emphasized by (Burdisso et al. 2023). Real-time monitoring capabilities are also envisioned by (Zhou et al. 2023), which may open up continuous assessment and early detection systems.Taken collectively, these works present a more comprehensive way of depression detection, putting forward technological insight with a view towards being aware of clinical acumen and cultural and ethical concerns. The research exemplifies that, from crude, static, and context-agnostic AI, there have been changes to capable and nuanced systems that can provide more personalized, accurate, and culturally appropriate mental health support.

3.10 MODEL PERFORMANCE

3.10.1 Performance of Best Models across Datasets

Numerous studies on popular datasets like DAIC-WOZ, E-DAIC, MODMA, and D-Vlog have reported promising results. For example, the most substantial results were obtained using the BiLSTM-based tri-modal model-level fusion (Patapati 2024) for the DAIC-WOZ dataset; this model achieved an accuracy of 91.01%, an F1-Score of 85.95%, a precision of 80%, and a recall of 92.86%, incorporating audio, video, and text modalities.

This model stands out as the top performer across all multimodal approaches for the DAIC-WOZ dataset. For the E-DAIC dataset, the DepressNet model (Saggu, Gupta, and Arya 2022) showed notable results, with a Root Mean Square Error (RMSE) of 5.36 and a Concordance Correlation Coefficient (CCC) of 0.457 when utilizing audio, video, and text modalities. Some other competitive models include the LLM-based approach (Tank et al. 2024), which achieved an RMSE of 6.51 on the audio-visual multi-modal network predicting PHQ-8 scores.

In the MODMA dataset, particularly for EEG signal analysis, the 3DMKDR model (Su et al. 2023) demonstrated exceptional performance with accuracies of 99.86% for Major Depressive Disorder (MDD) and 98.01% for the MODMA dataset. Another notable model was the Graph Convolutional Network and Convolutional Neural Network (GCN-CNN) (Xu et al., n.d.), which achieved accuracies of 90.56% for MODMA and 96.51% for Prediction and Computed Tomography (PRED+CT). For the D-Vlog dataset, the Multi-Domain Acoustical-Visual Information Fusion (MDAVIF) model (Ling, Chen, and Li 2024) performed particularly well, achieving a precision of 74.25% and an F1-Score of 75.25% using audio and video modalities. The DepMSTAT model (Tao et al. 2024) also showed competitive results with an accuracy of 71.53% across audio, video, and text modalities. These results highlight the significant advancements in multimodal depression detection, with different models excelling across various datasets and modality combinations.

3.10.2 Comparison Across Modalities

Multimodal approaches that integrate audio, video, and text have consistently outperformed unimodal systems. For the Audio Modality, the transformer-Convolutional Neural Network-Convolutional Neural Network regarded as Temporal Convolutional Network (CNN-CNN (TCC)) model (Yin et al. 2023) stands out with exceptional performance, achieving an impressive F1-score of 93.6% on the DAIC-WOZ dataset and 96.7% on the MODMA dataset.

Another notable alternative in the audio-based models' category is wav2vec 2.0 Long Short-Term Memory (LSTM) (Zhang et al. 2024), which scored 79% as the F1 score on the DAIC-WOZ dataset and 90.53% on the Cognitive and Mood Disorders Classification (CMDC) dataset. The (Domain Adaptive Audio Modeling) DAAMAudioTransformer also scored well with an F1 macro score of 0.72 [34].

For the video modality, the wav2vec 2.0 model (X. Huang et al. 2024) scored 0.9481% accuracy with an RMSE of 0.3810. The FPT-Former model (Li et al. 2024) achieved 79% accuracy with audio and video modalities on E-DAIC. An approach that was also competitive for video was the Pose and Motion-Based Feature Network (PMBFN) model (Hu et al. 2024), which yielded performance metrics on the DAIC-WOZ and E-DAIC datasets. For the text modality, the Graph Convolutional Network (GCN) (Burdisso et al. 2023) obtained a commendable F1 score of 0.84 on the DAIC-WOZ and E-DAIC datasets. Pre-trained BERT-based models (Abilkaiyrkyzy et al. 2024) achieved 69% in accuracy on the E-DAIC dataset. The Context-Aware Learning and Language Modeling (CALLM) framework (Wu et al. 2024) also performed strongly, achieving 0.77% accuracy, 0.70% F1, and 0.78% Area Under the Curve (AUC) on the augmented datasets.

These results underscore the differential performance of various models across modalities, highlighting the effectiveness of depression detection on the specific approach, dataset, and modality used under which estimates are made.

3.10.3 Insights from Novel Architectures

Novel architectures have demonstrated better capacity for capturing patterns in depression-related behaviors:

3.10.3.1 HiQuE (Hierarchical Question Embedding Network)

Jung et al. (2024) achieved a robust F1 score of 0.70, underlining the innovative approach to multimodal depression detection by leveraging hierarchical question embeddings in both the DAIC-WOZ and MIT interview datasets.

3.10.3.2 Transformer-based Models Like TCC (Transformer-CNN-CNN)

Yin et al. (2023) reported high results with F1 scores of 93.6% on DAIC-WOZ and 96.7% on MODMA, highlighting the power of transformer architectures in capturing complex emotional patterns.

3.10.3.3 DepressNet

Saggu, Gupta, and Arya (2022) introduced a novel multi-modal approach on the E-DAIC dataset, achieving an RMSE of 5.36 and a concordance correlation coefficient (CCC) of 0.457, demonstrating the potential of integrated neural network architectures in depression detection.

3.10.3.4 Self-Attention Graph Pooling with Soft Label (SGP-SL)

T. Chen et al. (2022) presented an innovative method for processing EEG signals, achieving an impressive 84.91% accuracy in detecting depression-related patterns through advanced graph-based neural networks.

3.10.3.5 DepMSTAT

Tao et al. (2024) proposed a multimodal approach on the D-Vlog dataset, achieving an accuracy of 71.53% by integrating audio, video, and text modalities in a novel feature extraction and fusion framework.

3.10.3.6 CALLM Framework

Wu et al. (2024) introduced a unique approach to text-based depression detection, utilizing advanced machine learning techniques to achieve an accuracy of 0.77, F1 of 0.70, and AUC of 0.78 on augmented datasets.

3.10.3.7 3DMKDR

Su et al. (2023) presented a groundbreaking method for EEG signal analysis, achieving near-perfect accuracies of 99.86% for major depressive disorder (MDD) and 98.01% for the MODMA dataset, demonstrating the potential of advanced signal processing techniques.

Indeed, these breakthroughs demonstrate the changing potential of deep learning and novel architectures in improving the capability of feature representation and detection in multimodal depression detection systems. The multiplicity of approaches, from hierarchical embedding networks to transformer-based models and sophisticated signal processing techniques, speaks to the swift evolution of this arena and the increasing sophistication of machine learning methodologies in mental health diagnostics. Overall, the trend is better toward complex multi-modal approaches to delineate the intricate and precise behavioral markers associated with depression with the promise of more accurate and comprehensive detection methods.

The multiple, different methodological approaches considered in this chapter are an indication of positive progress around automated depression detection. While standard datasets provide a powerful basis for model training, it is novel architectures and multimodal fusion that provide a better chance of achieving clinically meaningful outcomes. Continued addressing of ethical issues and improvements in generalizability will herald widespread implementation in mental healthcare systems.

3.11 CONCLUSION

Taking artificial intelligence, machine learning, and interdisciplinary collaboration as examples, current multimodal depression detection is being shaped in so many ways. From our review, we found an extraordinary movement away from conventional diagnoses toward advanced data-driven approaches that make smart use of multiple modalities from audio, video, text, and EEG signals. Such advancements include improved models, which achieved the highest rates of positive detection, with an unprecedented accuracy level of over 99% for experiments on several datasets by application of the transformer-based architectures and hierarchical embedding networks. The evolution of such technology goes beyond computational prowess; it proposes a change in the narrative of mental health diagnostics, potentially leading to more timely, accurate, and less invasive testing for depression. However, the path forward is complex and multifaceted. Critical challenges remain in areas of data quality, cross-cultural validation, and ethical implementation. There are large discrepancies in performance across demographic groups, warranting strong calls for developing standardized, inclusive, and culturally sensitive data collection protocols. A fundamental strategy for this is the combination of clinical expertise and AI technologies, with chances of interdisciplinary approaches yielding more dependable and context-aware detection systems. Privacy considerations, model interpretability, and seamless integration into an organization's existing healthcare infrastructure are other critical issues for consideration in a responsible approach to the development and deployment of these innovative technologies.

The future of diagnosing depression calls for all-around, modifying, and patient-center approaches. The most tremendous studies recommend real-time tracking frameworks, personal intervention approaches, and AI techniques to provide ongoing nuanced assessment and intervention of other mental health conditions. The convergence of advanced machine learning techniques, domain-specific expertise, and a deep understanding of human psychological complexity offers unprecedented opportunities for early intervention and personalized mental health support. As these technologies continue to evolve, they hold the potential to transform mental health diagnostics from reactive to initiative-taking models, improving patient outcomes, reducing diagnostic delays, and providing more targeted, compassionate care. The journey ahead requires ongoing collaboration between clinicians, data scientists, ethicists, and technology developers to realize the full transformative potential of these innovative approaches.

REFERENCES

Abilkaiyrkyzy, Akbobek, Fedwa Laamarti, Mufeed Hamdi, and Abdulmotaleb El Saddik. 2024. "Dialogue System for Early Mental Illness Detection: Toward a Digital Twin Solution." *IEEE Access* 12, 2007–2024. https://doi.org/10.1109/ACCESS.2023.3348783.

Aghaei, Alireza Afzal, and Nadia Khodaei. n.d. "Computational Mathematics and Computer Modeling with Applications Automated Depression Recognition Using Multimodal Machine Learning: A Study on the DAIC-WOZ Dataset." *Computational Mathematics and Computer Modeling with Applications (CMCMA)*, 2(1), 45–53. https://doi.org/10.48308/CMCMA.2.1.45.

Ahmed, Arfan, Sarah Aziz, Carla T. Toro, Mahmood Alzubaidi, Sara Irshaidat, Hashem Abu Serhan, Alaa A. Abd-alrazaq, and Mowafa Househ. 2022. "Machine Learning Models to Detect Anxiety and Depression through Social Media: A Scoping Review." *Computer Methods and Programs in Biomedicine Update* 2, 100066. https://doi.org/10.1016/j.cmpbup.2022.100066.

Ajmal, Sahar, and Muhammad Shoaib. n.d. "RSTFusionX: Leveraging Rhetorical Structure Theory and Ensemble Models for Depression Prediction in Social Media Posts." *IEEE Access*. https://doi.org/10.1109/ACCESS.2017.DOI.

Babu, Nirmal Varghese, and E. Grace Mary Kanaga. 2022. "Sentiment Analysis in Social Media Data for Depression Detection Using Artificial Intelligence: A Review." *SN Computer Science*, 3(1), 74. https://doi.org/10.1007/s42979-021-00958-1.

Burdisso, Sergio, Ernesto Reyes-Ramírez, Esaú Villatoro-Tello, Fernando Sánchez-Vega, Pastor López-Monroy, and Petr Motlicek. 2024. "DAIC-WOZ: On the Validity of Using the Therapist's Prompts in Automatic Depression Detection from Clinical Interviews." *arXiv preprint arXiv:2404.14463*.. https://doi.org/10.18653/v1/2024.clinicalnlp-1.8.

Burdisso, Sergio, Esaú Villatoro-Tello, Srikanth Madikeri, and Petr Motlicek. 2023. "Node-Weighted Graph Convolutional Network for Depression Detection in Transcribed Clinical Interviews." *arXiv preprint arXiv:2307.00920*. https://doi.org/10.21437/Interspeech.2023-1923.

Chen, Tao, Yanrong Guo, Shijie Hao, and Richang Hong. 2022. "Exploring Self-Attention Graph Pooling With EEG-Based Topological Structure and Soft Label for Depression Detection." *IEEE Transactions on Affective Computing* 13(4), 2106–2118. https://doi.org/10.1109/TAFFC.2022.3210958.

Chen, Xin, and Zhigeng Pan. 2021. "A Convenient and Low-Cost Model of Depression Screening and Early Warning Based on Voice Data Using for Public Mental Health." *International Journal of Environmental Research and Public Health* 18(12), 6441. https://doi.org/10.3390/ijerph18126441.

Devault, David, Ron Artstein, Grace Benn, Teresa Dey, Ed Fast, Alesia Gainer, Kallirroi Georgila, et al. n.d. "SimSensei Kiosk: A Virtual Human Interviewer for Healthcare Decision Support." Available at: http://simsensei.ict.usc.edu/.

Dia, Mamadou, Ghazaleh Khodabandelou, and Alice Othmani. n.d. "A Novel Stochastic Transformer-Based Approach for Post-Traumatic Stress Disorder Detection Using Audio Recording of Clinical Interviews." In *2023 IEEE 36th International Symposium on Computer-Based Medical Systems (CBMS)* (pp. 700–705). IEEE. https://doi.org/10.1109/CBMS58004.2023.00131.

Feng, Shutong, Guangzhi Sun, Nurul Lubis, Wen Wu, Chao Zhang, and Milica Gašić. 2023. "Affect Recognition in Conversations Using Large Language Models," September. Available at: http://arxiv.org/abs/2309.12881.

Gimeno-Gómez, David, Ana-Maria Bucur, Adrian Cosma, Carlos-David Martínez-Hinarejos, and Paolo Rosso. 2024. "Reading Between the Frames: Multi-Modal Depression Detection in Videos from Non-Verbal Cues," January. Available at: http://arxiv.org/abs/2401.02746.

Gratch, Jonathan, Ron Artstein, Gale Lucas, Giota Stratou, Stefan Scherer, Angela Nazarian, Rachel Wood, et al. n.d. "The Distress Analysis Interview Corpus of Human and Computer Interviews." Available at: http://www.biopac.com.

Gupta, Rohan Kumar, and Rohit Sinha. 2023. "Analyzing the Effect of Data Impurity on the Detection Performances of Mental Disorders," August. Available at: http://arxiv.org/abs/2308.05133.

Han, Zhuojin, Yuanyuan Shang, Zhuhong Shao, Jingyi Liu, Guodong Guo, Tie Liu, Hui Ding, and Qiang Hu. 2024. "Spatial-Temporal Feature Network for Speech-Based Depression Recognition." *IEEE Transactions on Cognitive and Developmental Systems* 16(1), 308–318. https://doi.org/10.1109/TCDS.2023.3273614.

Hu, Min, Lei Liu, Xiaohua Wang, Yiming Tang, Jiaoyun Yang, and Ning An. 2024. "Parallel Multiscale Bridge Fusion Network for Audio-Visual Automatic Depression Assessment." *IEEE Transactions on Computational Social Systems*. https://doi.org/10.1109/tcss.2024.3416029.

Huang, Xiangsheng, Fang Wang, Yuan Gao, Yilong Liao, Wenjing Zhang, Li Zhang, and Zhenrong Xu. 2024. "Depression Recognition Using Voice-Based Pre-Training Model." *Scientific Reports* 14(1), 12734. https://doi.org/10.1038/s41598-024-63556-0.

Huang, Zhaocheng, Julien Epps, Dale Joachim, Brian Stasak, James R. Williamson, and Thomas F. Quatieri. 2020. "Domain Adaptation for Enhancing Speech-Based Depression Detection in Natural Environmental Conditions Using Dilated CNNs." In *Proceedings of the Annual Conference of the International Speech Communication Association, INTERSPEECH*, 2020-October (pp. 4561–4565). International Speech Communication Association. https://doi.org/10.21437/Interspeech.2020-3135.

Ioannides, Georgios, Adrian Kieback, Aman Chadha, and Aaron Elkins. 2024. "Density Adaptive Attention-Based Speech Network: Enhancing Feature Understanding for Mental Health Disorders," August. Available at: http://arxiv.org/abs/2409.00391.

Iyortsuun, Ngumimi Karen, Soo Hyung Kim, Hyung Jeong Yang, Seung Won Kim, and Min Jhon. 2024. "Additive Cross-Modal Attention Network (ACMA) for Depression Detection Based on Audio and Textual Features." *IEEE Access* 12, 20479–20489. https://doi.org/10.1109/ACCESS.2024.3362233.

Jung, Juho, Chaewon Kang, Jeewoo Yoon, Seungbae Kim, and Jinyoung Han. 2024. "HiQuE: Hierarchical Question Embedding Network for Multimodal Depression Detection." In *International Conference on Information and Knowledge Management, Proceedings* (pp. 1049–1059). Association for Computing Machinery. https://doi.org/10.1145/3627673.3679797.

Kaya, Şuheda, and Burak Tasci. 2023. "Electroencephalogram-Based Major Depressive Disorder Classification Using Convolutional Neural Network and Transfer Learning." *Turkish Journal of Science and Technology* 18(1), 207–214. https://doi.org/10.55525/tjst.1242881.

Kumar, A., D. K. Rajak, P. Kumar, and A. Kumar (Eds.). 2024. *Advances in Sustainable Biomaterials: Bioprocessing 4.0, Characterizations, and Applications* (1st ed.). CRC Press. https://doi.org/10.1201/9781003434313.

Kumar, A., P. Kumar, S. Rathee, and B. Kumar (Eds.). 2025. *Mechatronics: Concepts, Tools, Applications, and New Trends* (1st ed.). CRC Press. https://doi.org/10.1201/9781003494478.

Li, Yifu, Xueping Yang, Meng Zhao, Zihao Wang, Yudong Yao, Wei Qian, and Shouliang Qi. 2024. "FPT-Former: A Flexible Parallel Transformer of Recognizing Depression by Using Audiovisual Expert-Knowledge-Based Multimodal Measures." *International Journal of Intelligent Systems* 2024(1), 1564574. https://doi.org/10.1155/2024/1564574.

Ling, Tianfei, Deyuan Chen, and Baobin Li. 2024. "MDAVIF: A Multi-Domain Acoustical-Visual Information Fusion Model for Depression Recognition from Vlog Data." In *ICASSP, IEEE International Conference on Acoustics, Speech, and Signal Processing - Proceedings* (pp. 8115–8119). Institute of Electrical and Electronics Engineers Inc. https://doi.org/10.1109/ICASSP48485.2024.10446491.

Ma, Jinping, Hai Zhou, Qinqin Fu, and Guohua Lu. 2023. "Facilitators and Barriers in the Development and Implementation of Depression Prevention and Treatment Policies in China: A Qualitative Study." *BMC Public Health* 23(1), 276. https://doi.org/10.1186/s12889-023-15201-0.

Mallol-Ragolta, Adria, Manuel Milling, and Björn Schuller. 2024. "Multi-Triplet Loss-Based Models for Categorical Depression Recognition from Speech." In *IberSPEECH 2024* (pp. 31–35). ISCA. https://doi.org/10.21437/IberSPEECH.2024-7.

Mei, Fuyu, and Zhidan Wang. 2024. "Trends in Mental Health: A Review of the Most Influential Research on Depression in Children and Adolescents." *Annals of General Psychiatry* 23(1), 36. https://doi.org/10.1186/s12991-024-00520-2.

Moon, Palash, and Pushpak Bhattacharyya. 2024. "We Care: Multimodal Depression Detection and Knowledge Infused Mental Health Therapeutic Response Generation," June. Available at: http://arxiv.org/abs/2406.10561.

Ogunleye, Bayode, Hemlata Sharma, and Olamilekan Shobayo. 2024. "Sentiment Informed Sentence BERT-Ensemble Algorithm for Depression Detection." *Big Data and Cognitive Computing* 8(9), 112. https://doi.org/10.3390/bdcc8090112.

Patapati, Santosh V. 2024. "Integrating Large Language Models into a Tri-Modal Architecture for Automated Depression Classification on the DAIC-WOZ," July. Available at: http://arxiv.org/abs/2407.19340.

Philip Thekkekara, Joel, Sira Yongchareon, and Veronica Liesaputra. 2024. "An Attention-Based CNN-BiLSTM Model for Depression Detection on Social Media Text." *Expert Systems with Applications* 249, 123834. https://doi.org/10.1016/j.eswa.2024.123834.

Praveen, S. V., Rajesh Ittamalla, and Gerard Deepak. 2021. "Analyzing Indian General Public's Perspective on Anxiety, Stress and Trauma during Covid-19 - A Machine Learning Study of 840,000 Tweets." *Diabetes and Metabolic Syndrome: Clinical Research and Reviews* 15(3), 667–671. https://doi.org/10.1016/j.dsx.2021.03.016.

Qin, Ruiyang, Ryan Cook, Kai Yang, Ahmed Abbasi, David Dobolyi, Salman Seyedi, Emily Griner, et al. 2024. "Language Models for Online Depression Detection: A Review and Benchmark Analysis on Remote Interviews." *ACM Transactions on Management Information Systems* 16(2), 1–35. https://doi.org/10.1145/3673906.

Ringeval, Fabien, Björn Schuller, Michel Valstar, Nicholas Cummins, Roddy Cowie, Leili Tavabi, Maximilian Schmitt et al. "AVEC 2019 workshop and challenge: state-of-mind, detecting depression with AI, and cross-cultural affect recognition." In Proceedings of the 9th International on Audio/visual Emotion Challenge and Workshop, pp. 3–12. 2019.

Ringeval, Fabien, Jonathan Gratch, Sharon Mozgai, Bjorn Schuller, Roddy Cowie, Nicholas Cummins, Maja Pantic, Michel Valstar, Stefan Scherer, and Maximilian Schmitt. 2017. "AVEC 2017 - Real-Life Depression and Affect Recognition Workshop and

Challenge." In *AVEC 2017 - Proceedings of the 7th Annual Workshop on Audio/Visual Emotion Challenge, Co-Located with MM 2017* (pp. 3–9). Association for Computing Machinery, Inc. https://doi.org/10.1145/3133944.3133953.

Saggu, Guramritpal Singh, Keshav Gupta, and K. V. Arya. 2022. "DepressNet: A Multimodal Hierarchical Attention Mechanism Approach for Depression Detection." *International Journal of Engineering Sciences* 15(1), 24–32. https://doi.org/10.36224/ijes.150104.

Schuller, Björn, Michel Valstar, Florian Eyben, Roddy Cowie, and Maja Pantic. 2012. "AVEC 2012 - The Continuous Audio/Visual Emotion Challenge." In *ICMI'12 - Proceedings of the ACM International Conference on Multimodal Interaction* (pp. 449–456). https://doi.org/10.1145/2388676.2388776.

Schuller, Björn, Michel Valstar, Florian Eyben, Gary Mckeown, Roddy Cowie, and Maja Pantic. n.d. "AVEC 2011-The First International Audio/Visual Emotion Challenge." Available at: http://sspnet.eu/avec2011.

Shen, Jian, Xiaowei Zhang, Xiao Huang, Manxi Wu, Jin Gao, Dawei Lu, Zhijie Ding, and Bin Hu. 2021. "An Optimal Channel Selection for EEG-Based Depression Detection via Kernel-Target Alignment." *IEEE Journal of Biomedical and Health Informatics* 25(7), 2545–2556. https://doi.org/10.1109/JBHI.2020.3045718.

Srimadhur, N. S., and S. Lalitha. 2020. "An End-to-End Model for Detection and Assessment of Depression Levels Using Speech." In *Procedia Computer Science* 171, 12–21. https://doi.org/10.1016/j.procs.2020.04.003.

Su, Yun, Zhixuan Zhang, Qi Cai, Bingtao Zhang, and Xiaohong Li. 2023. "3DMKDR: 3D Multiscale Kernels CNN Model for Depression Recognition Based on EEG." *Journal of Beijing Institute of Technology (English Edition)* 32(2), 230–241. https://doi.org/10.15918/j.jbit1004-0579.2022.096.

Tank, Chayan, Sarthak Pol, Vinayak Katoch, Shaina Mehta, Avinash Anand, and Rajiv Ratn Shah. 2024. "Depression Detection and Analysis Using Large Language Models on Textual and Audio-Visual Modalities," July. Available at: http://arxiv.org/abs/2407.06125.

Tao, Yongfeng, Minqiang Yang, Huiru Li, Yushan Wu, and Bin Hu. 2024. "DepMSTAT: Multimodal Spatio-Temporal Attentional Transformer for Depression Detection." *IEEE Transactions on Knowledge and Data Engineering* 36(7), 2956–66. https://doi.org/10.1109/TKDE.2024.3350071.

Tejaswini, Vankayala, Korra Sathya Babu, and Bibhudatta Sahoo. 2024. "Depression Detection from Social Media Text Analysis Using Natural Language Processing Techniques and Hybrid Deep Learning Model." *ACM Transactions on Asian and Low-Resource Language Information Processing* 23(1), 1–20. https://doi.org/10.1145/3569580.

Valstar, Michel, Björn Schuller, Kirsty Smith, Florian Eyben, Bihan Jiang, Sanjay Bilakhia, Sebastian Schnieder, Roddy Cowie, and Maja Pantic. 2013. "AVEC 2013 - The Continuous Audio/Visual Emotion and Depression Recognition Challenge." In *AVEC 2013 - Proceedings of the 3rd ACM International Workshop on Audio/Visual Emotion Challenge* (pp. 3–10). Association for Computing Machinery. https://doi.org/10.1145/2512530.2512533.

Valstar, Michel, Björn Schuller, Kirsty Smith, Timur Almaev, Florian Eyben, Jarek Krajewski, Roddy Cowie, and Maja Pantic. 2014. "AVEC 2014 - 3D Dimensional Affect and Depression Recognition Challenge." In *AVEC 2014 - Proceedings of the 4th International Workshop on Audio/Visual Emotion Challenge, Workshop of MM 2014* (pp. 3–10). Association for Computing Machinery. https://doi.org/10.1145/2661806.2661807.

Valstar, Michel, Jonathan Gratch, Björn Schuller, Fabien Ringevaly, Denis Lalanne, Mercedes Torres Torres, Stefan Scherer, Giota Stratou, Roddy Cowie, and Maja Panticz. 2016. "AVEC 2016 - Depression, Mood, and Emotion Recognition Workshop and Challenge." In *AVEC 2016 - Proceedings of the 6th International Workshop on Audio/Visual Emotion Challenge, Co-Located with ACM Multimedia 2016* (pp. 3–10). Association for Computing Machinery, Inc. https://doi.org/10.1145/2988257.2988258.

Vandana, Nikhil Marriwala, and Deepti Chaudhary. 2023. "A Hybrid Model for Depression Detection Using Deep Learning." *Measurement: Sensors* 25, 100587. https://doi.org/10.1016/j.measen.2022.100587.

Wang, Jinhan, Vijay Ravi, and Abeer Alwan. 2023. "Non-Uniform Speaker Disentanglement for Depression Detection From Raw Speech Signals." In *Proceedings of the Annual Conference of the International Speech Communication Association, INTERSPEECH*, 2023-August (pp. 2343–2347). International Speech Communication Association. https://doi.org/10.21437/Interspeech.2023-2101.

Wang, Zhiyong, Longxi Chen, Lifeng Wang, and Guangqiang Diao. 2020. "Recognition of Audio Depression Based on Convolutional Neural Network and Generative Antagonism Network Model." *IEEE Access* 8, 101181–101191. https://doi.org/10.1109/ACCESS.2020.2998532.

Wu, Yuqi, Kaining Mao, Yanbo Zhang, and Jie Chen. 2024. "CALLM: Enhancing Clinical Interview Analysis through Data Augmentation with Large Language Models." *IEEE Journal of Biomedical and Health Informatics*. https://doi.org/10.1109/JBHI.2024.3435085.

Xu, Chenyang, Feiyi Fan, Jianfei Shen, Hanguang Wang, Zhongyi Zhang, and Qinghao Meng. n.d. "UC Merced Proceedings of the Annual Meeting of the Cognitive Science Society Title An EEG-Based Depressive Detection Network with Adaptive Feature Learning and Channel Activation Permalink Publication Date An EEG-Based Depressive Detection Network with Adaptive Feature Learning and Channel Activation." Vol. 46. Available at: https://escholarship.org/uc/item/76b4c4qw.

Yin, Faming, Jing Du, Xinzhou Xu, and Li Zhao. 2023. "Depression Detection in Speech Using Transformer and Parallel Convolutional Neural Networks." *Electronics (Switzerland)* 12(2), 328. https://doi.org/10.3390/electronics12020328.

Yirmiya, Raz. 2024. "The Inflammatory Underpinning of Depression: An Historical Perspective." *Brain, Behavior, and Immunity*. https://doi.org/10.1016/j.bbi.2024.08.048.

Yoo, Hyewon, and Hayoung Oh. 2023. "Depression Detection Model Using Multimodal Deep Learning." https://doi.org/10.20944/preprints202305.0663.v1.

Yoon, Jeewoo, Chaewon Kang, Seungbae Kim, and Jinyoung Han. 2022. "D-Vlog: Multimodal Vlog Dataset for Depression Detection." Available at: https://sites.google.com/view/jeewoo-yoon/dataset.

Zhang, Xu, Xiangcheng Zhang, Weisi Chen, Chenlong Li, and Chengyuan Yu. 2024. "Improving Speech Depression Detection Using Transfer Learning with Wav2vec 2.0 in Low-Resource Environments." *Scientific Reports* 14(1), 9543. https://doi.org/10.1038/s41598-024-60278-1.

Zhou, Li, Zhenyu Liu, Zixuan Shangguan, Xiaoyan Yuan, Yutong Li, and Bin Hu. 2023. "TAMFN: Time-Aware Attention Multimodal Fusion Network for Depression Detection." *IEEE Transactions on Neural Systems and Rehabilitation Engineering* 31, 669–679. https://doi.org/10.1109/TNSRE.2022.3224135.

Zulfiker, Md Sabab, Nasrin Kabir, Al Amin Biswas, Tahmina Nazneen, and Mohammad Shorif Uddin. 2021. "An In-Depth Analysis of Machine Learning Approaches to Predict Depression." *Current Research in Behavioral Sciences* 2, 100044. https://doi.org/10.1016/j.crbeha.2021.100044.

4 Machine Learning and Natural Language Processing Strategies for Fake News Detection

An Empirical Study

Ananya Gupta, Akashvi Bhardwaj, and Ashish Kumar

4.1 INTRODUCTION

Nowadays, content on the internet plays a significant part in influencing users' opinions. E-commerce customers rely only on reviews to know if the product is good [1 –3]. The problem with fake news has increased in recent days because it acts like spam that misleads people. Fake news can have a major impact on the reader's mind, leading them to make poor decisions [4]. The primary objective of this chapter is to present an in-depth evaluation of existing methodologies used in the detection of fake news. This study aims to give both researchers and practitioners an extensive understanding of the present status of the field by evaluating the strengths and limits of various approaches. The term "fake news" refers to deliberately misleading or deceiving readers by presenting false information as genuine news [5, 6]. It is not only articles that usually disseminate fake news, but social media also has recently emerged as the primary source for the rapid and widespread propagation of fake news [7]. In the method proposed here, NLP techniques and ML are used. In this chapter, we demonstrate that fake news can be detected by analyzing the entire text, and new opportunities can be discovered that would otherwise remain hidden. ML is a subfield of artificial intelligence (AI) that refers to the development of computer systems that study data and improve their performance without the need for explicit programming [8–11]. Natural language processing (NLP) is a branch of computer science that studies how computers can interpret and process human language in a similar way as humans. It involves teaching computers to understand and respond to speech by analyzing, interpreting, and generating human language. To put it in simple words, NLP enables computers to interpret and interact with human language.

 DOI: 10.1201/9781032724508-5

The process of preparing data for analysis is always a time-consuming step in developing a ML model. The objective of fake news detection research is to develop innovative techniques, algorithms, and methods that can identify and classify fake news in real time from genuine information. The chapter is structured as follows: Section 4.2 provides a literature review and highlights key research contributions in fake news detection. Section 4.3 outlines the proposed algorithm employed in this study, including data collection, data exploration, NLP techniques, and a detailed analysis of various ML algorithms. Section 4.4 discusses the evaluation metrics employed in assessing the performance of detection models. Finally, Section 4.5 summarizes the key findings and proposes potential avenues for future research. Figure 4.1 presents the flowchart of the proposed algorithm.

4.2 LITERATURE REVIEW

In the domain of fake news detection NLP and ML have gained popularity in recent years. As false news continues to threaten the integrity of online information, researchers have focused their efforts on researching various approaches and techniques to identify and detect fake news.

Raja et al. discovered that employing ML approaches can considerably increase the effectiveness of fake news detection. The outcomes of their investigation show that SVM utilizing TD-IDF attained the best results with an accuracy of 93% [1]. Kishwar and Zafar applied ML techniques by getting an F1 score and accuracy of 89%, and their findings demonstrate that KNN is the best-performing ML model. It is important to understand if the KNN algorithm performs best on different types of fake news [4]. Sudhakar et al. presented two ML algorithms, namely logistic regression and Naïve Bayes classifier; out of these, logistic regression provided higher accuracy (98.70%). Their study demonstrated the potential of ML to prevent false news and advance fake news techniques [8]. Granik M. et al.'s research showed that the Naïve Bayes algorithm can successfully categorize news as authentic or fraudulent. The outcome could identify fake news with reasonable accuracy [12]. Capuano et al. focus on approaches and strategies for identifying fake news-based content, providing an overview of existing research in the area [13]. Khan et al. compared several ML algorithms for spotting fake news online. They emphasized the need for standardized metrics [14]. Pal and Pradhan used a decision tree model with the highest accuracy of 99.609% [15]. According to Sharma et al., the highest accuracy of 91% was achieved by the passive aggressive model [16]. In the study of Probierz et al. best results we obtained by the SVM algorithm with an accuracy of 94% [17, 18, 19]. These papers demonstrated that ML algorithms such as the Naïve Bayes classifier accurately identify and reduce the spread of fake news. They emphasize the significance of applying these strategies to stop the spread of false information on social media and other platforms. In conclusion, each author showcased their most preferred and appropriate model but it is important to give a detailed analysis of the dataset used and whether the model will work in different situations. The rapid of artificial intelligence (AI) and its potential influence on society has opened up new

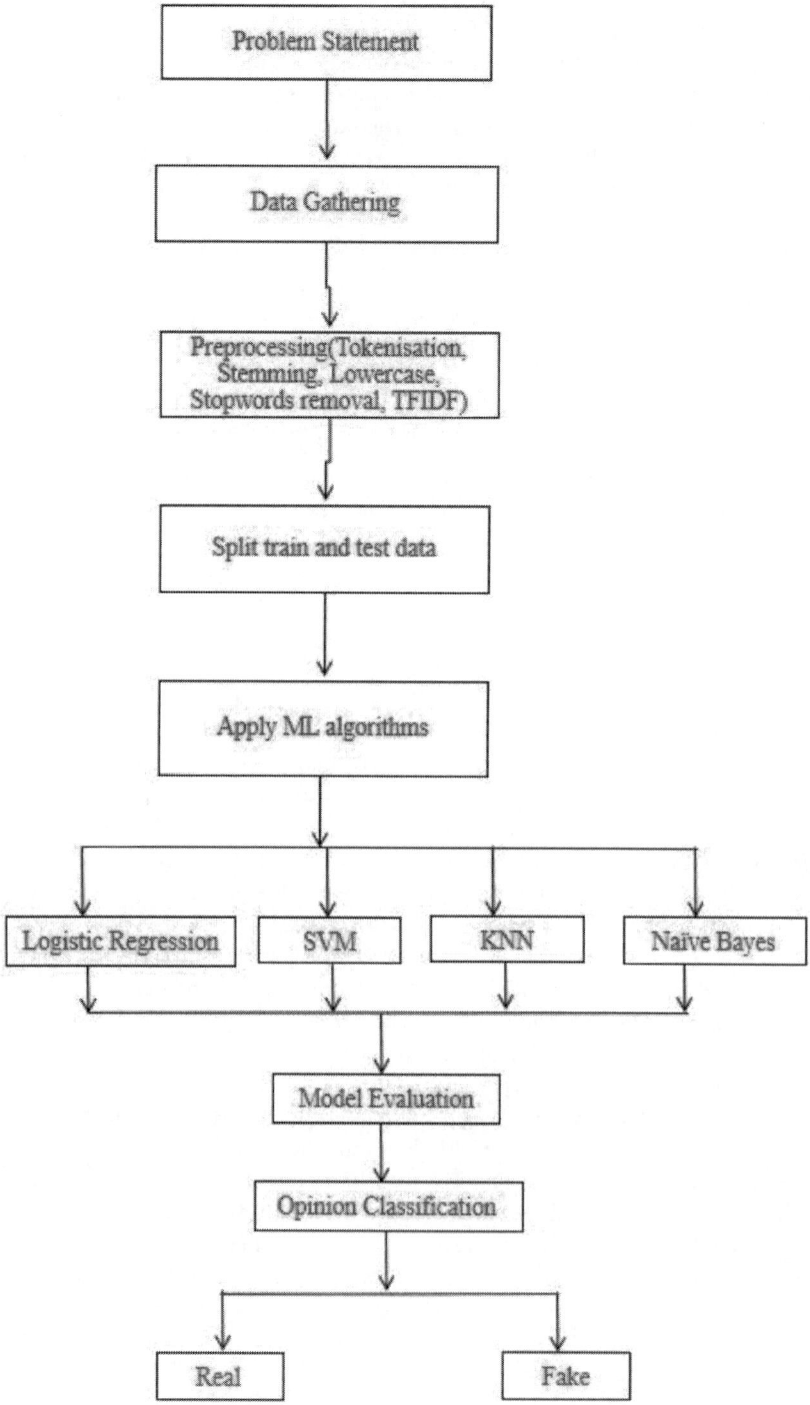

FIGURE 4.1 Flowchart of the proposed algorithm

avenues for addressing the critical issue of fake news. Using AI with ML and NLP techniques improves our capacity to determine the reliability of news sources and content [20, 21–23].

Table 4.1 summarizes various studies on fake news detection using different ML models, including their focus, algorithms used, and outcomes achieved.

4.3 PROPOSED METHODOLOGY

4.3.1 Dataset Collection:

In our project, we utilized a dataset that is specifically curated for fake news detection.

Figure 4.2 depicts a partial snapshot of the complete dataset, offering a glimpse into the richness of the data available in our study. The dataset consists of 44,898

TABLE 4.1
Overview of Studies on Fake News Detection

Study Focus	Machine Learning Models	Outcome
Raj et al. [1]	SVM	Successfully applied SVM with TF-IDF
Kishwer & Zafar [4]	KNN	KNN is the best algorithm based on accuracy and F1 Score
Sudhakar et al. [8]	Logistic Regression	Logistic Regression achieved the highest accuracy i.e. 98%
Granik et al. [12]	Naïve Bayes	Various approaches were used out of which Naïve Bayes was able to categorize between fake and real news
Capuano et al. [13]	Multiple approaches	Highlighted existing research in fake news detection
Khan et al. [14]	Multiple ML algorithm	Compared multiple machine learning model
Pal & Pradhan [15]	Decision Tree	Decision Tree achieved the highest accuracy of 99.6%
Sharma et al. [16]	Passive Aggressive	The passive-aggressive model is assessed and analyzed based on real-time datasets
Probierz et al. [19]	SVM	SVM gave the best performance in different types of fake news

	title	text	subject	date	target
0	Commerce Secretary says Trump-Xi talks will ad...	BEIJING (Reuters) - Meetings between U.S. Pres...	worldnews	November 8, 2017	true
1	Senate confirms Callista Gingrich as U.S. Amba...	WASHINGTON (Reuters) - The U.S. Senate on Mond...	politicsNews	October 16, 2017	true
2	BREAKING: Trump Tops Off INSANE Weekend By An...	Donald Trump s most dangerous and offensive ca...	News	March 4, 2017	fake
3	LEFTIST MEDIA EXPOSES Democrat Party For Ignor...	Hollywood producer kingpin and mega-Democrat P...	left-news	Nov 14, 2017	fake
4	John McCain's Son Really Hates Ignorant Racis...	After racists threw a temper tantrum over Old ...	News	May 4, 2016	fake

FIGURE 4.2 Dataset

instances and 5 attributes, namely title, text, subject, and target, which are gathered from: https://www.kaggle.com/datasets/clmentbisaillon/fake-and-real-news-dataset.

Each instance provides valuable information for analysis, aiding in the identification of fake news through comprehensive news analysis. The curated results help in easily identifying fake news through news analysis.

4.3.2 Data Exploration

Figure 4.3 represents the distribution of news articles based on subject categories in our research on fake news detection. The x-axis enlists various subjects such as political news, world news, the Middle East, and US news. The y-axis shows the frequency or percentage of articles in each category. The "politics news" and the "worldnews" had the highest occurrences, while the "Middle East" and US_news had the lowest occurrences. This distribution is crucial for fake news research as it helps us to understand the diversity and focus of the dataset. Analyzing the

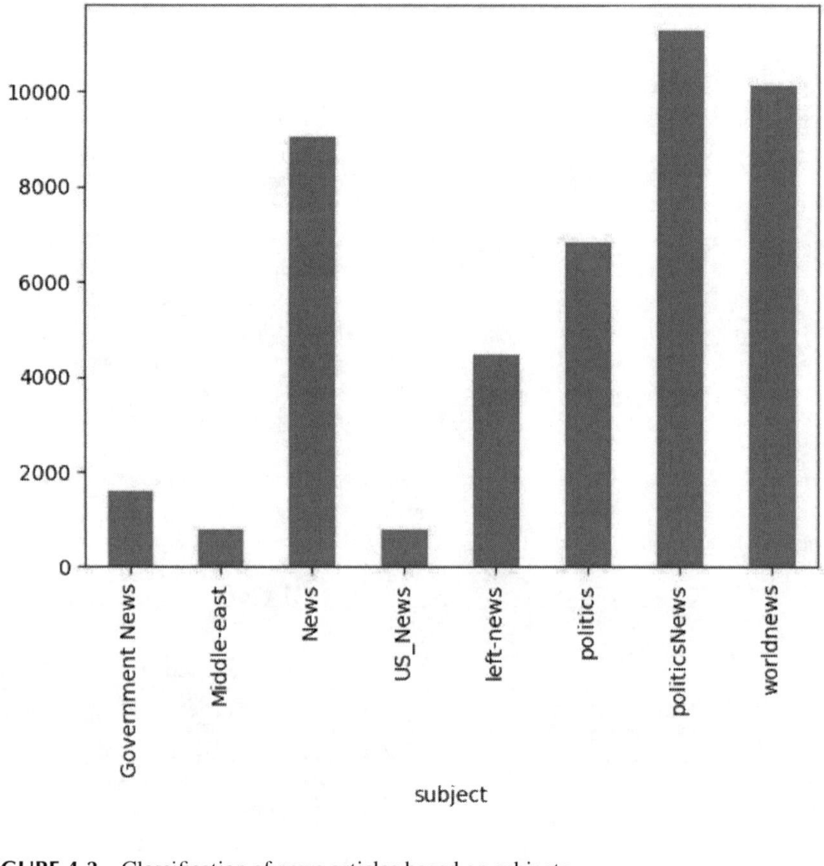

FIGURE 4.3 Classification of news articles based on subjects

distribution helps identify potential issues and ensures a robust and effective fake news detection system.

Figure 4.4 illustrates the analysis of distribution of tokens within the text for both fake news and true news articles that shows frequency of allocation of words. The bar graph shows 20 most frequent words in fake and true news articles, allowing a comparison for token usage. This helps identify patterns and differences in language use between the two types of news. Understanding the token distribution helps refine

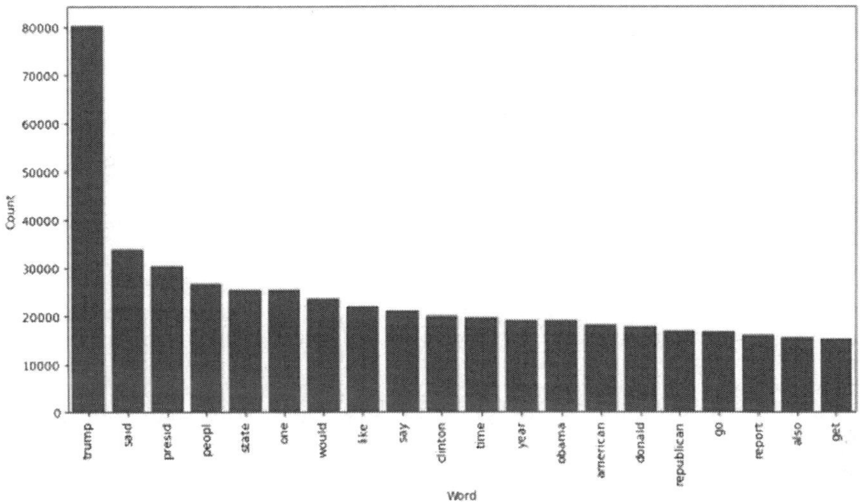

a)

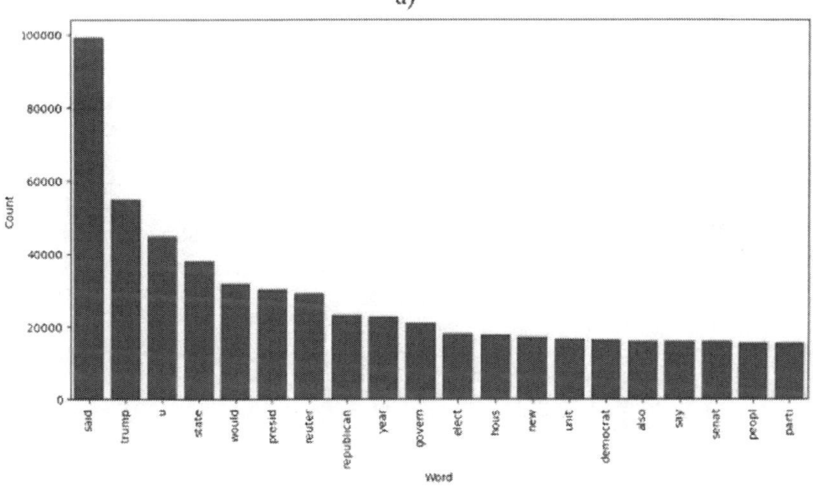

b)

FIGURE 4.4 Tokens details existing in text. (a) Fake news bar graph. (b) True news bar graph

the fake news detection system, improving accuracy and reliability of the classification process.

4.3.3 Preprocessing of the Dataset Using NLP Techniques

a) Tokenization

Tokenization breaks the sentence into different components called tokens. It is an important step in NLP because it allows computers to interact and interpret individual text components. In this study, a sentence-level tokenization approach was adopted using the NLTK (Natural Language Processing Toolkit) library in Python.

b) Stop Word Removal

Stop words are the common words that are filtered out during natural language data pre-processing because they don't carry significant meaning and add nothing to comprehending the context of the phrase. They are removed to help focus on the crucial words that convey more significance and increase the accuracy and efficiency of the language processing techniques.

c) Stemming

The process of reducing a word to its stem which affixes to suffixes and prefixes or the roots of the word is known as "stemming." It is crucial in text analysis as it help in simplifying and standardizing word representations. It can impact the semantics and context of news items, necessitating further investigation and debate due to its potential to alter the subtle meaning of certain words in different situations.

d) Lowercase

When we lowercase the text, we change all capital letters into small letters. This helps to ensure the words are treated consistently regardless of whether they are written in uppercase or lowercase. It prevents the news from being categorized as fake or real from being influenced by the use of capital letters.

e) Vectorization

When we analyze text data, there are instances of words that appear with high frequency. Such words do not contain any valuable or special information. So, too much redundancy in such data leads to a high-level computational burden on the effort of the learning process. Here, we have examined two feature selection methods, Term Frequency (TF) and Term Frequency-Inverse Document Frequency (TF IDF). They are used to measure the frequency of words in a document.

TF measures how often a word appears in a document relative to the total number of words, while IDF measures the rarity or uniqueness of a word across a collection of documents.

A higher TF-IDF score indicates that the word is a relatively unique and potentially important feature.

$$tf - idf(t,d) = tf(t,d) * idf(t)$$

4.3.4 Model Evaluation Using Several ML Algorithms

Numerous ML algorithms are well known for their excellent performance on classification problems. The classification techniques used for this model are logistic regression, decision tree, random forest, and Naïve Bayes. The amount of data used for training and testing data is around 80% and 20%, respectively, in each approval. After using four ML classification algorithms, different results are generated and compared.

Logistic Regression is one of the ML algorithms that are used for solving classification problems. Logistic regression forecasts the result of a dependent variable that is categorical. Therefore, the output must be discrete in nature. It fails to provide the exact values of 0 and 1, instead providing the probabilistic values ranging between 0 and 1. It can be either yes or no, 0 or 1, and so on. It is suitable for classifying new data using both continuous and discrete datasets.

K-nearest Neighbor is a simple ML method based on supervised learning. It is used for solving classification and regression problems but is mostly used for classification problems. It operates in the categorization context by the location of the k closest neighbors to a given data point based on the similarity of their features. In addition, it is also referred to as a lazy-learner algorithm, as it does not learn from the training set immediately but rather stores it and performs a classification once it has analyzed it.

A **support vector machine, or SVM,** is a supervised learning algorithm that is widely used for classification and regression problems. The technique is mostly used for classification problems in ML. A basic objective of the SVM algorithm is to find the optimal line or decision boundary that distinguishes the n-dimensional space into classes in order to use it as a tool to automatically categorize new data points in the future.

Naïve Bayes is a supervised learning technique that solves classification problems using Bayes theorem. It is mostly used in text categorization with a large training set. It makes predictions based on the probability of an object.

To implement all the algorithms described here, the Python language was employed via commonly available libraries such as pandas (for loading and reading data), re (for conversion), nltk (for stemming, stop words removal), matplotlib, sklearn (for vectorization and splitting of data, also for learning and evaluating algorithm).

4.4 RESULTS AND EXPERIMENTS

This model uses Jupyter Notebook's most recent version. It is the most useful and reliable ML library. Then we imported other libraries as required by the program.

We enter the dataset into the program, and the model is trained and tested using various NLP techniques and ML algorithms.

For a confusion matrix, the following abbreviations are commonly used:

True Positives (TP): The number of instances correctly identified as positive by the model. In the context of fake news detection, TP represents the number of fake news articles correctly classified as fake news.

True Negatives (TN): The number of instances correctly identified as negative by the model. In fake news detection, TN represents the number of genuine news articles correctly classified as genuine news.

False Positives (FP): The number of instances incorrectly identified as positive by the model. In fake news detection, FP represents the number of genuine news articles mistakenly classified as fake news.

False Negatives (FN): The number of instances incorrectly identified as negative by the model. In fake news detection, FN represents the number of fake news articles mistakenly classified as genuine news.

Accuracy: The overall classification accuracy is calculated as the ratio of correctly classified instances to the total number of instances.

$$accuracy=TP+TN/TP+TN+FP+FN$$

Precision: The proportion of true positive predictions (correctly labeled fake news) out of all positive predictions.

$$precision=TP/TP+FP$$

Recall: The proportion of true positive predictions out of all actual fake news instances.

$$recall=TP/TP+FN$$

F1 Score: The harmonic mean of precision and recall, providing a balanced measure between the two.

$$F1\text{-}score =TP+TP/FN+FP+TP+TP$$

We can understand which classifier performs the most efficiently using the confusion matrix.

Figures 4.5–4.8 represent a confusion matrix, which summarizes the performance of the classification model by presenting counts of true positives, true negatives, false positives, and false negatives.

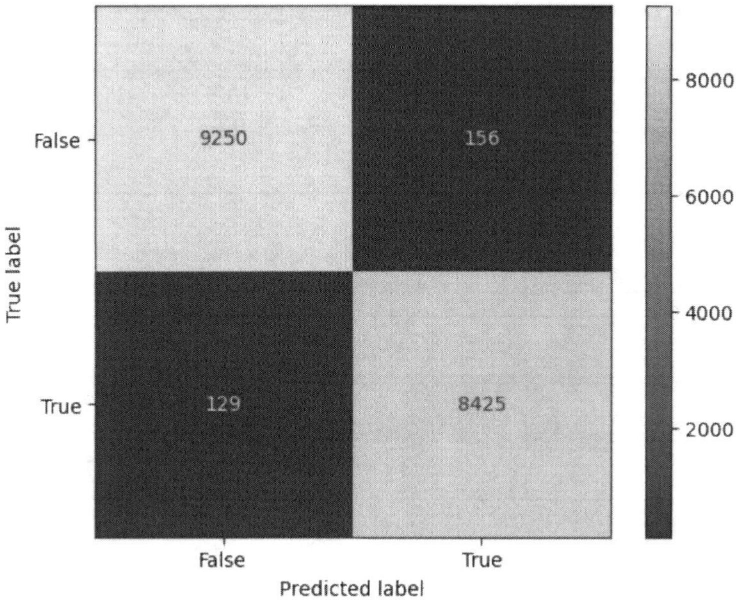

FIGURE 4.5 Confusion matrix logistic regression

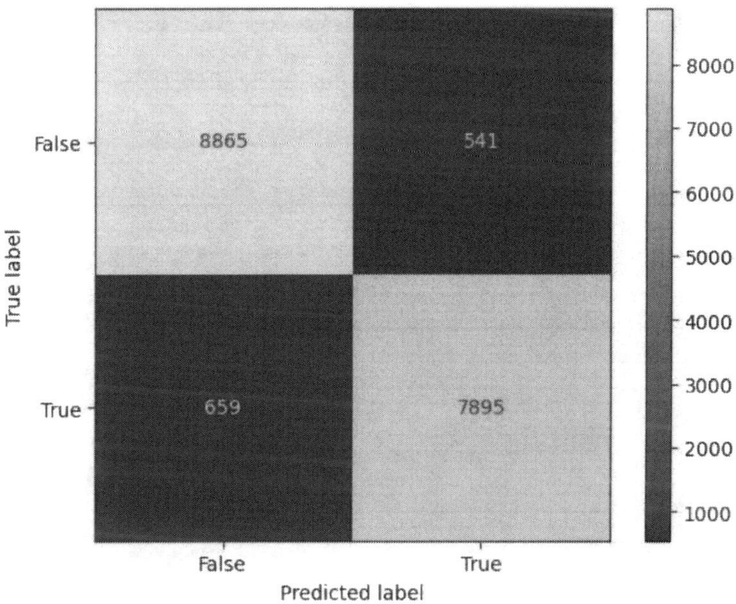

FIGURE 4.6 Confusion matrix naïve Bayes

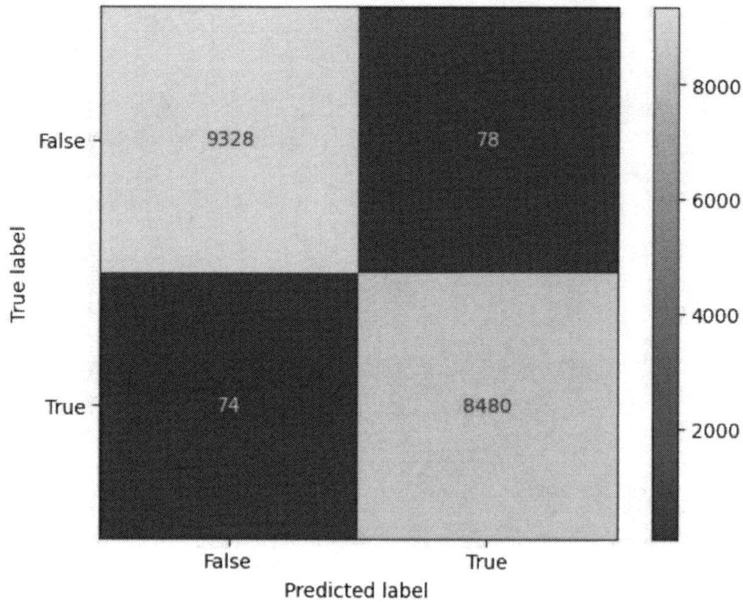

FIGURE 4.7 Confusion matrix SVM

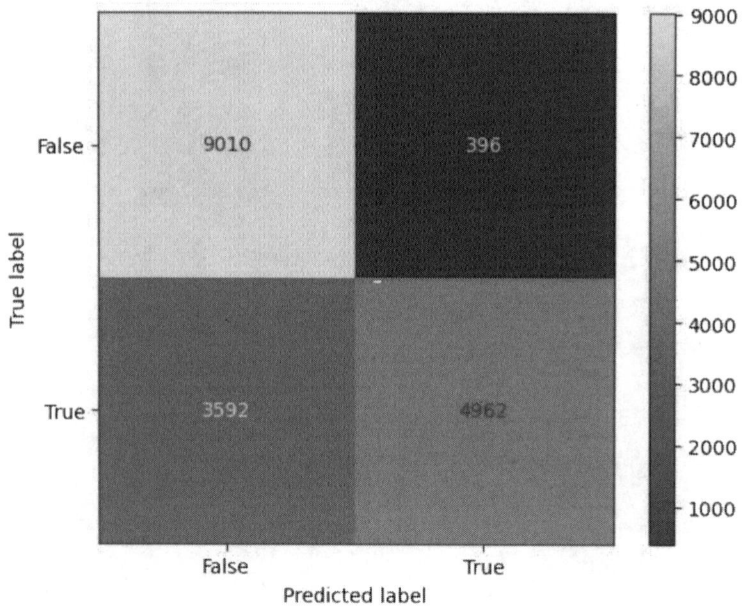

FIGURE 4.8 Confusion matrix

In Figure 4.5 we see a TP count of 8,425, indicating the correct identification of 8,425 fake news articles. The TN count of 9,250 demonstrates that 9,250 real news articles were correctly classified. The FP count of 156 suggests that 156 real news articles were mistakenly classified as fake news, while the FN count of 129 reveals that 129 fake news articles were misclassified as real news.

In Figure 4.6 we see a TP count of 7,895, indicating the identification of 7,895 fake news articles. The TN count of 8,865 demonstrates that 8,865 real news articles were correctly classified. The FP count of 541 suggests that 541 real news articles were mistakenly classified as fake news, while the FN count of 659 reveals that 659 fake news articles were misclassified as real news.

In Figure 4.7 we see a TP count of 8,480, indicating the identification of 8,480 fake news articles. The TN count of 9,328 demonstrates that 9,328 real news articles were correctly classified. The FP count of 78 suggests that 78 real news articles were mistakenly classified as fake news, while the FN count of 74 reveals that 74 fake news articles were misclassified as real news.

In Figure 4.8 we see a TP count of 4,962, indicating the identification of 4,962 fake news articles. The TN count of 9,010 demonstrates that 9,010 real news articles were correctly classified. The FP count of 396 suggests that 396 real news articles were mistakenly classified as fake news, while the FN count of 3,592 reveals that 3,592 fake news articles that misclassified as real news.

Table 4.2 presents a thorough evaluation and comparison of the performance of each technique in the analysis. It is emphasized that the table is comprehensive, includes multiple techniques, and covers accuracy, precision, recall, and F1 score.

Figure 4.9 shows the performance of four popular ML algorithms, namely logistic regression, Naïve Bayes, K-nearest neighbor, and SVM in the form of a bar graph. We evaluate their effectiveness in terms of accuracy, precision, recall, and F1 score.

The evaluation metrics showed that all four algorithms performed well in detecting fake news. The SVM has achieved the highest accuracy (99%) and F1 score (99%), indicating their strong overall performance and also its efficiency in differentiating fake news from genuine news. It also exhibited high precision (99%) and recall (99%), indicating its ability to correctly classify both fake and real news articles. Our results are in perfect accordance with the other studies, notably Raja et al.'s work, which shows an impressive accuracy of 93% when combining SVM with TFIDF [1]. By demonstrating SVM robustness it becomes a more reliable tool for

TABLE 4.2
Comparative Analysis

Techniques	Accuracy	Precision	Recall	F1 Score
Logistic Regression	98%	98%	98%	98%
Naïve Bayes	94%	94%	94%	94%
KNN	80%	83%	79%	79%
SVM	99%	99%	99%	99%

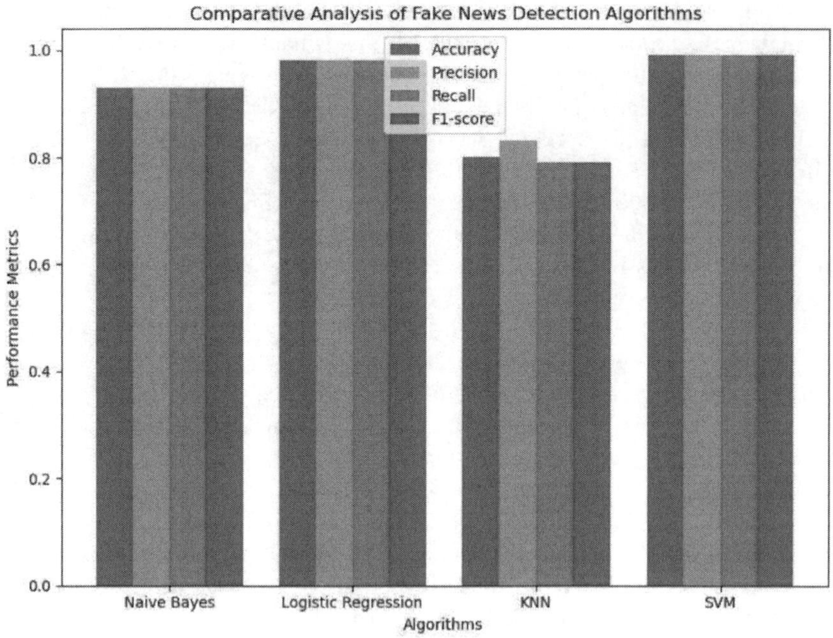

FIGURE 4.9 Comparative analysis of classifier models

detecting fake news, thereby enhancing its reliability as a detection method. Using SVM, Probierz et al. also achieved 94% accuracy, which is consistent with the findings presented here [19]. SVM has been viewed as a critical component of fake news detection due to the similarity of results between these studies.

In addition to performing efficiently, Naïve Bayes achieved an accuracy of 94% with an F1 score of 94%, which indicates that it has a strong performance. However, the results are consistent with those of Granik et al., who showed that Naïve Bayes is capable of classifying news as real or fake [12]. Naïve Bayes can also be used as a useful tool in the fight against fake news because of the stunning consistency of the results.

The logistic algorithm also performed well, with an accuracy of 98% and an F1 score of 98%.

KNN, though slightly outperformed by the other algorithms, still achieved respectable results. It achieved an accuracy of 80% and an F1 score of 79%, indicating its capability in distinguishing between fake and real news. Our findings are consistent with those from a research by Kishwar and Zafar, who showed an impressive F1 score and accuracy of 89% when using KNN [4].

The results of our study indicate that ML algorithms can effectively detect fake news. It shows that SVM with TFIDF have shown the best performance on the dataset with an accuracy of 99%.

4.5 RESULTS

- The goal of this chapter is to develop a model for quickly discovering fake news based on the news text analysis. In this study, the data was analyzed using four different ML algorithms.
- The proposed strategy obtains the maximum accuracy when combined with TF-IDF and SVM, reaching a remarkable level of accuracy of 99%.
- Our study stands out by demonstrating the impressive performance of the SVM model, supported by TF-IDF.
- Detecting fake news is a crucial societal need, and a significant process has been established in this area, but still there is still a lot of work to be done.
- Fake news challenges us to become more informed, robust, and resilient through the refinement and development of detection methodology, collaboration, knowledge sharing, and promoting digital literacy.
- The chapter suggests extending the project for real-time fake news detection involving designing algorithms and systems that can process and analyze social media posts, videos, and images in real time, providing immediate feedback to users to help prevent the spread of fake news.
- In future studies, we will rigorously validate the claim that SVM achieves 99% accuracy. We will use comprehensive data validation techniques and cross-validation methods to ensure the robustness and reproducibility of the results.

REFERENCES

1. Raja, M. S., & Raj, L. A. Fake news detection on social networks using Machine learning techniques. *Materials Today: Proceedings*, *62*, 4821–4827 (2022).
2. Huang, Z., & Benyoucef, M. From e-commerce to social commerce: A close look at design features. *Electronic Commerce Research and Applications*, *12*(4), 246–259 (2013).
3. Rosário, A., & Raimundo, R. Consumer marketing strategy and E-commerce in the last decade: a literature review. *Journal of Theoretical and Applied Electro Commerce Research*, *16*(7), 3003–3024 (2021).
4. Kishwar, A., & Zafar, A. Fake news detection on Pakistani news using machine learning and deep learning. *Expert Systems With Applications*, *211*, 118558 (2023).
5. Egelhofer, J. L., & Lecheler, S. Fake news as a two-dimensional phenomenon: A framework and research agenda. *Annals of the International Communication Association*, *43*(2), 97–116 (2019).
6. Rodríguez-Ferrándiz, R. An Overview of the Fake News Phenomenon: From Untruth-Driven to Post-Truth-Driven Approaches. *Media and Communication*, *11*(2), 15–29 (2023).
7. Zhang, X., & Ghorbani, A. A. An overview of online fake news: Characterization, detection, and discussion. *Information Processing & Management*, *57*(2), 102025 (2020).
8. Sudhakar, M., & Kaliyamurthie, K. P. Effective prediction of fake news using two machine learning algorithms. *Measurement: Sensors*, *24*, 100495 (2022).

9. Aggarwal, K., Mijwil, M. M., Al-Mistarehi, A. H., Alomari, S., Gök, M., Alaabdin, A. M. Z., & Adulrhman, S. H. Has the future started? The current growth of artificial intelligence, machine learning, and deep learning. *Iraqi Journal for Computer Science and Mathematics*, *3*(1), 115–123 (2022).
10. De Mauro, A., Sestino, A., & Bacconi, A. Machine learning and artificial intelligence use in marketing: a general taxonomy. *Italian Journal of Marketing*, *2022*(4), 439–457 (2022).
11. Theodosiou, A. A., & Read, R. C. Artificial intelligence, machine learning and deep learning: potential resources for the infection clinician. *Journal of Infection*, *87*(4), 287–294 (2023).
12. Granik, M., & Mesyura, V. Fake news detection using naive Bayes classifier. In *2017 IEEE first Ukraine conference on electrical and computer engineering (UKRCON)* (pp. 900–903). IEEE (2017).
13. Capuano, N., et al., Content-based fake news detection with machine and deep learning: a systematic review. *Neurocomputing*, *530*, 91–103 (2023).
14. Khan, J. Y., Khondaker, M. T. I., Afroz, S., Uddin, G., & Iqbal, A. A benchmark study of machine learning models for online fake news detection. *Machine Learning with Applications*, *4*, 100032 (2021).
15. Pal, A., & Pradhan, M. Survey of fake news detection using machine intelligence approach. *Data & Knowledge Engineering*, *144*, 102118 (2023).
16. Sharma, A., Sharma, M. K., & Dwivedi, R. K. Exploratory data analysis and detection in news articles on social media using machine learning classifiers. *Ain Shams Engineering Journal*, *14*(10), 102166 (2023).
17. Shu, K., Wang, S., & Liu, H. Beyond news contents: The role of Social context for fake news detection. In *Proceedings of the Twelfth ACM International Conference of Web Search and Data Mining* (pp. 312–320) (2019, January).
18. Shu, K., Sliva, A., Wang, S., Tang, J., & Liu, H. Fake news detection on social media: A data mining perspective. *ACM SIGKDD Explorations Newsletter*, *19*(1), 22–36 (2017).
19. Probierz, B., Stefański, P., & Kozak, J. Rapid detection of fake news based on machine learning methods. *Procedia Computer Science*, *192*, 2893–2902 (2021).
20. Rajendra, P., Kumari, M., Rani, S., Dogra, N., Boadh, R., Kumar, A., & Dahiya, M. Impact of artificial intelligence on civilization: Future perspectives. *Materials Today: Proceedings*, *56*, 252–256 (2022).
21. Rani, S., Kumar, A., Bagchi, A., Yadav, S., & Kumar, S. RPL based Routing protocols for load balancing in IoT network. In *Journal of Physics: Conference Series* (Vol. 1950, No. 1, p. 012073). IOP Publishing (2021, August).
22. Mridha, M. F., Keya, A. J., Hamid, M. A., Monowar, M. M., & Rahman, M. S. A comprehensive review on fake news detection with deep learning. *IEEE Access*, *9*, 156151–156170 (2021).
23. Ali, I., Ayub, M. N. B., Shivakumara, P., & Noor, N. F. B. M. Fake news detection techniques on social media: A survey. *Wireless Communications and Mobile Computing*, *2022*(1), 6072084 (2022).

5 Exploring the Nexus of AI, the Metaverse, and Quality of Life

Amandeep Dhaliwal

5.1 INTRODUCTION

The metaverse has introduced a transformative shift in the realm of social interaction, prompting social media platforms to incorporate virtual reality (VR) and augmented reality (AR) technologies to offer immersive experiences that blur the boundaries between online and offline engagement [1–3]. Within this altered landscape, VR chatrooms and virtual events have emerged, facilitating interactions in three-dimensional spaces, thereby redefining the manner in which individuals connect [4].

Coined by Neal Stephenson in his 1992 science-fiction novel *Snow Crash*, the concept of the metaverse denotes a collective virtual shared space that integrates elements of the physical and digital realities. Recent advancements in technologies such as AR, VR, blockchain, and others have propelled the metaverse from the realms of science fiction into the realm of feasibility. It encompasses immersive environments, social interactions, digital economies, and user-generated content [5, 6]. The evolution of the metaverse can be traced from its early conceptualization in virtual worlds to its contemporary manifestations on digital platforms such as Fortnite, Decentraland, and Roblox [6]. While its inception was rooted in science-fiction literature, the present-day metaverse is firmly anchored in the progress of technology, where VR enables individuals to immerse themselves in digital realms, AR overlays digital elements onto the physical world, and blockchain secures transactions and digital asset ownership within this expanding digital milieu [3, 7, 8].

The growing significance of the metaverse extends across diverse sectors, exerting transformative influence. Economically, it has given rise to virtual economies, facilitating the generation, exchange, and possession of digital assets with tangible real-world value. Additionally, the metaverse holds profound implications for domains such as education, entertainment, healthcare, and social interaction [9]. This dynamic convergence of technology and human experience heralds a new era in our perception and navigation of reality.

The metaverse has revolutionized the realm of e-commerce, where retailers employ AR to enable customers to visualize products within their personal environments

DOI: 10.1201/9781032724508-6

before making purchases. Virtual showrooms and immersive shopping experiences empower users to explore products comprehensively, fostering informed decision-making and enhancing the online shopping experience [7]. This aligns with the trend towards experiential retail, offering customers immersive product interactions prior to purchase. Similarly, the metaverse is increasingly integrated into educational contexts, featuring virtual classrooms and training simulations that provide dynamic and engaging learning experiences. Educational institutions explore VR for practical training, medical simulations, and virtual field trips, democratizing access to quality education while transcending geographical constraints [10].

The entertainment industry seamlessly embraces the metaverse, with video games evolving into immersive worlds where users engage in intricate narratives and social interactions. Platforms such as Fortnite and Roblox have metamorphosed into metaverse-like environments, hosting concerts, events, and creative communities [11, 12]. These platforms blur the boundary between gaming and social interaction, offering users multifaceted experiences. Additionally, the metaverse has made significant inroads in real estate and architecture, giving rise to virtual property markets and digital real estate ownership within blockchain-based metaverse platforms such as Decentraland [2]. Architects and designers leverage VR to visualize and present their creations, enabling clients to experience spaces before the commencement of construction [67].

This chapter undertakes an examination of the potential impact of the metaverse on individuals' quality of life in an increasingly digitized world. As the digital landscape continues to evolve, it becomes imperative to scrutinize its potential ramifications on individuals' quality of life. This chapter offers an in-depth exploration of the multifaceted effects of the metaverse on various dimensions of quality of life, considering both its potential advantages and the challenges it presents.

5.2 METHODOLOGY

The research technique included a comprehensive literature analysis to synthesize existing research findings and provide theoretical answers to the concerns. This chapter investigated the metaverse from the standpoint of its possible usefulness, its implications on enhancing quality of life, potential misuse, and privacy problems. Given that this area and the metaverse domain are still evolving, there were many new types of research coming out, and our review found that this research topic had tremendous promise to be further explored. Thus, secondary sources were chosen from peer-reviewed worldwide sources and established journals and databases, with a concentration on research articles published after 2015. The research sources include studies published in well-known indexes as well as other white papers and other research available online. After removing the repetitious article listings and non-English language, only 65 papers from trustworthy sources and top-tier publications were further shortlisted and studied in depth.

5.2.1 Research Model and Questions

The current study focused on three main aspects of the metaverse to further investigate as given in Figure 5.1.

The research questions framed for the study were:

RQ1: What are the potential applications of the metaverse, and what are their implications for quality of life in general?

RQ2: What are the challenges and dangers of the metaverse that may have a negative impact on society?

5.3 METAVERSE AND ITS IMPLICATIONS ON QUALITY OF LIFE

The metaverse exerts a multifaceted impact on various facets of quality of life, necessitating a comprehensive examination of its implications, both positive and negative.

5.3.1 Social Connectivity and Interaction

The metaverse has introduced unprecedented opportunities for augmenting social connectivity. It enables individuals to partake in virtual gatherings, collaborative endeavors, and shared experiences, transcending geographical boundaries. The formation of digital communities within the metaverse fosters a sense of belonging and mitigates feelings of isolation, aligning with Putnam's social capital theory, which underscores the contribution of robust social networks to psychological well-being and community resilience [13]. Consequently, the metaverse holds the potential to positively influence the social dimension of quality of life.

The metaverse's capacity to overcome geographical constraints allows individuals to establish connections irrespective of physical distance [3, 4]. This phenomenon is notably evident in virtual events and gatherings, where participants from around the world converge. Virtual conferences, concerts, and workshops hosted within the metaverse democratize access to knowledge and entertainment, affording individuals the opportunity to engage with experts, artists, and thought leaders they may not have encountered in the physical realm.

Collaboration experiences a revival within the immersive environments of the metaverse [1]. Shared virtual spaces serve as platforms for creative projects, collaborative problem-solving, and co-creation. Instances such as virtual art galleries, collaborative design workshops, and educational simulations exemplify how the metaverse nurtures shared encounters that transcend traditional constraints. The

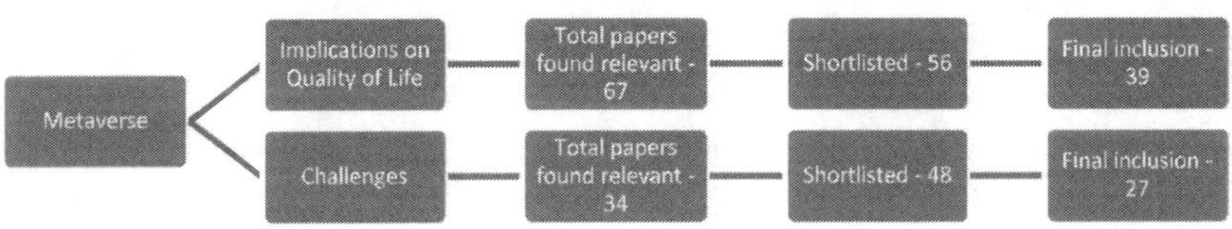

FIGURE 5.1 Research model

metaverse has given rise to vibrant virtual communities [6], enabling individuals with common interests to establish connections through online games, social VR platforms, and digital forums. These communities cultivate a sense of belonging and camaraderie, frequently resulting in meaningful relationships that bridge geographical and cultural divides [3, 6].

The metaverse's impact on social connectivity aligns with the principles of social capital as articulated by Putnam (2000). Social capital underscores the significance of social networks and communal bonds. The metaverse enhances social capital by facilitating connections that traverse conventional boundaries, allowing individuals to forge bonds over shared interests, experiences, and values, thereby fostering a rich tapestry of social interactions [14–16].

However, the heightened social connectivity facilitated by the metaverse engenders both advantageous and apprehensive elements [17, 18]. On the positive side, it broadens individuals' horizons, promoting cultural exchange and exposure to diverse perspectives. Furthermore, it provides a lifeline for individuals confronting social isolation or physical constraints. Nevertheless, concerns surrounding the authenticity of digital interactions and the potential for depersonalization also loom, echoing concerns articulated by Turkle [19].

5.3.2 Psychological Well-Being and Escapism

The advent of the metaverse, characterized by the fusion of virtual and physical domains, has instigated profound transformations in the realms of individuals' psychological well-being and their capacity for escapism [20]. The metaverse has notably elevated psychological well-being and facilitated escapism, affording individuals innovative channels for self-expression, creative exploration, and temporary respite from the exigencies of reality [21].

The metaverse serves as a canvas upon which individuals can navigate and articulate their thoughts, emotions, and identities in ways that transcend the constraints of the physical world [15, 22, 23]. This avenue for self-expression has been correlated with enhanced psychological well-being, as it offers an outlet for catharsis, emotional release, and the cultivation of a positive self-concept [24]. Avatars, virtual personas, and digital environments empower individuals to experiment with diverse facets of their identities, potentially fostering self-esteem and self-assuredness [21].

Furthermore, the metaverse provides an opportunity for individuals to engage in activities that align with their passions and interests. This engagement, typified by the concept of "flow," has been associated with heightened psychological well-being [22]. The immersive and interactive attributes of the metaverse can induce a state of flow, wherein individuals experience deep concentration, satisfaction, and a sense of personal growth [25].

Escapism, characterized by the yearning for temporary respite from the pressures and challenges of reality, finds a profound outlet within the metaverse [26]. Virtual environments, distinct from the physical world, furnish a sanctuary for individuals in search of solace and refuge. Virtual realms offer opportunities to disengage from real-world pressures, providing interludes of relaxation and serenity [20, 21].

Engaging in creative pursuits, traversing imaginative landscapes, or simply interacting with others within the metaverse can serve as a therapeutic intermission from the intricacies of daily life.

Additionally, the metaverse empowers individuals to explore fantastical realms and narratives that offer respite from the mundane aspects of reality. This form of narrative escapism, wherein users immerse themselves in intricate storylines, empowers individuals to temporarily inhabit alternate worlds and experiences [25]. These encounters possess the potential to alleviate stress, foster emotional well-being, and rekindle a sense of wonder and awe [27].

5.3.3 Economic Implications

Digital economies flourishing within the metaverse, underpinned by the utilization of cryptocurrencies and blockchain technology, offer fresh avenues for economic engagement [9]. Virtual employment opportunities, digital assets, and the emergence of non-fungible tokens (NFTs) present prospects for income generation. However, the presence of economic disparities and the potential for exploitation within virtual labor markets necessitate a conscientious exploration of ethical dimensions. Through the provision of innovative platforms for entrepreneurial endeavors, employment opportunities, and the fostering of inventive pursuits, the metaverse has undergone a transformative impact on various sectors, enabling individuals to participate in novel forms of economic enterprise [28].

The metaverse has emerged as a fertile terrain for entrepreneurial activities and innovative ventures. Virtual marketplaces within the metaverse, exemplified by platforms such as Decentraland and CryptoVoxels, have afforded entrepreneurs the capacity to engage in the acquisition, sale, and exchange of virtual assets via blockchain technology [29]. These platforms have ushered in a burgeoning market for digital real estate, artistic creations, and collectibles, thereby generating substantial economic value [30]. Furthermore, the metaverse has served as an incubator for startups dedicated to crafting immersive VR and AR experiences, spanning diverse domains from entertainment to education. These enterprises harness the metaverse's immersive potential to offer distinctive products and services tailored to the discerning preferences of consumers [31].

Content creators have devised inventive methods for monetizing their creative output within the metaverse. The advent of non-fungible tokens (NFTs) has wrought a revolution in the notions of ownership and value within the digital sphere [32]. Artists and content creators can tokenize their work as NFTs, granting them the ability to directly vend and trade their creations within virtual marketplaces. This development has inaugurated novel revenue streams, affording creators the opportunity to directly reap the benefits of their content's value in ways that were hitherto unattainable.

5.3.4 Employment and Remote Work

The advent of the metaverse has ushered in a transformation in the realm of employment, leading to a restructuring of conventional employment paradigms through its capacity to facilitate remote work and thereby transcending geographical constraints [33]. The metaverse introduces virtual offices and collaborative environments, enabling employees to engage in work-related tasks, meetings, and interactions within a shared digital realm [34]. This shift carries implications for both businesses and the workforce, offering heightened flexibility and opportunities for integrating work with personal life. As the metaverse blurs the boundaries between physical and digital presence, it has given rise to an innovative employment landscape that necessitates the acquisition of new skill sets and adaptability.

The emergence of the metaverse has engendered a fundamental redefinition of remote work and global collaboration. Virtual offices and shared digital spaces have enabled professionals to participate in collaborative endeavors regardless of geographic limitations [33, 34]. This transition expands the talent pool available to businesses and organizations, surmounting traditional constraints associated with physical proximity. Consequently, employment opportunities have transcended borders, affording individuals the capability to contribute to projects from any corner of the globe [35].

The rise of the metaverse has given birth to a diverse array of novel job roles and specialized positions. Industries such as virtual reality development, AR content creation, and metaverse architecture have flourished, necessitating a workforce equipped with technical proficiency in these domains [31]. Professionals adept at designing immersive experiences, orchestrating virtual events, and curating virtual environments are in growing demand, underscoring the metaverse's potential to create fresh employment avenues [36]. Employment within the metaverse demands a high degree of skill diversification and adaptability. The metaverse landscape is in a constant state of evolution, compelling professionals to remain abreast of technological advancements and emerging trends [33, 37]. In this dynamic milieu, the cultivation of ongoing skill development becomes imperative, motivating individuals to enhance their proficiency in domains such as virtual reality programming, blockchain technology, and digital marketing. Adaptability and the capacity to swiftly acquire new skills have emerged as pivotal assets for thriving in the job market driven by the metaverse [35].

5.3.5 Entertainment and Content Creation

The metaverse has emerged as a pivotal force in shaping the landscape of entertainment. It has redefined the paradigms of consumer engagement, artistic expression, and the very essence of content dissemination. The metaverse has revolutionized entertainment by affording unparalleled opportunities for immersive virtual experiences. Virtual concerts, conferences, and events conducted within the metaverse have garnered substantial global audiences, instilling a profound sense of presence among participants [33, 38, 39]. Attendees are empowered to engage with performers, navigate virtual venues, and partake in distinctive social interactions, transcending the confines of physical space. This heightened engagement has redefined the concept of live entertainment, ushering in an entirely novel dimension of audience participation [39].

Content creation within the metaverse transcends conventional modes of storytelling. Creators now possess the capacity to craft interactive narratives that beckon users to actively partake in and influence the trajectory of the storyline [40]. AR and VR experiences immerse users in dynamic environments where they can shape the narrative's progression [31, 41]. This participatory narrative approach blurs the demarcation between consumer and creator, fostering a deeper level of engagement and personalized connection with the content. The metaverse has become a canvas for artists and creators to push the boundaries of artistic expression [42]. Virtual art exhibitions and installations provide artists with a platform to showcase their work to a global audience, frequently experimenting with new mediums and forms [30, 40]. Additionally, the metaverse facilitates collaboration among artists hailing from diverse backgrounds, enabling them to co-create immersive experiences that meld visual art, music, and interactive elements [41]. This collaborative milieu has sparked fresh manifestations of cross-disciplinary creativity.

5.3.6 Retail and Consumer Engagement

Retail experiences have undergone profound transformation owing to the influence of the metaverse. Virtual storefronts and showrooms empower businesses to exhibit their products and engage with customers within immersive digital environments [43, 44]. Consumers can interact with products prior to making purchases, elevating their online shopping encounters [45]. Notably, virtual reality try-on experiences have revolutionized the fashion industry, enabling customers to visualize clothing and accessories within a virtual context, thereby mitigating concerns regarding fit and appearance [43]. These innovations underscore the metaverse's potential to bridge the gap between e-commerce and conventional retail, stimulating economic growth within these sectors.

The metaverse has fundamentally redefined the notion of virtual storefronts, enabling retailers to construct immersive shopping experiences that transcend the confines of traditional e-commerce [44]. Virtual showrooms and interactive three-dimensional environments empower consumers to explore products within lifelike settings, fostering a deeper connection with merchandise [32]. This heightened sensory engagement results in enhanced consumer confidence and a more informed decision-making process [18, 46].

A particularly impactful contribution of the metaverse to consumer engagement lies in its ability to offer AR try-on experiences. In sectors such as fashion and beauty, consumers can virtually test clothing, accessories, and cosmetics prior to purchase. This personalized interaction heightens consumer satisfaction by alleviating concerns related to fit and appearance, ultimately reducing the likelihood of product returns and exchanges.

Furthermore, the metaverse has seamlessly bridged the gap between online and offline shopping experiences. Retailers can integrate physical store locations with virtual counterparts, delivering a consistent and comprehensive brand experience to consumers [47]. For instance, shoppers can visit a virtual store to explore products, receive personalized recommendations, and then seamlessly transition to an offline store for a tactile experience [7]. This omnichannel approach augments consumer engagement by catering to diverse shopping preferences.

The notion of virtual economies within the metaverse has unlocked new channels for consumer engagement and revenue generation [47]. Digital collectibles, represented as non-fungible tokens (NFTs), have emerged as a distinctive avenue for brands to connect with consumers through limited-edition virtual merchandise and exclusive content. These digital collectibles resonate with consumers' desires for ownership and exclusivity, cultivating a sense of belonging within virtual communities [44].

5.3.7 Education and Skill Development

The metaverse has exerted a discernible influence on the domains of education and skill development. Virtual classrooms and training simulations have been instrumental in delivering dynamic learning experiences, catering to the diverse spectrum of learning styles [48]. Enterprises have harnessed the metaverse to furnish employee training in authentic settings, augmenting the acquisition and retention of skills. Additionally, educational institutions have ventured into the realm of immersive learning platforms within the metaverse, fostering heightened engagement and knowledge acquisition among students. This confluence of education and technology underscores the metaverse's capacity to nurture human capital development, thereby contributing substantively to economic advancement.

The metaverse has introduced immersive learning environments that transcend the confines of conventional classroom settings [49]. Smartphones have greatly contributed to big data, which increases the efficiency of machine learning [50]. Through the utilization of virtual reality (VR) and AR technologies, students can engage with intricate concepts and scenarios in dynamic and interactive ways [51]. This pedagogical approach to learning amplifies comprehension, retention, and engagement, as students actively partake in the learning process, rather than passively absorbing information [10].

The metaverse's impact on education extends to the provision of real-world simulations and experiential learning opportunities. Disciplines such as medicine, engineering, and aviation have derived notable benefits from VR-based simulations that afford students the opportunity to practice skills within secure and controlled environments [52, 53]. These simulations faithfully replicate real-world scenarios, enabling learners to cultivate practical skills and hone problem-solving capabilities without exposure to associated risks. This experiential learning approach is especially effective in nurturing critical thinking and decision-making proficiencies [53].

The metaverse surmounts geographical constraints, extending global access to education and knowledge. Students from diverse backgrounds can participate in virtual classrooms, workshops, and lectures, irrespective of their physical location [34]. This inclusivity fosters cross-cultural exchange, intercultural collaboration, and the sharing of varied perspectives, thereby enriching the educational experience [49]. The global accessibility facilitated by the metaverse holds the potential to alleviate educational disparities and offer equitable learning opportunities.

In the metaverse, education can be meticulously tailored to individual learning styles and preferences. Adaptive learning technologies embedded within the metaverse are capable of scrutinizing students' progress and learning patterns, permitting the formulation of personalized learning trajectories [48, 52]. Students can engage with educational content at their own pace, receiving targeted support and interventions as needed. This personalized pedagogical approach heightens engagement, motivation, and overall learning outcomes [42].

Some of the prominent applications of Metaverse are given below in Table 5.1.

5.4 CHALLENGES OF METAVERSE ON LIFE QUALITY

As the metaverse unfolds, it brings forth a multitude of prospects and challenges across various domains, necessitating meticulous examination and consideration. In the realm of human interaction, commerce, and entertainment, the convergence of physical and digital realities is poised to reshape fundamental aspects of society. This transformation is accompanied by a spectrum of complex and multifaceted future considerations and policy implications, drawing insights from contemporary discourse and scholarly works.

a) Identity in Flux: The metaverse's profound impact on human identity is a central consideration [19]. As individuals seamlessly traverse the boundaries between physical and virtual realms, traditional notions of selfhood, privacy, and authenticity are increasingly blurred. These shifts have the potential to reshape individual well-being, mental health, and societal cohesion. Therefore, a nuanced examination of how these changes influence human experiences is essential [54].
b) *Economic Transformation:* The metaverse's implications for the labor force and the economy are poised to be transformative. Virtual workspaces, remote collaboration, and digital entrepreneurship have the potential to redefine conventional employment models [35]. However, the metaverse's global reach may also exacerbate issues related to income inequality and labor rights. Striking a balance between economic growth and ensuring equitable working conditions becomes imperative [55].
c) *Digital Rights and Privacy:* As users increasingly engage with the metaverse, their personal data becomes a valuable asset. In the recent decade, there has been a dramatic increase in the development of computer malware. Malicious software (malware) is now commonly employed by cybercriminals to target computer systems [55, 56]. Policy measures must be proactive in safeguarding users' personal information, ensuring transparency in data collection, and empowering individuals to retain control over their digital

TABLE 5.1
Various Applications of Metaverse

Application Area	Description
Gaming and Entertainment	Creating virtual worlds, immersive gaming experiences, and interactive entertainment.
Social Interaction	In virtual areas, we can facilitate social gatherings, meetings, and communication.
Education	Virtual classrooms, training simulations, and collaborative learning environments are available.
Work and Collaboration	Remote work, virtual offices, and collaborative workplaces are made possible for teams and enterprises.
Healthcare	Offering virtual medical simulations, telemedicine, and therapy sessions.
Art and Creativity	Creating platforms for artists, musicians, and makers to promote and collaborate on their work.
Commerce and Retail	Purchase goods and services through virtual storefronts, marketplaces, and shopping experiences.
Real Estate	Virtual property tours, architectural design, and property development planning are all available.
Travel and Tourism	Experiences in virtual tourism, travel planning, and location exploration.
Sports and Events	Virtual attendance, live broadcasting of athletic events, and interactive sports simulations are all options.
Research and Science	Simulations and collaborative research settings for a wide range of scientific disciplines.
NFTs and Digital Collectibles	Digital assets and non-fungible tokens (NFTs) are traded and displayed.
Governance and Civic Engagement	Virtual town halls, assemblies, and political participation in cyberspace.
Environmental Conservation	Ecosystem simulation and research, climate change, and wildlife conservation.
Mental Health and Therapy	Offering secure and confidential locations for virtual therapy and mental health help.
Fashion and Personal Expression	Virtual fashion displays, avatar customization, and digital style expression are all possibilities.

identities. A comprehensive approach to data protection is essential to prevent issues related to surveillance capitalism [58].

d) Accessibility and Inclusion: Policies must address the digital divide by providing affordable access to metaverse technologies and fostering digital literacy through training programs [56]. Additionally, inclusive design practices should be encouraged to ensure that the metaverse remains accessible to individuals with disabilities, promoting a diverse and inclusive user community [59].

e) Ethical Dilemmas: The expansion of the metaverse raises ethical questions surrounding content moderation, virtual property rights, and freedom of expression. Policymakers need to collaborate with stakeholders to develop guidelines that balance creative freedom with the prevention of harmful or offensive content [38, 60]. Virtual property rights also present complex issues related to ownership, inheritance, and intellectual property that policy frameworks should address [61].

f) Cultural Preservation: Consideration should be given to the metaverse's potential to reshape cultural expressions and experiences [62]. Collaborations between technologists, cultural institutions, and policymakers can ensure that Indigenous knowledge, historical artifacts, and traditional practices are respected and preserved within the digital landscape [63, 64].

g) International Collaboration: Given the borderless nature of the metaverse, international collaboration is paramount. Policymakers and stakeholders should engage in cross-border discussions to establish common standards for security, data governance, and user protection [65]. The creation of international regulatory bodies can provide a platform for addressing global challenges and harmonizing policy approaches [40].

Thus, the metaverse's emergence ushers in a new era of possibilities and complexities, requiring a proactive and multifaceted policy approach. The problems given in Table 5.2 highlight the importance of careful study and responsible metaverse development to ensure that it improves, rather than degrades, the quality of life for individuals and communities. Policymakers must navigate the intricate interplay of technology, society, and ethics to harness the metaverse's potential while mitigating its associated challenges and risks [29, 66].

5.5 CONCLUSION

In conclusion, the metaverse has emerged as a potent driver of expanded economic prospects spanning diverse sectors. Its capacity to stimulate entrepreneurial endeavors, redefine labor paradigms, mold entertainment paradigms, revolutionize retail practices, and reformulate educational methodologies has unveiled its transformative potential in shaping economic landscapes. However, unlocking these potentials mandates a judicious response to existing challenges, necessitating the establishment of robust regulatory frameworks. As technology perpetually advances, the metaverse is poised to assume an increasingly prominent role in augmenting economic opportunities, providing a glimpse into the future of interconnected digital economies.

The metaverse has effectively transcended conventional boundaries, particularly in the realm of virtual and physical commerce, ushering in immersive shopping experiences and AR try-on functionalities. Its profound impact on employment is evident through the reconfiguration of work modalities, fostering novel approaches to work, collaboration, and income generation. Additionally, the metaverse has left

TABLE 5.2
Various Challenges of Metaverse to Quality of Life

Challenge	Description
Physical Health Concerns	Excessive screen time and continuous usage of VR/AR systems can cause eye strain, motion sickness, and sedentary behavior, all of which can have a negative impact on physical health.
Mental Health and Addiction	Overuse of the metaverse, social isolation, and addiction to virtual settings can all lead to mental health problems such as sadness, anxiety, and addiction.
Real-world Disconnection	Excessive immersion in the metaverse may result in a detachment from real-life social interactions, potentially harming relationships and emotional well-being.
Privacy Concerns	Increased digital presence poses privacy concerns, such as data security, spying, and the potential misappropriation of personal information.
Work-Life Balance	Work and personal life integration in virtual environments may blur boundaries and make it difficult to maintain a healthy work-life balance.
Economic Disparities	Economic inequities can arise as those with access to advanced technology have greater possibilities and benefits in the workplace.
Social Inequality	Discrimination, harassment, and exclusion in virtual communities can reflect real-world socioeconomic inequities and reduce users' quality of life.
Content and Cyberbullying	Cyberbullying, harassment, and improper content in the metaverse can have a negative impact on users' mental health and well-being.
Addiction to Virtual Worlds	The immersive aspect of the metaverse can lead to addictive habits that detract from real-world experiences and quality of life.
Environmental Impact	Metaverse infrastructure's energy usage can have an environmental impact, contributing to carbon footprints and climate change.
Ethical and Moral Dilemmas	The creation of AI beings and debates about the moral accountability of digital actions may inject ethical quandaries into the metaverse.
Disconnection from Nature	People who rely too heavily on virtual experiences may become disconnected from nature and environmental concerns, affecting their general well-being.
Loss of Physical Spaces	Physical locations such as shopping districts may suffer as a result of the metaverse, hurting local communities and urban life.

an indelible mark on the landscape of entertainment and content creation, offering innovative avenues to engage diverse audiences, narrate compelling stories, and monetize creative pursuits.

The multifaceted influence of the metaverse on the quality of life is undeniable, with the potential to enrich social connectedness, stimulate creative expression, and expand economic horizons. However, the realization of these affirmative potentials hinges upon the judicious resolution of ethical quandaries, psychological considerations, and socioeconomic disparities. These issues must be earnestly addressed to harness the metaverse's positive contributions fully.

Through a holistic and interdisciplinary approach coupled with continuous research endeavors, society stands poised to navigate the intricate tapestry of challenges and opportunities woven by the metaverse. In doing so, it can mold a digital future that not only enriches but also sustains human well-being.

As the metaverse combines psychology, technology, ethics, and other disciplines, research on the influence of the metaverse on quality of life is innovative. It investigates the impact of emerging technologies, such as VR and AR, on mental health, work, relationships, and economics. Researchers are investigating the metaverse's social and psychological ramifications, ethical quandaries, and environmental impact. This study also looks at the implications for privacy, security, and digital identity. As the metaverse evolves, it presents new problems regarding policy, inclusivity, and cultural transitions, further enhancing its originality in impacting human well-being and society dynamics.

5.6 FUTURE RESEARCH DIRECTIONS

Some of the future study directions on how the metaverse impacts quality of life can diversify into many interdisciplinary fields, providing insights into this expanding digital universe. Some of the crucial research directions can be in the area of the long-term effects of the metaverse, where the researchers can study the long-term social and psychological effects of metaverse use, both positive (improved connectivity, collaboration) and negative (digital addiction, isolation). Another important area could be focusing on developing and evaluating the ethical norms for the metaverse, particularly in areas such as AI behavior, digital consent, and moral duties. Another aspect could be the rules and regulations of governments, as well as their effectiveness in addressing data governance and digital rights in the metaverse, which must be scrutinized, and the metaverse's environmental impact, which should be studied in order to reduce its carbon footprint and promote sustainability. On the societal aspect, future researchers can investigate inclusivity and accessibility as well as human-AI interactions and their impact on cultural and artistic expressions. Thus, as the metaverse is an evolving technology, it offers many promising future directions of research.

REFERENCES

1. Amicis R, Fiore F, Perini A. Immersive co-design and co-creation in the metaverse. In: *Extended Abstracts of the 2021 CHI Conference on Human Factors in Computing Systems*; 2021.
2. Anderson J, Rainie L. The metaverse in 2040. *Pew Research Centre* 2022;30:1–203.
3. Gupta A, George D. The Metaverse: A paradigm shift in virtual social spaces. *Frontiers in Psychology* 2021;12:667508.
4. Qin J, Liu Y, Zhang D, Fu H. An exploratory study of the impact of virtual reality on social interaction. *Frontiers in Psychology* 2020;11:59.
5. Narula H. Virtual Society: The Metaverse and the New Frontiers of Human Experience. Random House; 2022.

6. Wohn DY, Freeman D, Gutierrez-Osuna R. Metaverse social spaces: Understanding user experiences and network structures. *ACM Transactions on Computer-Human Interaction (TOCHI)* 2019;26(6):1–36.
7. Chen J, Wang L, Ma L. Virtual reality try-on: A new opportunity in the fashion industry. *Journal of Fashion Marketing and Management* 2020;24(4):605–19.
8. Rani S, Tripathi K, Arora Y, Kumar A. A Machine Learning Approach to Analyze Cloud Computing Attacks. In: *2022 5th International Conference on Contemporary Computing and Informatics (IC3I)* (pp. 22–26); 2022. Available from: https://ieeexplore.ieee.org/document/10073468 (cited 2023 Oct 25).
9. Schmitt M. *Metaverse: Bibliometric review, building blocks, and implications for business, government, and society.* Building Blocks, and Implications for Business, Government, and Society (July 21, 2022) 2022.
10. Mikropoulos TA, Natsis A. Educational virtual environments: A ten-year review of empirical research (1999–2009). *Computers & Education* 2011;56(3):769–80.
11. Evans L, Frith J, Saker M. *From microverse to metaverse: Modelling the future through today's virtual worlds.* Emerald Publishing Limited; 2022.
12. Jackson B. Into the metaverse; 2023. Available from: https://www.infotech.com/research/ss/into-the-metaverse (cited 2023 Jul 10).
13. Putnam RD. *Bowling alone: The collapse and revival of American community.* Simon & Schuster; 2000.
14. Amichai-Hamburger Y. Technology and social isolation. In: Zalta EN, editor. *The Stanford Encyclopedia of Philosophy.* Stanford University; 2013.
15. Christensen JF, Satchell C. Social robots and human-like conversational artificial intelligence. In: *Social Robots.* Springer; 2018. pp. 1–16.
16. Jin SAA. From parasocial to social interaction: The mediating role of perceived social interaction in the relationships between parasocial interaction and social media users' psychological outcomes. *Health Communication* 2020;35(1):49–59.
17. Kuss DJ, Griffiths MD. Internet gaming addiction: A systematic review of empirical research. *International Journal of Mental Health and Addiction* 2012;10(2):278–96.
18. Adams D. Virtual retail in the metaverse: Customer behavior analytics, extended reality technologies, and immersive visualization systems. *Linguistic and Philosophical Investigations* 2022;(21):73–88.
19. Turkle S. *Reclaiming conversation: The power of talk in a digital age.* Penguin Books; 2015.
20. Lee EJ, Lee W, Bae I. What is the draw of the metaverse? Personality correlates of zepeto use motives and their associations with psychological well-being. *Cyberpsychology, Behavior, and Social Networking* 2023;26(3):161–168.
21. Han DID, Bergs Y, Moorhouse N. Virtual reality consumer experience escapes: preparing for the metaverse. *Virtual Reality* 2022;26(4):1443–1458.
22. Csikszentmihalyi M. *Flow: The Psychology of Optimal Experience.* Harper & Row; 1990.
23. Gaggioli A. Transformative experience design. In: Calvo R, D'Mello S, editors. *New Perspectives on Affect and Learning Technologies.* Springer; 2016. pp. 225–239.
24. O'Connor BP, Jewell MJ. Hope, optimism, and college student well-being. *Journal of College Student Development* 2002;43(3):363–368.
25. Ud Din I, Almogren A. Exploring the psychological effects of Metaverse on mental health and well-being. *Information Technology & Tourism* 2023;25(3):367–389.
26. Cheung CK. Online social networks: Why do students use Facebook? *Computers in Human Behavior* 2011;27(4):1337–43.
27. Hamari J, Koivisto J, Sarsa H. Does gamification work? A literature review of empirical studies on gamification. In: *2017 50th Hawaii international conference on system sciences (HICSS).* IEEE; 2017. pp. 3025–3034.

28. De Felice F, De Luca C, Di Chiara S, Petrillo A. Physical and digital worlds: implications and opportunities of the metaverse. *Procedia Computer Science* 2023;217:1744–1754.
29. Polona C, André M, Maria N. Metaverse: Opportunities, risks and policy implications, *EPRS: European Parliamentary Research Service* 2022. Belgium. Retrieved from https://coilink.org/20.500.12592/gjscr8 on 30 Apr 2025. COI: 20.500.12592/gjscr8.
30. Johnson W, Seltzer A, Thelwall M. *The Market for Digital Art NFTs*; 2020.
31. Smith D, Anderson CA. The Virtual Reality Gold Rush: Entrepreneurial Opportunities and Challenges. *California Management Review* 2019;61(1):51–76.
32. Chen J, Wang L, Ma L. Virtual reality try-on: A new opportunity in the fashion industry. *Journal of Fashion Marketing and Management* 2020;24(4):605–619.
33. Kral P, Janoskova K, Dawson A. Virtual skill acquisition, remote working tools, and employee engagement and retention on blockchain-based metaverse platforms. *Psychosociological Issues in Human Resource Management* 2022;10(1):92–105.
34. Wagner D, López JDJ, Guiry MD. Virtual Reality and Augmented Reality in the Workplace: A Comprehensive Review and Research Agenda. *International Journal of Management Reviews* 2021;23(3):464–496.
35. Carter D. Immersive employee experiences in the Metaverse: Virtual work environments, augmented analytics tools, and sensory and tracking technologies. *Psychosociological Issues in Human Resource Management* 2022;10(1):35–49.
36. Martin M, Rickman A. The challenge of understanding artificial intelligence and its effect on employment. *Labour Economics* 2020;65:101851.
37. Huang MH, Rust RT. Engaged to a robot? The role of AI in service. *Journal of Service Research* 2021;24(1):30–41.
38. Faraboschi P, Frachtenberg E, Laplante P, Milojicic D, Saracco R. Virtual worlds (Metaverse): From skepticism, to fear, to immersive opportunities. *Computer* 2022;55(10):100–106.
39. Ko P, Kim D, Kim B. Evaluating the user experience of virtual reality live concerts. *Journal of Business Research* 2022;150:722–732.
40. Wang W, Li X, Wang X. AI personalization in e-commerce: A systematic mapping study. *Electronic Commerce Research and Applications* 2020;40:100909.
41. Narin NG. A content analysis of the metaverse articles. *Journal of Metaverse* 2021;1(1):17–24.
42. Lin H, Wan S, Gan W, Chen J, Chao HC. *Metaverse in Education: Vision, Opportunities, and Challenges*. IEEE; 2022. pp. 2857–2866.
43. Ismail AR. Virtual Shopping in The Metaverse: How Brand Engagement in Self-Concept Influence Brand Loyalty. Available at *SSRN* 2023;4494222.
44. Popescu GH, Valaskova K, Horak J. Augmented reality shopping experiences, retail business analytics, and machine vision algorithms in the virtual economy of the metaverse. *Journal of Self-Governance and Management Economics* 2022;10(2):67–81.
45. Chen X, Li Y, Su Z. A study of chatbots in financial services. In: *International Conference on Service Systems and Service Management (ICSSSM)*. IEEE; 2018. pp. 1–5.
46. Abidin C. Visibility Labour: Engaging with Influencers' Fashion Brands and #OOTD Advertisements on Instagram. *Media International Australia* 2016;161(1):86–100.
47. Dawson A. Data-driven consumer engagement, virtual immersive shopping experiences, and blockchain-based digital assets in the retail metaverse. *Journal of Self-Governance and Management Economics* 2022;10(2):52–66.
48. Huang W, Huang WH, Chen DZ, Lin CM. Exploring the acceptance of immersive virtual reality for learning: The roles of presence and personality traits. *Computers & Education* 2021;166:104155.
49. Preston J. Facebook, the metaverse and the monetisation of higher education. *Impact of Social Sciences Blog* 2021.

50. Rajendra P, Kumari M, Rani S, Dogra N, Boadh R, Kumar A, et al. Impact of artificial intelligence on civilization: Future perspectives. *Materials Today: Proceedings* 2022;56:252–256.
51. Huang MH, Rust RT. Artificial intelligence in service. *Journal of Service Research* 2018;21(2):155–72.
52. Arifin A, Djumat I, Nicolas DG, Syam ASM, Saputra N. Metaverse in education; innovation strategy, learning acceleration, and optimization. *Journal of Namibian Studies: History Politics Culture* 2023;34:1470–1485.
53. Azoury N, Hajj C. *Perspective Chapter: The Metaverse for Education*; 2023.
54. Spence EH. Meta ethics for the metaverse: The ethics of virtual worlds. *Current Issues in Computing and Philosophy* 2008;175(3).
55. Wells R. Navigating privacy and security in the metaverse. *Communications of the ACM* 2023;66(1):44–47.
56. Kour M, Rani K. Challenges and opportunities to the media and entertainment industry in metaverse. *Applications of Neuromarketing in the Metaverse* 2023;88–102.
57. Rani S, Tripathi K, Arora Y, Kumar A. Analysis of anomaly detection of malware using KNN. In: *2022 2nd International Conference on Innovative Practices in Technology and Management (ICIPTM)*; 2022. pp. 774–779. Available from: https://ieeexplore.ieee.org/document/9754044?denied= (cited 2023 Oct 25).
58. Fernandez CB, Hui P. *Life, the Metaverse and Everything: An Overview of Privacy, Ethics, and Governance in Metaverse*. IEEE; 2022. pp. 272–277.
59. Bonales-Daimiel G. New Formulas of interactive entertainment: from advergaming to metaverse as an advertising strategy. In: *Handbook of Research on the Future of Advertising and Brands in the New Entertainment Landscape*. IGI Global; 2023. pp. 76–103.
60. Dignum V, al-Rifaie MM. Responsible AI: From ethical design to ethical behavior. In: *IEEE/ACM International Conference on Ethics of Autonomous Systems and Robots*. IEEE; 2018. pp. 1–5.
61. Choi HY. Working in the metaverse: Does telework in a metaverse office have the potential to reduce population pressure in megacities? Evidence from young adults in Seoul, South Korea. *Sustainability* 2022;14(6):3629.
62. Di Pietro R, Cresci S. *Metaverse: Security and Privacy Issues*. IEEE; 2021. pp. 281–288.
63. Consalvo M. Convergence Culture and the Metaverse. In: Huntemann N, Payne M, editors. *Joystick Soldiers: The Politics of Play in Military Video Games*. Routledge; 2019. pp. 167–182.
64. Łach K, Ćwiek-Karpowicz J. Unmasking the "Right to Explanation": Current challenges for algorithmic accountability. *Computer Law & Security Review* 2021;40:105483.
65. Angwin J, Larson J, Mattu S, Kirchner L. Machine Bias: There's Software Used Across the Country to Predict Future Criminals. And it's Biased Against Blacks; 2016. Available from: https://www.propublica.org/article/machine-bias-risk-assessments-in-criminal-sentencing.
66. Hof S, Kosta E. Privacy concerns and privacy behavior: How do they relate? *Journal of Consumer Policy* 2015;38(2):155–79.
67. Kumar A, Kumar P, Rathee S, Kumar B (Eds.). *Mechatronics: Concepts, Tools, Applications, and New Trends* (1st ed.). CRC Press; 2025. https://doi.org/10.1201/9781003494478.

6 A Literature Survey on AI-Driven Code-Mixed Text Analysis and Normalization

Poonam Gupta and Parvinder Singh

6.1 INTRODUCTION

The expansion of the internet and other technological innovations has sparked interest in researching social computing in several different fields, including anthropology, languages, and literature, as well as virtual worlds and e-culture. India is a culturally diverse country, and each of its regions has a distinctive culture of its own. People from all around the country, most of whom reside in rural areas, utilize online forums such as Twitter, Facebook, and YouTube to share their thoughts and suggestions. Code-mixed text is a prevalent practice in many cultures, especially in bilingual groups. For instance, due to the country's linguistic diversity, code-mixed writing is common on social networking sites in India. In their online communications, people frequently combine regional dialects, Hindi, and English. They communicate on social media sites in their regional dialects, which are not understood by others who belong to other cultures or who speak a different language.

Social media activity in code-mixed native languages has considerably increased over the past few years thanks to affordable internet and rising smartphone usage. Social media has dramatically changed how we communicate and engage with one another. By allowing us to connect with people around the globe and share thoughts, ideas, and experiences, technology now plays a major role in how we go about living our daily lives. One interesting aspect of social media is the phenomenon of code-mixed text, where people seamlessly blend multiple languages within a single conversation or post. Code-mixed text refers to the practice of using multiple languages, often within the same sentence or phrase, to express oneself on social media platforms. This linguistic fusion occurs due to various factors, including the diversity of users, multicultural environments, and the need to cater to a specific audience.

DOI: 10.1201/9781032724508-7

6.2 CODE-MIXING AND CODE-SWITCHING OVERVIEW

The act of juggling between two or more languages in a single discourse is known as code-switching. There are three types of code-switching:

1. Inter-Sentential Switching
2. Intra-Sentential Switching
3. Tag Switching or Extra Sentential.

Inter-sentential switching is the process of switching between sentences at their beginnings or ends, such as when transitioning from English to Hindi [1]. When a change is made within a single sentence, there are no pauses, hesitations, or breaks to indicate the change, and then it is known as "intra-sentential switching." When a word, a phrase, or both are translated between two languages, it is referred to as "tag switching." On social media sites such as Facebook, Twitter, and WhatsApp, code-mixing is the practice of sending a single message in multiple languages. In code-mixing, people used Roman script in combination with the English language as well as their native language to express their thoughts and views. Code-mixing techniques include when a foreign word or phrase is used in a sentence or sentence fragment that is largely composed of the local language; this is known as "intra-sentential code-mixing." Inter-sentential code-mixing is the act of switching between languages during a single speech or discussion. Code-mixing that occurs within a single word is referred to as "intra-word codemixing." In Indian languages, code-mixing is used in various fields such as information extraction (IE), information retrieval (IR), POS tagging, question answering systems, language identification, etc. [1, 2].

6.3 RELATED WORK

The field of NLP is very old and has been studied scientifically for close to fifty years. Although there is evidence of activity from earlier times, NLP history generally starts in the 1950s. Making machines understand human languages was the primary goal of early NLP research. Machine Translation (MT), a straight dictionary look-up system created in 1954 to translate Russian sentences into English, is considered the first use of NLP that is still widely used today [3, 4]. In 1965, a young American linguist named Noam Chomsky first proposed the idea of syntactic structure description using generative grammar rules. Since then, he has had a significant impact on almost all NLP activity [5]. For the analysis of code-mixed data, NLP approaches and machine learning (ML) algorithms dominate the literature study. Table 6.1 shows the summary of related work in the area of code-mixed text analysis, normalization, and natural language processing. Table 6.2 shows limitations of code-mixed text.

TABLE 6.1
Related Work

S.No.	Authors	Year	Description
1	Muhammad Ramzan et al. [6]	2021	They investigated the variations in the usage of code-mixing and coding switching. The number of words (noun, adjective, verb and adverb), phrases (verb, noun), and clauses and sentences (sentence insertion and clause) are all counted.
2	Shree Harsh et al. [7]	2015	They focused on identifying infected noun forms in both Hindi and English before translating them into the desired target language.
3	Gurpreet Singh Josan et al. [8]	2008	They wanted to know if higher order n-gram models improved word sense in Punjabi. Entropy has been shown to be a valid indicator of an n-gram model's fitness for the process of word sense disambiguation.
4	Grigori Sidorov et al. [9]	2014	Sn-grams use in authorship attribution was also covered in detail. As a starting point, the conventional n-grams of words, characters, and POS tags were classified using three classifiers: SVM, NB, and J48. SVM classifier performs better when using sn-grams.
5	Vishal Gupta [10]	2014	All Punjabi words, including all adjectives, adverbs, nouns, verbs, pronouns, and proper names, are fully automatically stemmed, was the author's main goal. The proposed Punjabi stemming algorithm has an efficiency of 87.37%.
6	Amandeep Kaur et al. [11]	2015	The NER System for Punjabi is assessed using a variety of feature combinations, including as the context word window, different digital features, and unusual word length features. Among all these, feature set with word window 5 performs better with f-score, precision and recall values 87.46, 90.99% and 84.19% respectively.
7	Amandeep Kaur et al. [12]	2017	A hybrid system that combines N-gram modeling, subjective lexicon and support vector machines is built for the sentiment analysis of Punjabi text. This hybrid system performs more effectively in terms of precision, recall and f-score.
8	Melvin Johnson et al. [13]	2017	It is demonstrated that it is possible to train multilingual NMT models that using a single model with shared parameters, translate between different languages with the added advantage of raising the quality of translation for the languages that are mixed in that have limited resources, using a simple solution for multilingual Neural Machine Translation (NMT).
9	Vaibhav Aggarwal et al. [14]	2018	IAN (Interactive Analyser) tool and the X-Bar technique is used to demonstrate how Punjabi Natural Language can be UNLized for the UGO-A1, UC-A1 and AESOP-A1.

(*Continued*)

TABLE 6.1 (CONTINUED)
Related Work

S.No.	Authors	Year	Description
10	Anupam Jamatia, et al. [15]	2019	Code-mixing between Bengali-English and Hindi-English as well as a mixture of these two language pairings is studied in Facebook, Twitter, and WhatsApp postings from an Indian social media corpus. In terms of language identification performance, bidirectional LSTM demonstrated the best, with the deep learners outperforming the CRF.
11	Anab Maulana et al. [16]	2019	A task that normalizing code-mixed Twitter data, particularly in the Indonesian-English language, was the authors' main concern. The pipeline's output has a WER score of 31.89 and a BLEU score of 54.07.
12	Jasmeet Singh et al. [17]	2019	A flexible stemming technique has been developed that may be applied to applications other information retrieval, such as classification of text, sentiment analysis, removing inflections, etc.
13	Randa Zarnoufi et al. [18]	2020	They elucidated how to leverage the tools and techniques already in place to normalize the usage of dialects in SMC (Social Media Communication) writing.
14	Jagroop Kaur et al. [19]	2020	Every Roman/English word is transliterated to its Gurmukhi equivalent in order to provide a standard for Roman to Gurmukhi conversion of CSMT in Punjabi. The developed model's CER & BLEU are, respectively, 0.268112 and 0.680793.
15	Ali Saeed et al. [20]	2019	The widely used language of Urdu is given a brand-new, openly available All-Words WSD (Word Sense Disambiguation) dataset, despite the fact that there are very few NLP research resources available for it. Word 4-gram yields the best results for word n-gram techniques with accuracy = 57.71%. Character 10-gram approach get the best results for character n-gram approaches with accuracy = 56.89%.
16	H. Sintayehu et al. [21]	2021	Graph-based label propagation approach for the Amharic NER problem is investigated. It is a straightforward semi supervised, iterative algorithm that propagates labels throughout the dataset.
17	Kamal Deep Singh et al. [22]	2020	A technique in this article for bilingual NMT from Punjabi to English is developed. The Neural Machine Translation system's accuracy is increased by the usage of Byte Pair Encoding and Word Embedding. The reported BLEU rating for the NMT from English to Punjabi is 36.96, while the Punjabi to English value is 38.30.

(Continued)

TABLE 6.1 (CONTINUED)
Related Work

S.No.	Authors	Year	Description
18	Sitender et al. [23]	2021	In this work, a hybridized form of rule based and direct MT is used to create a translation system for the conversion of Sanskrit to English language. The suggested system is assessed using Python's Natural Language Toolkit API, along with BLEU ratings of 0.7606, 3.72 for adequacy and 3.63 for fluency.
19	Sitender et al. [24]	2022	SANSUNL, Sanskrit text expanded to the Universal Networking language, was discussed by upgrading POS(Parts of Speech) tagging, Sanskrit Language parsing and processing. The system showed an overall efficiency of 95.375% with an BLEU score of 0.81 and fluency score of 3.705.
20	Sitender et al. [25]	2021	A machine translation method for Sanskrit language to Universal Networking Language is discussed. The suggested approach receives a 0.85 BLEU rating and 93.18% success rate in successfully resolving UNL relations.
21	Sreebha Bhaskaran et al. [26]	2021	The study focused on short phrases that lasted between two and six words in 15 Indian languages, categorizing each text line based on specific language. It recognizes the proper Indian languages when a document is found to be multilingual.
22	Saurabh Sharma et al. [27]	2021	A variety of unsupervised keyword extraction methods are presented that are not dependent on the size, structure, or subject matter of the texts. The suggested technique is built on a novel set of cognitively inspired phrase, common, word embedding, and external information source properties. The suggested approach improves classification performance by 8% and 6% using SVM classifier, and by 15% and 25% using KNN classifier for the IMDB movie review and WebKB datasets, respectively.
23	Daniel W. Otter et al. [28]	2020	This provides a concise description of deep learning architectures and methodologies are provided. After that, it sifts through the voluminous recent research and aggregates a wide range of pertinent contributions.
24	Navdeep Singh et al. [29]	2022	The extraction of a named entity from the data written in Punjabi Gurmukhi is proposed using deep learning.
25	Rahul Manohar Samant et al. [30]	2022	Various language models in the domain of NLU are discussed. The evolution and significance of deep learning based models are also discussed.

(*Continued*)

TABLE 6.1 (CONTINUED) Related Work

S.No.	Authors	Year	Description
26	Tanveer Singh et al. [31]	2022	Using finite automata, it suggested a technique for getting rid of stop words in Punjabi. Comparisons are made between the performance of the suggested method and the traditional stop-word removal method.
27	Arun Vallarasi et al. [32]	2022	The concept of Cultural Computing was discussed. Also elaborated that cultural computing was based upon on Kansei Mediation.
28	Seema Sharma et al. [33]	2016	The effects of invention have been highlighted after a detailed analysis of personal, cooperative, and social computing. Also illustrated the development of cultural computing as a technological paradigm.
29	Lou Grimal et al. [34]	2021	Various types of human computer interactions available for next generation engineers who will use new technologies are discussed.
30	Seemu Sharma et al. [35]	2018	The necessity for cultural computing is addressed and thoroughly examined various cultural computing methodologies and paradigms.
31	K Sreelakshi et al. [36]	2020	A machine learning model that combines English and Hindi coding to detect hate speech in data is developed. When the suggested methodology's performance is contrasted with word2vec and doc2vec features, FastText features offer enhanced feature representation with SVM-Radial Basis Function (RBF) classifier.
32	Amitava Das et al. [37]	2015	Along with instances of mixed Bengali-English and Hindi-English Facebook communications, the approach created to identify language borders in code-mixed text.
33	Utsab Barman et al. [38]	2014	The social media's issue: language's automatic language identification is discussed.
34	Vishal Gupta [39]	2014	An analysis of different natural language processing methods is presented.
35	Amruta Godase et al. [40]	2015	Various machine translation initiatives carried out in India are focused, along with their characteristics and application areas.
36	Mallama V Reddy et al. [41]	2014	Ad-Hoc multilingual Task created English-Kannada and English-Telugu CLIR(Cross Language Information Retreival) system is shown.
37	R. Mahesh K.Sinha et al. [42]	2005	A approach for automatically converting Hinglish into versions of pure (standard) Hindi and pure English is presented.

(Continued)

TABLE 6.1 (CONTINUED)
Related Work

S.No.	Authors	Year	Description
38	K. Pavan et al. [43]	2010	RoLI(Romanized text language identification system) is proposed for the identification of Romanized text for several Indian languages. In tests conducted across five Indian language web sites with a combination of different languages, RoLI attained a high accuracy of 98.3%.
39	Eleanor Clark et al. [44]	2011	The challenges of automatically normalizing social media English are examined and possible uses for it.
40	Shubhangi Sharma et al. [45]	2012	Phrase-based SMT is used to perform machine transliteration for an English-Hindi language pair that includes Indian names utilizing two alternative character encodings (i.e., in UTF and wx-notation). The target side test set's results demonstrate that wx- notation is more accurate than UTF.
41	Shashank Sharma et al. [46]	2015	Numerous approaches are offered to normalizing the text and used a variety of sentiment resources to determine whether the statement was positive or negative. 85% accuracy can be attained using our model.
42	Royal Denzil Sequiera et al. [47]	2014	The language of a word is determined using the N-grams methodology, and a rule-based method is used to return the original character of a word in Kannada language that has been Romanized.
43	Sukanya Dutta et al. [48]	2015	A work on Romanized transliteration of Bangla words as well as text normalization of English terms is demonstrated in code-mixed social media language that contains words from English language. The word-level language recognition as a whole was 90.5% accurate.
44	Prakash Ranjan et al. [49]	2016	The results of code-mixed data with those from regular text are compared.
45	Vijay Kumar Sharma et al. [50]	2016	A query translation method based on the Wikipedia API is suggested. Queries are tokenized, and the N-gram approach is used to construct multi-word query keywords. The query translation uses Wikipedia's title and inter-wiki link features. Without utilizing any linguistic resources, the recommended technique obtained very good MAP (Mean Average Precision). A maximum of 0.2685 MAP was attained using solely Wikipedia's title and inter-wiki link features.
46	Akshata Phadte et al. [51]	2017	For CMST in Konkani and English, a word-level language identification technique is presented.

(Continued)

TABLE 6.1 (CONTINUED)
Related Work

S.No.	Authors	Year	Description
47	B S Sowmya Lakshmi et al. [52]	2017	Different supervised classification techniques are researched for LID(Language Identification), and discovered that word-level code mixing can be more accurately identified when a dictionary module is added. Results from TF-IDF, and Bigram features outperform CRF.
48	Anupam Jamatia et al. [53]	2018	Using language tags at the word level, utterance breaks, and parts of speech, this work gathered corpus of code-mixed social media content from tweets and posts on Facebook and Twitter that were published in English, Hindi, and Bengali using a coarse-grained and fine-grained tagset.
49	Deepthi Mave et al. [54]	2018	A variety of word-level language recognition methods for coding switching, an outline of the proposed research compares Hindi-English data with Spanish-English dataset. For Spanish-English and Hindi-English, the CRF model performs better than neural network-based model.
50	Shashi Shekhar et al. [55]	2019	To compare the difficulties of language recognition at the word level and to clarify how the index in code mixed data is used in Indian social media posts.
51	S Nagesh Bhattu et al. [56]	2020	The sequential application of CRF and BiLSTM is proposed as a paradigm for POS tagging in code-mixed text. The F1-score with embeddings outperforms the CRF-based baseline in various languages.
52	M.K. Vathshala et al. [57]	2020	A particular neural network which is recurrent (RNN) is LSTM Network is utilized to assess social media data for code-switching and transliteration to English.
53	Suman Dowlagar et al. [58]	2021	The combined BERT and CNN model is introduced for POS tagging and Language identification.
54	Devansh Gautam et al. [59]	2021	They proved that mBART is capable of translating sentences which are code-mixed in Hindi and English into English, and Hindi-English translation improves mBART's performance on code-mixed translation.

6.4 DISCUSSIONS

The consequences of code-mixing for NLP tasks such as language detection, sentiment analysis, and machine translation were examined in the chapter. It emphasized how crucial it is for NLP models to take into account the difficulties of code-mixing in order to process code-mixed material more accurately and efficiently. The review focused on how code-mixing enables speakers to more effectively convey their emotions, cultural allusions, and identity. It promotes inclusivity in various language

TABLE 6.2
Limitations of Code-Mixed Text

Sr. No.	Category	Limitation	Description
1.	**Language Identification**	Complexity of Language Boundaries	Difficulty in distinguishing where one language ends and another begins, especially with co\de-switching at the word or morpheme level.
2.	**Lack of Resources**	Limited Datasets	Fewer annotated datasets available for training robust models.
3.	**Linguistic Diversity**	Dialects and Variants	Regional dialects, slang, and colloquial expressions not well-represented.
4.	**Contextual Understanding**	Contextual Dependencies	Need for understanding the context is amplified with multiple languages.
5.	**Normalization Challenges**	Standardization	Difficulty in creating a standardized form due to variability in language mixing.
6.	**Model Performance**	Bias and Fairness	Monolingual-trained models may show bias or reduced performance with code-mixed text.
7.	**Cultural and Social Factors**	User Preferences	Specific ways of mixing languages preferred by users are hard to capture.

communities and improves communication's authenticity. The analysis emphasized how code-mixing is becoming more common on digital platforms and social media. It showed how the use of code-mixing has increased in the digital age, giving rise to hybrid languages and linguistic advances in the online sphere.

6.5 CONCLUSION

Code-mixing is a typical and normal occurrence in multilingual societies, and as we have seen throughout the review, due to the extensive use of social media and digital communication technologies in the digital era, it has become more common. This research study concluded with a thorough examination of related work in the area of code-mixed text. The importance of code-mixing in text analysis was made clear in a number of ways. It is indispensable for sentiment analysis, language identification, and natural language processing.

6.6 FUTURE SCOPE

The future scope of these fields is vast, with several promising directions for advancement. Improved language models specifically trained on code-mixed datasets are expected to enhance the understanding and generation of multilingual text, making them more context-aware and capable of capturing nuanced meanings. Cross-lingual understanding will benefit from improved transfer learning and deeper semantic and pragmatic comprehension, enabling models to handle code-mixed sentences more effectively. Furthermore, the integration of ethical considerations will ensure that models are fair and inclusive, reducing biases against any language or dialect. Overall, advancements in code-mixing and normalization are poised to make technology more accessible and effective for diverse, multilingual populations, fostering better communication and understanding across languages.

REFERENCES

1. Thara, S., and Prabaharan Poornachandran. "Code-mixing: A brief survey." In *2018 International Conference on Advances in Computing, Communications and Informatics (ICACCI)*, pp. 2382–2388. IEEE, 2018.
2. Kumar, A., P. Kumar, S. Rathee, and B. Kumar (Eds.). *Mechatronics: Concepts, Tools, Applications, and New Trends* (1st ed.). CRC Press. 2025. https://doi.org/10.1201/9781003494478.
3. Khurana, Diksha, Aditya Koli, Kiran Khatter, and Sukhdev Singh. "Natural language processing: State of the art, current trends and challenges." *Multimedia Tools and Applications* 82, no. 3 (2023): 3713–3744.
4. Johri, Prashant, Sunil K. Khatri, Ahmad T. Al-Taani, Munish Sabharwal, Shakhzod Suvanov, and Avneesh Kumar. "Natural language processing: History, evolution, application, and future work." In *Proceedings of 3rd International Conference on Computing Informatics and Networks: ICCIN 2020*, pp. 365–375. Springer Singapore, 2021.
5. Chomsky, Noam. *Aspects of the Theory of Syntax*. No. 11. MIT Press, 2014.
6. Ramzan, Muhammad, Aamir Aziz, and Maimoona Ghaffar. "A study of code-mixing and code-switching (Urdu and Punjabi) in children's early speech." *Journal of Language and Linguistic Studies* 17, no. 2 (2021): 869–881.
7. Harsh, A. S., T. V. Prasad, and G. Ramakrishna. "Identification and Translation of Noun in Bilingual Code Mixed Language into Pure Form." *International Journal of Applied Engineering Research* 10, no. 9 (2015): 21591–21604.
8. Josan, Gurpreet Singh, and Gurpreet Singh Lehal. "Size of N for word sense disambiguation using N-gram model for Punjabi language." *International Journal of Translation* 20, no. 1–2 (2008): 47–56.
9. Sidorov, Grigori, Francisco Velasquez, Efstathios Stamatatos, Alexander Gelbukh, and Liliana Chanona-Hernández. "Syntactic n-grams as machine learning features for natural language processing." *Expert Systems with Applications* 41, no. 3 (2014): 853–860.
10. Gupta, Vishal. "Automatic stemming of words for Punjabi language." In *Advances in signal processing and intelligent recognition systems*, pp. 73–84. Springer International Publishing, 2014.
11. Kaur, Amandeep, and Gurpreet Singh Josan. "Evaluation of named entity features for Punjabi language." *Procedia Computer Science* 46 (2015): 159–166.
12. Kaur, Amandeep, and Vishal Gupta. "A Novel Approach for Sentiment Analysis of Punjabi Text using SVM." *International Arab Journal of Information Technology (IAJIT)* 14, no. 5 (2017).
13. Johnson, Melvin, Mike Schuster, Quoc V. Le, Maxim Krikun, Yonghui Wu, Zhifeng Chen, Nikhil Thorat et al. "Google's multilingual neural machine translation system: Enabling zero-shot translation." *Transactions of the Association for Computational Linguistics* 5 (2017): 339–351.

14. Agarwal, Vaibhav, and Parteek Kumar. "UNLization of Punjabi text for natural language processing applications." *Sādhanā* 43 (2018): 1–23.
15. Jamatia, Anupam, Amitava Das, and Björn Gambäck. "Deep learning-based language identification in English-Hindi-Bengali code-mixed social media corpora." *Journal of Intelligent Systems* 28, no. 3 (2019): 399–408.
16. Barik, Anab Maulana, Rahmad Mahendra, and Mirna Adriani. "Normalization of Indonesian-English code-mixed Twitter data." In *Proceedings of the 5th Workshop on Noisy User-generated Text (W-NUT 2019)*, pp. 417–424. 2019.
17. Singh, Jasmeet, and Vishal Gupta. "A novel unsupervised corpus-based stemming technique using lexicon and corpus statistics." *Knowledge-Based Systems* 180 (2019): 147–162.
18. Zarnoufi, Randa, Hamid Jaafar, and Mounia Abik. "Machine normalization: Bringing social media text from non-standard to standard form." *ACM Transactions on Asian and Low-Resource Language Information Processing (TALLIP)* 19, no. 4 (2020): 1–30.
19. Kaur, Jagroop, and Jaswinder Singh. "Roman to gurmukhi social media text normalization." *International Journal of Intelligent Computing and Cybernetics* 13, no. 4 (2020): 407–435.
20. Saeed, Ali, Rao Muhammad Adeel Nawab, Mark Stevenson, and Paul Rayson. "A sense annotated corpus for all-words Urdu word sense disambiguation." *ACM Transactions on Asian and Low-Resource Language Information Processing (TALLIP)* 18, no. 4 (2019): 1–14.
21. Sintayehu, H., and G. S. Lehal. "Named entity recognition: a semi-supervised learning approach." *International Journal of Information Technology* 13 (2021): 1659–1665.
22. Deep, Kamal, Ajit Kumar, and Vishal Goyal. "Punjabi to English Bidirectional NMT System." In *Proceedings of the 17th International Conference on Natural Language Processing (ICON): System Demonstrations*, pp. 7–9. 2020.
23. Sitender and Seema Bawa. "A Sanskrit-to-English machine translation using hybridization of direct and rule-based approach." *Neural Computing and Applications* 33, no. 7 (2021): 2819–2838.
24. Sitender and Seema Bawa. "Sanskrit to universal networking language EnConverter system based on deep learning and context-free grammar." *Multimedia Systems* 28, no. 6 (2022): 2105–2121.
25. Sitender and Seema Bawa. "Sansunl: a Sanskrit to UNL enconverter system." *IETE Journal of Research* 67, no. 1 (2021): 117–128.
26. Bhaskaran, Sreebha, Geetika Paul, Deepa Gupta, and J. Amudha. "Indian language identification for short text." In *Advances in Computational Intelligence and Communication Technology: Proceedings of CICT 2019*, pp. 47–58. Springer Singapore, 2021.
27. Sharma, Saurabh, Vishal Gupta, and Mamta Juneja. "Diverse feature set based Keyphrase extraction and indexing techniques." *Multimedia Tools and Applications* 80 (2021): 4111–4142.
28. Otter, Daniel W., Julian R. Medina, and Jugal K. Kalita. "A survey of the usages of deep learning for natural language processing." *IEEE Transactions on Neural Networks and Learning Systems* 32, no. 2 (2020): 604–624.
29. Singh, Navdeep, Munish Kumar, Bavalpreet Singh, and Jaskaran Singh. "DeepSpacy-NER: an efficient deep learning model for named entity recognition for Punjabi language." *Evolving Systems* 14, no. 4 (2023): 673–683.
30. Samant, Rahul Manohar, Mrinal R. Bachute, Shilpa Gite, and Ketan Kotecha. "Framework for deep learning-based language models using multi-task learning in natural language understanding: A systematic literature review and future directions." *IEEE Access* 10 (2022): 17078–17097.

31. Kochhar, Tanveer Singh, and Gulshan Goyal. "Design and implementation of stop words removal method for Punjabi language using finite automata." In *Advances in Data Computing, Communication and Security: Proceedings of I3CS2021*, pp. 89–98. Singapore: Springer Nature Singapore, 2022.
32. Vallarasi, Arun and S. Regi. "Introduction to Cultural Computing." *International Advanced Research Journal in Science, Engineering and Technology*, 9, no. 3 (2022): 69–74.
33. Sharma, Seemu, Seema Bawa, and Hitashi Lomash. "Proliferation of social computing: cultural computing paradigm." *International Journal of Computer Applications* 137, no. 9 (2016): 27–30.
34. Grimal, Lou, Ines Di Loreto, and Nadège Troussier. "Exploratory research on cultural computing for human-computer interaction. The case of sustainability." *Proceedings of the Design Society* 1 (2021): 1471–1480.
35. Sharma, Seemu, Seema Bawa, and Hitashi Lomash. "Approaches in cultural computing: a survey and inference from social computing with dynamics of mind." *Wireless Personal Communications* 103 (2018): 2693–2713.
36. Sreelakshmi, K., B. Premjith, and K. P. Soman. "Detection of hate speech text in Hindi-English code-mixed data." *Procedia Computer Science* 171 (2020): 737–744.
37. Das, Amitava, and Björn Gambäck., "Code-Mixing in Social Media Text The Last Language Identifier Frontier?" *Traitement Automatique des Langues* 54 (2015): 41–64.
38. Barman, Utsab, Amitava Das, Joachim Wagner, and Jennifer Foster. "Code mixing: A challenge for language identification in the language of social media." In *Proceedings of the First Workshop on Computational Approaches to Code Switching*, pp. 13–23. 2014.
39. Gupta, Vishal. "A survey of natural language processing techniques." International *Journal of Computer Science & Engineering Technology* 5, no. 1 (2014): 14–16.
40. Godase, Amruta, and Sharvari Govilkar. "Machine translation development for Indian languages and its approaches." *International Journal on Natural Language Computing* 4, no. 2 (2015): 55–74.
41. Reddy, Mallamma V., and M. Hanumanthappa. "NLP challenges for machine translation from English to Indian languages." *International Journal of Computer Science and Informatics* 3, no. 1 (2013): 35.
42. Sinha, R. Mahesh K., and Anil Thakur. "Machine translation of bi-lingual hindi-english (hinglish) text." In *Proceedings of Machine Translation Summit X: Papers*, pp. 149–156. 2005.
43. Pavan, Kosuru, Niket Tandon, and Vasudeva Varma. "Addressing challenges in automatic language identification of romanized text." In *Proceedings of 8th International Conference on Natural Language Processing (ICON-2010)*. 2010.
44. Clark, Eleanor, and Kenji Araki. "Text normalization in social media: progress, problems and applications for a pre-processing system of casual English." *Procedia-Social and Behavioral Sciences* 27 (2011): 2–11.
45. Sharma, Shubhangi, Neha Bora, and Mitali Halder. "English-hindi transliteration using statistical machine translation in different notation." *Training* 20000, no. 297380 (2012): 20000.
46. Sharma, Shashank, P. Y. K. L. Srinivas, and Rakesh Chandra Balabantaray. "Text normalization of code mix and sentiment analysis." In *2015 international conference on advances in computing, communications and informatics (ICACCI)*, pp. 1468–1473. IEEE, 2015.
47. Sequiera, Royal Denzil, Shashank S. Rao, and B. R. Shambavi. "Word-level language identification and back transliteration of Romanized text." In *Proceedings of the 6th Annual Meeting of the Forum for Information Retrieval Evaluation*, pp. 70–73. 2014.

48. Dutta, Sukanya, Tista Saha, Somnath Banerjee, and Sudip Kumar Naskar. "Text normalization in code-mixed social media text." In *2015 IEEE 2nd International Conference on Recent Trends in Information Systems (ReTIS)*, pp. 378–382. IEEE, 2015.
49. Ranjan, Prakash, Bharathi Raja, Ruba Priyadharshini, and Rakesh Chandra Balabantaray. "A comparative study on code-mixed data of Indian social media vs formal text." In *2016 2nd International Conference on Contemporary Computing and Informatics (IC3I)*, pp. 608–611. IEEE, 2016.
50. Sharma, Vijay Kumar, and Namita Mittal. "Exploiting Wikipedia API for Hindi-English cross-language information retrieval." *Procedia Computer Science* 89 (2016): 434–440.
51. Phadte, Akshata, and Ramrao Wagh. "Word level language identification system for Konkani-English code-mixed social media text (CMST)." In *Proceedings of the 10th Annual ACM India Compute Conference*, pp. 103–107. 2017.
52. Lakshmi, BS Sowmya, and B. R. Shambhavi. "An automatic language identification system for code-mixed English-Kannada social media text." In *2017 2nd International Conference on Computational Systems and Information Technology for Sustainable Solution (CSITSS)*, pp. 1–5. IEEE, 2017.
53. Jamatia, Anupam, Björn Gambäck, and Amitava Das. "Collecting and annotating indian social media code-mixed corpora." In *International Conference on Intelligent Text Processing and Computational Linguistics*, pp. 406–417. Cham: Springer International Publishing, 2016.
54. Mave, Deepthi, Suraj Maharjan, and Thamar Solorio. "Language identification and analysis of code-switched social media text." In *Proceedings of the Third Workshop on Computational Approaches to Linguistic Code-Switching*, pp. 51–61. 2018.
55. Shekhar, Shashi, Dilip Kumar Sharma, and M. M. Sufyan Beg. "An effective cybernated word embedding system for analysis and language identification in code-mixed social media text." *International Journal of Knowledge-based and Intelligent Engineering Systems* 23, no. 3 (2019): 167–179.
56. Bhattu, S. Nagesh, Satya Krishna Nunna, Durvasula VLN Somayajulu, and Binay Pradhan. "Improving code-mixed POS tagging using code-mixed embeddings." *ACM Transactions on Asian and Low-Resource Language Information Processing (TALLIP)* 19, no. 4 (2020): 1–31.
57. Vathsala, M. K., and Ganga Holi. "RNN based machine translation and transliteration for Twitter data." *International Journal of Speech Technology* 23, no. 3 (2020): 499–504.
58. Dowlagar, Suman, and Radhika Mamidi. "A pre-trained transformer and CNN model with joint language ID and part-of-speech tagging for code-mixed social-media text." In *Proceedings of the International Conference on Recent Advances in Natural Language Processing (RANLP 2021)*, pp. 367–374. 2021.
59. Gautam, Devansh, Kshitij Gupta, and Manish Shrivastava. "Translate and Classify: Improving Sequence Level Classification for English-Hindi Code-Mixed Data." In *Proceedings of the Fifth Workshop on Computational Approaches to Linguistic Code-Switching*, pp. 15–25. 2021.

7 Reviewing Different Artificial Intelligence Algorithms for Sarcasm Detection

Rahul Yadav and Jaswinder Singh

7.1 INTRODUCTION

Sarcasm is a way of communication in which a person oppositely states their way of expression. Sarcasm has no precise definition. The significant obstacle that arises in sarcasm is its ambiguous nature and expanding number of languages. Daily, numerous numbers of new slang words are being created and used on various social networking sites and other platforms. Day by day new developments are made in social networking sites, such as the use of emoticons has the power to change the context meaning differently or can make the text sarcastic. The various factors mentioned here make sarcasm a very critical issue that needs to be addressed. The detection of sarcastic content employs natural language processing (NLP) and sentiment analysis, with the aim of identifying the polarity of text through various techniques. "Sarcastic Invader" is an analytic tool that is developed by a French company that can identify sarcasm in comments on social media platforms like Facebook with up to 80% accuracy [1]. The amount of data created, captured, copied, and consumed globally is increasing exponentially as it touched 62.4 zettabytes in 2020 and is expected to grow to more than 180 zettabytes in 2025. So lately, the preprocessing and processing of data have increased exponentially, and this has emerged the bigger picture of machine learning. This work examines research questions that would allow us to find the latest trends in machine learning and sarcasm detection. These research questions are as follows:

RQ1: What are the machine learning algorithms used for sarcasm detection in previous works?
RQ2: What are the tools used for sarcasm detection by the researchers for the implementation of their models?
RQ3: What are the performance parameters used by the researchers for sarcasm detection to evaluate their models?

 DOI: 10.1201/9781032724508-8

There are three sections in this chapter. Section 7.1 contains an introduction and literature review, from which we get a clear idea of previous work done on sarcasm detection. Section 7.2 lists the types and steps of the sarcasm detection model and the challenges that arise in sarcasm detection. This section gives a view of how sarcasm detection is done and the challenges involved in it. Section 7.3 presents the concluding remarks.

7.2 TYPES OF SARCASM

Different types of sarcasm are shown in Figure 7.1.

- **Self-Deprecating**: Self-deprecating sarcasm is a form of self-awareness. In this type, the individual points out to the negative or positive things that they do not like about themselves but comments on them in a loud, funny, and joking way. In self-deprecating sarcasm, people make fun of themselves mostly; for example, "I have been single ever seems I mingled."

FIGURE 7.1 Types of sarcasm

- **Brooding Sarcasm**: In brooding sarcasm, the speaker says something in a polite tone with irritation and bitterness. The speaker expresses pity for their situation and also for themselves; for example, "'At hotel' Receptionist – Would you like a room? Customer – No, Not at all, I came here to dance, stage Please."
- **Deadpan Sarcasm**: In deadpan sarcasm, the speaker shows complete seriousness in his wording. The word "deadpan" means expressionless. The speaker who is using deadpan sarcasm makes the statement sarcastic without any emotions. This makes it very difficult for the listener or reader to recognize whether the speaker is using sarcasm or not. i.e. "I am not saying I love you, I am just saying I love you."
- **Polite sarcasm**: In polite sarcasm, the speaker appears to be sincere but is insincere. In this kind of sarcasm, the listener understands the meaning of the words but slowly realizes the context of sarcasm; for example, "I'm not snarky I'm selectively polite."
- **Obnoxious sarcasm**: In obnoxious sarcasm, the speaker uses obvious sarcasm and directly offends the listener at the face. In this kind of sarcasm, the listener intends to literally punch in the face off the listener. This type of sarcasm is not funny; for example, "I got called 'Pretty' today! Well, the full statement was 'You're pretty annoying' but I only focused on positioning things."
- **Manic sarcasm**: In this kind of sarcasm the speaker sounds happy but is unhappy. In manic sarcasm, the sounds are so unnatural that the speaker appears to be manic in a mental state; for example, "When a person is seen stressed by saying 'I am fine'."
- One employee to another: "If a company is 'always hiring', it means they are also 'always firing'."
- **Raging sarcasm**: In this type of sarcasm, the speaker expresses rage and anger and also uses a lot of exaggeration most of the time. This kind of sarcasm relies on hyperbole and threats of violence.

For example, I have a raging headache.

Other countries: "Here, take an Advil"
India: "Sara din phone mein ghusa rahega toh aur kya hoga?"

7.3 MACHINE LEARNING

7.3.1 Why Machine Learning?

Sarcasm means being sad by being funny at the same time and contradictory to it. It has been part of every human being's expression for many years. Nowadays, it is also used in news headlines and on various social media platforms and is gaining ground day by day. Sarcasm detection is an NLP and binary classification task that

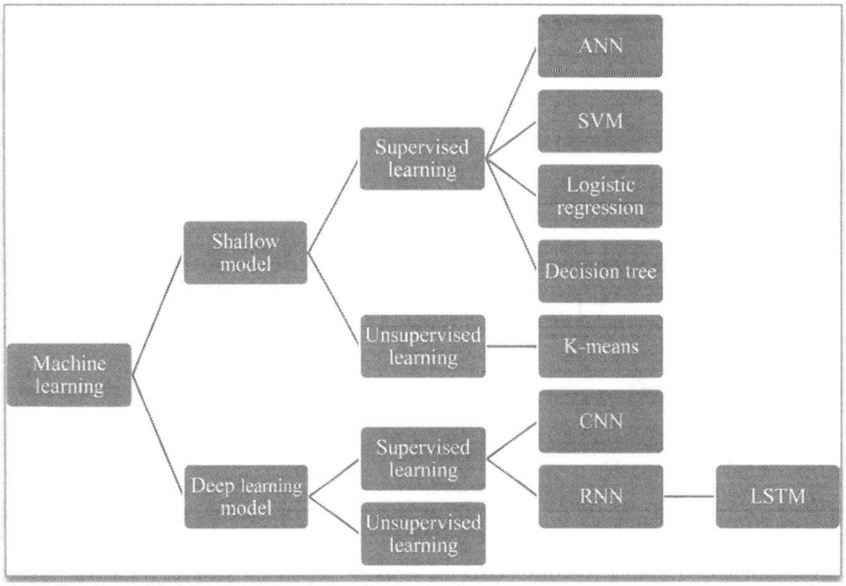

FIGURE 7.2 Taxonomy of machine learning

can be easily achieved by training a machine learning algorithm to detect whether the sentence is sarcastic or not.

Machine learning is a developing field of technology that allows computers to use past data to analyze. A range of methods are used by machine learning to create mathematical models and make predictions based on recently learned methods, information, and data. The taxonomy of machine learning is shown in figure 7.2.

7.3.2 Machine Learning Classifiers

1. **SVM:** It is a binary classification type of searching algorithm that searches for a prime hyperplane that divides the dataset into classes considering optimal features in the negative and positive with maximum margin. By using different kernel functions linear and non-linear classification problems can be handled by SVM. Some common features of SVM for sarcasm detection are "N-gram features", which uses sequences of *n* words or characters as features to capture the lexical and syntactic patterns of sarcasm, such as punctuations and interjections. "Sentiment features" use polarity or intensity of the text as a feature to capture the contrast. "Contextual features" utilize the information from preceding or surrounding texts to capture the situational or pragmatic uses of sarcasm.
2. **KNN:** K- Nearest Neighbor in machine learning for sarcasm detection is a supervised method that uses K- Nearest Neighbor as a classifier to distinguish between sarcastic and non-sarcastic texts based on the similarity or

distance between the text and its neighbors. KNN is a machine learning algorithm that assigns a label to a new text based on the majority vote of its K closest neighbors in the training data. KNN can handle both numerical and categorical features. Some important features used in KNN for sarcasm detection are N-gram features, sentiment features, contextual features, etc.

3. **Logistic Regression:** This is a supervised machine learning method used for sarcasm detection. Regression and classification are two different types of logistic regression. When input is provided to the binary classification model of logistic regression, it gives output probability in response to that given input. The output probability was evaluated based on training data, and the output probability gives a result that the context of the text is sarcastic or not.
4. Machine learning algorithms are used for text classification (sarcasm detection), as it works on various algorithms such as LSTM and SVM that process training data and learn the pattern associated with the text. When the model is applied to unseen data (test data), it classifies the data established on the learned patterns from the training data, based on which the proposed model is trained on.

NLP is concerned with allowing the computer to understand the text and spoken words of human beings with the same efficiency as humans. Computers perform computational linguistics of human language using machine learning, statistical, and deep learning models, in conjunction with NLP. The above-mentioned technologies enable computers to understand the speaker's or writer's intended meaning. However, NLP is sometimes unable to show or understand the context of the sentence.

7.4 LITERATURE REVIEW

D. K. Nayak et al. in [2] focused on detecting sarcasm using a supervised learning approach in news headlines. This type of model consists of a total of six different supervised learning models, including the Naive Bayes, logistic regression, SVM, DistilBERT, RoBERT and (BERT). L. Xu et al. (2017) in [3] employed both background and reply texts to enhance the detection of sarcasm in replies. SVMs and LSTMs with attention were the primary models used. S. M. Sarcasm et al. (2020) in [4] concluded that the SVM was used with AMLA for detecting sarcastic tweets on Twitter, which gave the best result. In addition, when SVM is combined with convolutional neural network (CNN), it was found that it offers a high level of accuracy. S. Razall et al. (2021) in [5] focused on sarcasm in various tweets by extracting features combined with contextual handcrafted features using deep learning. A. Ben Meriem et al. (2021) in [6] presented work on a fuzzy method for identifying sarcasm on social media. A binary categorization system was used in which each text is sarcastic or not sarcastic. K. Sentamilselvan et al. (2021) in [7] detected the sarcasm model and used machine learning classifiers and a rule-based approach. S. K. Bharti et al. (2022) in [8] introduced a strategy based on deep learning sarcastic remarks posted on social media that has become popular in the present world. This study presented a new method for sarcasm detection in conversations by using both textual and audible parameters. A. M. A. Barhoom et al. (2022) in [9] provided work on machine learning and deep learning algorithms for detecting sarcasm in top news stories by utilizing one deep learning method and 21 machine learning algorithms. R. Anan et al. (2023) in [10] detected sarcasm using the BERT model with stratified k-fold cross-validation on the Bangla language because regional languages are not explored as much as English. W. Q. Al-Jamal et al. (2022) in [11] detected sarcasm in Arabic text using deep learning. In this work, two different transformation-based models are analyzed, namely MARBERT and AraBERT. Here, the AraBERT model outperformed the MARBERT model in the F1 score. Y. Salini et al. (2023) in [12] focused on methods and approaches used for sarcasm detection by employing comparisons of different algorithms and techniques used and also suggested the types of algorithms and techniques best suited for different kinds of data and types of sarcasm. B. Singh et al. (2023) in [13] detected sarcasm based on a survey of sarcasm detection in NLP. The study concentrated on the difficulties and advancements in the field of sarcasm detection, as well as the necessity of sarcasm detection.

7.4.1 Summary of Literature Review of Sarcasm Detection

A summary of some more papers is described in Table 7.1, which is based on the different factors such as data source, model, year of paper, and parameters such as "accuracy and precision."

7.5 RESEARCH GAP

Addressing the research gap not only advances the field of sarcasm detection but also enables the creation of more dependable and accurate models. Almost all sarcasm detection models are binary-oriented and differentiate sarcasm, such as negative sarcasm, positive sarcasm, mild sarcasm, sarcastic sarcasm, humble sarcasm, and many others, which give more concrete ideas about sarcasm. Sarcasm detection focuses on audio and video rather than on only text format. Also, Sarcasm detection should explore more areas of data acquisition rather than focusing only on Twitter (As of right present, Twitter data is used in nearly all sarcasm detection research projects). Researchers also focus on implementing the model in real life with the help of chatbot customer care service. In future, these gaps need to be explored gradually from time to time.

7.6 STEPS OF SARCASM DETECTION

The whole process of sarcasm detection is explained in figure 7.3 and each step is also explained below explicitly.

TABLE 7.1
Some Research papers on sarcasm detection

Sr. No.	Dataset source	Model	Year	Parameter	Reference
1	Forum and Twitter	SVM and Naïve Bayes	2017	Accuracy (65.2) Precision (96.5)	[14]
2	Twitter	LSTM	2017	Accuracy (85.5) Precision (85.5)	[15]
3	Twitter	CASCADE	2018	Accuracy (79)	[16]
4	Twitter	MIARN	2018	Accuracy (72.8) Precision (86.1)	[17]
5	Twitter	Multimodal ANN	2019	Accuracy (83.4) Precision (76.6)	[18]
6	Twitter	Multimodal ANN	2020	Accuracy (86.1) Precision (80.9)	[19]
7	Twitter	Recurrent CNN RoBERTA	2020	Accuracy (82) Precision (81)	[20]
8	Twitter	Feature-based statistical model	2020	Accuracy (93.5) Precision (93.8)	[21]
9	Twitter	BiLSTM-CNN	2020	Accuracy (85) Precision (78.8)	[22]
10	Twitter	BERT	2021	Accuracy (70.6) Precision (68.7)	[23]
11	Twitter	PCA-K-mean-SVM	2023	Accuracy (91.9)	[24]
12	News headlines and comments	LSTM	2023	Accuracy (91.4)	[25]
13	Twitter	MHSA-GRU	2023	Accuracy (97.6)	[26]

7.6.1 Data Acquisition

Acquiring data for sarcasm detection is very challenging due to the subjective nature of sarcasm and the need for labeled data that indicates sarcastic intent. However, here are a few approaches that can be considered for data acquisition:

- **Existing Datasets:** Some publicly available datasets have been created specifically for sarcasm detection. Examples include the Twitter Sarcasm Corpus, the Reddit Sarcasm Dataset, and the News Headlines Dataset. These datasets can serve as a starting point for training and evaluating sarcasm detection models.
- **Crowdsourcing:** One effective approach is to crowd-source data labeling by presenting users with text samples and asking them to classify whether the text is sarcastic or not. Platforms such as Amazon Mechanical Turk

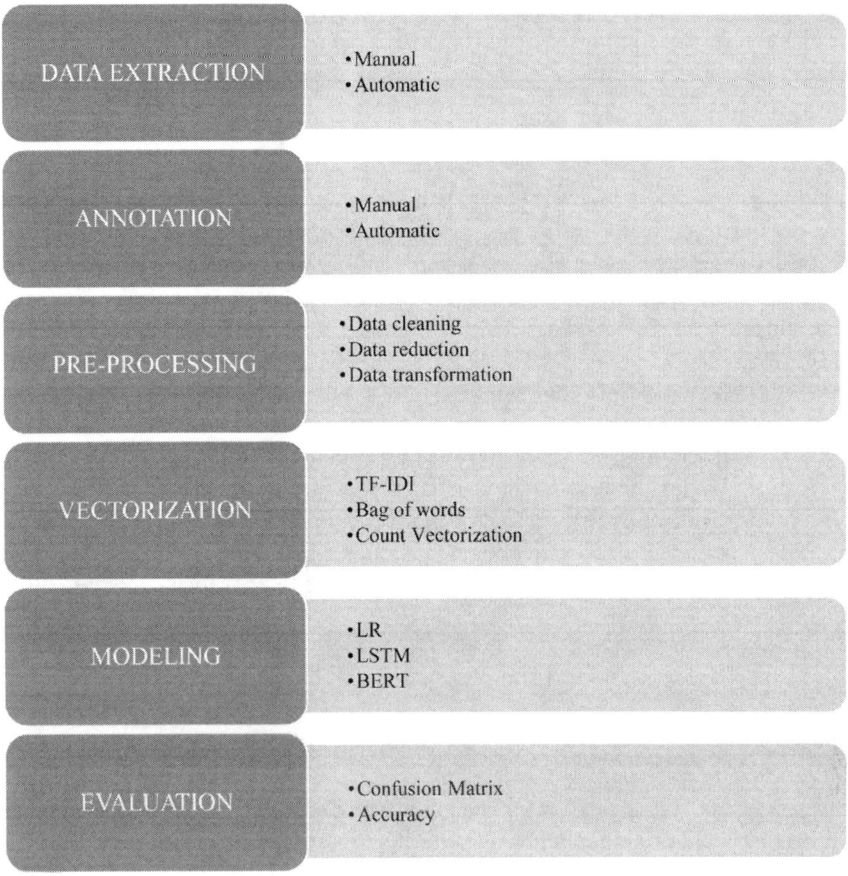

FIGURE 7.3 Steps of sarcasm detection

or Figure Eight can be used to gather labeled data from a large number of human annotators.

- **Domain-Specific Data:** If interested in sarcasm detection in a specific domain, such as customer reviews or movie dialogues, we can acquire data from sources that specialize in that domain. For example, scraping product reviews from e-commerce websites or obtaining movie scripts can provide domain-specific data for training sarcasm detection models.

7.6.2 Annotation

The technique of marking data or text examples to indicate whether or not they include sarcasm is known as annotation in sarcasm detection. To train machine learning models to recognize and comprehend sarcasm in textual data, this annotation is essential. Sarcasm detection annotation usually entails human annotators going over text samples and classifying them as sardonic or non-sarcastic. To identify sarcasm, the annotators may employ particular standards or criteria, such as linguistic signals, context, tone, and sentiment.

7.6.3 Data Preprocessing

Preprocessing is a part of NLP [27]. It is done because of unwanted weight and biases in the process. Regarding this various kinds of preprocessing techniques were used as data set into lowercase, stop words are removed, removal of punctuation signs, and change into their root form [5]. Some important data preprocessing techniques for sarcasm detection are:

- **Tokenization:** Splitting the text into smaller units such as words to sub-words.
- **Normalization:** Converting the text into a standard form such as lower casing, removing punctuation, and expanding contractions.
- **Stop word removal:** Eliminating words that are too common or irrelevant for the task, such as articles and conjunctions.
- **Lemmatization:** Reducing words to their base or root form, such as cats to cat.
- **Embedding:** Converting words into vector numbers that snatch their semantic and syntactic features such as word2vac and GloVe.

7.6.4 Vectorization

In the context of sarcasm detection, the practice of encoding textual data into numerical vectors appropriate for input into machine learning models is known as vectorization. By turning words, phrases, or sentences into numbers that reflect their structure and meaning, vectorization helps machine learning algorithms analyze and understand the text better in sarcasm detection. Several techniques for vectorizing the text data are as follows:

- **Bag-of-Words (BoW):** BoW uses numerical vectors to represent each document. A unique word in the corpus vocabulary corresponds to each dimension, and the value of the vector indicates how often the word appears in the text. This method is straightforward and effective, but it ignores context and word order.
- **TF-IDF:** Similar to BoW, TF-IDF gives words weights according to how frequently they occur in the document (TF) and how uncommonly they occur throughout the corpus (IDF). This aids in the prioritization of terms that are more discriminative or informative in detecting sarcasm.
- **Word Embedding:** Word embedding captures the semantic relationships between words by representing them as low-dimensional, dense vectors. Based on the context of words in a large corpus, techniques such as Word2Vec, GloVe, and FastText learn distributed representations of words.

These embeddings can be taught from scratch on task-specific datasets, or they can be pre-trained on extensive text corpora.

7.6.5 Modeling

The perspectives used for the detection of sarcasm have developed numerously in past. The characteristics of machine learning (such as logistic regression and long short-term memory) were applied in the past, but currently these are transferred to deep learning techniques (such as BERT) for the detection of sarcasm. Also, nowadays researchers give high importance to deep learning features with NLP. An advantage of using deep learning is that it can automatically detect the optimum feature for the given input [28].

7.6.6 Evaluation

Here in finding results regarding sarcasm detection, we need to evaluate various parameters such as accuracy, F1 measure, precision, and recall by evaluating the confusion matrix as the matrix indicates predicate values and actual values in the shape of False positive, True negative, True positive and False negative.

7.7 CHALLENGES IN SARCASM DETECTION

Contextual Understanding: Sarcasm heavily relies on context, tone of voice, and non-verbal cues. In written text, sarcasm detection can be difficult because it needs these additional cues. High dependency on the context of the text in sarcasm detection makes it a very difficult challenge for an automated system to accurately interpret it [12].

Ambiguity: Sarcasm often involves the use of ambiguous language and statements that can be interpreted both literally and sarcastically. Distinguishing between literal and sarcastic meanings can be challenging, as it needs to understand the speaker's intention also with the underlying irony.

Cultural and Semantic Variations: Sarcasm can differ greatly across cultures and languages. What may be considered sarcastic in one culture may not be perceived as such in another. Sarcasm can also involve wordplay, puns, or cultural references that are difficult to capture without a comprehensive understanding of the specific language or cultural context.

Evolution of Language: Sarcasm evolves, and new forms of sarcasm emerge as language and culture change. Keeping up with the latest trends and patterns in sarcasm can be a significant challenge for automated systems, as they may not have access to real-time data and linguistic updates.

Subjectivity and Subject Matter: Sarcasm can be highly subjective and dependent on the topic or subject matter being discussed. What one person finds sarcastic, another may perceive as genuine or non-sarcastic. Recognizing sarcasm accurately requires a deep understanding of the specific domain and the nuances associated with it [12].

Data Limitations: Training sarcasm detection models requires large amounts of annotated data that indicate sarcastic intent. However, such datasets are often limited, making it challenging to build robust models that can handle various instances of sarcasm effectively.

7.8 RESULT AND DISCUSSION

By analyzing and reading various research papers on sarcasm detection in previous work, we have found the results of our various research questions as follows:

RQ1: In Sarcasm detection, various trends come from time to time which shaped the NLP research. The 1st algorithm used for sarcasm detection was the "semi-supervised" sarcasm detection identification algorithm (SASI) in 2010 concentrated on machine learning and feature extraction models. As the research progressed more algorithms were explored from time to time such as SVM, CASCADE, and Naïve Bayes in 1st decade. Then it shifts toward Convolution Neural Networks and LSTM. Now at present sarcasm detection algorithms are more focused on BERT and RoBERTA for new advancements.

RQ2: Various tools used for sarcasm detection models are NLP, Python, WEKA (Waikato Environment for Knowledge Analysis), IBM tone analyzer etc.

RQ3: In sarcasm detection parameters used for evaluating the proposed system are accuracy, precision, recall, F1 score etc. and these factors can be evaluated with the help of a confusion matrix by predicted values and actual values.

7.9 CONCLUSION

In conclusion, our review chapter on machine learning techniques for sarcasm detection has provided an inclusive examination of the existing conditions of knowledge in this field of sarcasm detection. We have addressed various types of sarcasm, the sources that are targeted, the importance of detecting sarcasm, the evaluation metrics to evaluate the performance factors of sarcasm detection algorithms, and the challenges and limitations of using machine learning algorithms for sarcasm detection. Our findings have shown that various machine learning algorithms have been widely utilized for sarcasm detection. Earlier, the focus was more on machine learning approaches, but to achieve more accuracy and to train the model better, the focus shifted toward deep learning approaches. Tools used for detecting sarcasm are Python, NLP, MATLAB, etc. However, there are still some challenges associated with the utilization of machine learning and deep learning algorithms for sarcasm detection, such as the sense of humor of every person being of a different kind and it being hard to train a machine according to everyone's mind. Machines also need to be trained to identify negative and positive sarcasm to analyze good opinions about data. Also, the practical accuracy achieved by various researchers for the detection of sarcasm is minimal and needs to be improved.

REFERENCES

1. M. Bouazizi and T. Ohtsuki, "Sarcasm Over Time and Across Platforms: Does the Way We Express Sarcasm Change?" *IEEE Access*, 10, 55958-55987, 2022. doi: 10.1109/ACCESS.2022.3174862.
2. D. K. Nayak and B. K. Bolla, *Efficient Deep Learning Methods for Sarcasm Detection of News Headlines*, Vol. 269, no. February. Springer Singapore, 2022. doi: 10.1007/978-981-16-7996-4_26.
3. L. Xu and V. Xu, "Project Report: Sarcasm Detection," 2019. https://web.stanford.edu/class/archive/cs/cs224n/cs224n.1194/reports/custom/15791781.pdf
4. S. M. Sarsam, H. Al-Samarraie, A. I. Alzahrani, and B. Wright, "Sarcasm detection using machine learning algorithms in Twitter: A systematic review," *Int. J. Mark. Res.*, 62(5), 578–598, 2020. doi: 10.1177/1470785320921779.
5. M. S. Razali, A. A. Halin, L. Ye, S. Doraisamy, and N. M. Norowi, "Sarcasm Detection Using Deep Learning with Contextual Features," *IEEE Access*, 9, 68609–68618, 2021. doi: 10.1109/ACCESS.2021.3076789.
6. A. Ben Meriem, L. Hlaoua, and L. Ben Romdhane, "A fuzzy approach for sarcasm detection in social networks," *Procedia Comput. Sci.*, 192, 602–611, 2021. doi: 10.1016/j.procs.2021.08.062.
7. K. Sentamilselvan, P. Suresh, G. K. Kamalam, S. Mahendran, and D. Aneri, "Detection on sarcasm using machine learning classifiers and rule-based approach," *IOP Conf. Ser. Mater. Sci. Eng.*, 1055(1), 012105, 2021. doi: 10.1088/1757-899x/1055/1/012105.
8. S. K. Bharti, R. K. Gupta, P. K. Shukla, W. A. Hatamleh, H. Tarazi, and S. J. Nuagah, "Multimodal Sarcasm Detection: A Deep Learning Approach," *Wirel. Commun. Mob. Comput.*, 2022, 2022, 1653696. doi: 10.1155/2022/1653696.
9. A. M. A. Barhoom, B. S. Abunasser, and S. S. Abu-Naser, "Sarcasm Detection in Headline News using Machine and Deep Learning Algorithms," *Int. J. Eng. Inf. Syst.*, 6(4), 66–73, 2022. Available at: www.ijeais.org/ijeais
10. R. Anan, T. S. Apon, Z. T. Hossain, E. A. Modhu, S. Mondal and M. G. R. Alam, "Interpretable Bangla Sarcasm Detection using BERT and Explainable AI," in *2023 IEEE 13th Annual Computing and Communication Workshop and Conference (CCWC)*, Las Vegas, NV, USA, 2023, pp. 1272–1278. doi: 10.1109/CCWC57344.2023.10099331.
11. W. Q. Al-Jamal, A. M. Mustafa and M. Z. Ali, "Sarcasm Detection in Arabic Short Text using Deep Learning," in *2022 13th International Conference on Information and Communication Systems (ICICS)*, Irbid, Jordan, 2022, pp. 362–366. doi: 10.1109/ICICS55353.2022.9811153.
12. Y. Salini and J. HariKiran, "Sarcasm Detection: A Systematic Review of Methods and Approaches," in *2023 3rd International Conference on Smart Data Intelligence (ICSMDI)*, Trichy, India, 2023, pp. 15–22. doi: 10.1109/ICSMDI57622.2023.00012.
13. B. Singh and D. K. Sharma, "A Survey of Sarcasm Detection Techniques in Natural Language Processing," in *2023 6th International Conference on Information Systems and Computer Networks (ISCON)*, Mathura, India, 2023, pp. 1–6. doi: 10.1109/ISCON57294.2023.10112176.
14. S. Saha, J. Yadav, and P. Ranjan, "Proposed Approach for Sarcasm Detection in Twitter." *Indian J. Sci. Technol.*, 10, 1–8, 2017.
15. B. Felbo, A. Mislove, A. Søgaard, I. Rahwan, S. Lehmann, "Using millions of emoji occurrences to learn any-domain representations for detecting sentiment, emotion and sarcasm," in *Proceedings of the 2017 Conference on Empirical Methods in Natural Language Processing*, Copenhagen, Denmark, 7–11 September 2017.

16. D. Hazarika, S. Hazarika, S. Gorantla, E. Cambria, R. Zimmermann, and R. Mihalcea, "CASCADE: Contextual Sarcasm Detection in Online Discussion Forums," in *Proceedings of the 27th International Conference on Computational Linguistics*, Santa Fe, NM, USA, 20–26 August 2018.
17. Y. Tay, A. Luu, S. Hui, and J. Su, "Reasoning with sarcasm by reading in in-between," in *Proceedings of the 56th Annual Meeting of the Association for Computational Linguistics*, Melbourne, Australia, 15–20 July 2018.
18. Y. Cai, H. Cai, and X. Wan, "Multi-Modal Sarcasm Detection in Twitter with Hierarchical Fusion Model," in *Proceedings of the 57th Annual Meeting of the Association for Computational Linguistics*, Florence, Italy, 28 July–2 August 2019.
19. H. Pan, Z. Lin, Y. Qi, P. Fu, and W. Wang, "Modeling Intra and Inter-modality Incongruity for Multi-Modal Sarcasm Detection," in *Findings of the Association for Computational Linguistics: EMNLP 2020*; Association for Computational Linguistics: Stroudsburg, PA, USA, 2020; Vol. 2020.
20. R. Potamias, G. Siolas, and A. Stafylopatis, "A transformer-based approach to irony and sarcasm detection," *Neural Comput. Appl.*, 32, 17309–17320, 2020.
21. S. Sonawane, and S. Kolhe, "TCSD: Term Co-occurrence Based Sarcasm Detection from Twitter Trends," *Procedia Comput. Sci.*, 167, 830–839, 2020.
22. D. Jain, A. Kumar, and G. Garg, "Sarcasm detection in mash-up language using soft-attention based bi-directional LSTM and feature-rich CNN," *Appl. Soft Comput. J.*, 91, 106198, 2020.
23. M. Shrivastava, and S. Kumar, "A pragmatic and intelligent model for sarcasm detection in social media text," *Technol. Soc.*, 64, 101489, 2021.
24. S. Modak, and A. Mondal, "Sentiment Analysis of Twitter Data Using Clustering and Classification," in *Proceedings of Third International Conference on Computing, Communications, and Cyber-Security*, Vol. 421, 2023, pp. 651–664.
25. M. Pal, and R. Prasad, "Sarcasm Detection followed by Sentiment Analysis for Bengali Language: Neural Network & Supervised Approach," in *Proceedings of the International Conference on Advances in Intelligent Computing and Applications*, Kochi, India, 1–3 February 2023, pp. 1–7.
26. A. R. W. Sait, and M. K. Ishak, "Deep Learning with Natural Language Processing Enabled Sentimental Analysis on Sarcasm Classification," *Comput. Syst. Sci. Eng.*, 44, 2553–2567, 2023.
27. A. Ghosh and T. Veale, "Fracking sarcasm using neural network," in *Proc. 7th Workshop Comput. Approaches Subjectivity, Sentiment Social Media Anal.*, 2016, pp. 161–169.
28. A.-C. Băroiu, and Ș. Trăușan-Matu, "Automatic sarcasm detection: Systematic literature review," *Information* 13(8), 399, 2022.

8 Advances in Automatic Question Generation by AI-based Algorithms

A Review

Nikita Singhal, Nikhil Yadav, Madhukar, Navneet Das, and Krishna Hirve

8.1 INTRODUCTION

Manually creating questions can be challenging and time-consuming, especially if the literature or subject under study is lengthy or complex. It requires a thorough understanding of the subject matter, the ability to identify the key concepts and ideas, and the ability to formulate compelling questions based on this knowledge. There are various benefits of using an automatic question generator, including the following:

1. Efficiency: Automatic question generators can rapidly and easily generate a huge number of questions based on a given text or topic. This can save time and effort compared to generating questions manually and can serve to support learning and research activities.
2. Relevance: Automatic question generators can generate questions that are suited to the exact content and context of the material or topic being studied. This can help to ensure that the questions are relevant and interesting and can give a more successful and engaging learning experience.
3. Adaptability: Automatic question generators can be simply altered and customized to match the individual demands and requirements of diverse audiences and circumstances. This might help to produce questions that are suited for a wide range of purposes and situations.

Automatic question generation (AQG) is a rapidly developing field with the potential to revolutionize the way we learn and assess. AQG has a number of potential benefits, including saving time and effort, personalizing learning, promoting active learning, and providing feedback to learners. MathXL uses AQG to generate personalized practice questions for students based on their individual performance data.

DOI: 10.1201/9781032724508-9

This helps students to focus on the areas where they need the most practice and to make steady progress. Duolingo, a popular language learning app, utilizes AQG to generate personalized lessons for users. Quizizz is a gamified learning platform that uses AQG to generate quizzes and other learning activities for students. These are just a few examples of how AQG is being used today. As AQG systems become more sophisticated and accurate, they are likely to be used in even more ways in the future.

The general procedure of AQG is depicted in Figure 8.1. Preprocessing the text to get rid of any stopwords or to perform word sense disambiguation is the first step in the procedure. Then, crucial phrases are chosen from the corpus using ontology or subject-matter expertise. Following is the selection of key terms, which can be accomplished using statistical techniques such as term frequency (TF). Then, using key terms, any predefined rules or templates, or any trained model, questions are created. Distractors can be generated as alternatives to the correct answer in multiple-choice questions (MCQs) or gap-filling questions (GFQs), respectively. The user is then shown the questions and distractions that were developed. There will not be a need for a distractor to be generated if the inquiry type is subjective.

There are various elements that can make question generation (QG) challenging. For example, it might be tough to recognize the important concepts and ideas in a document, especially if the content is vast or complex. Additionally, it might be challenging to produce questions that are clear, relevant, and fascinating while simultaneously avoiding ambiguity or confusion. Furthermore, formulating questions that are appropriate for a certain audience or setting might sometimes be difficult. For example, producing questions for a quiz or test that are suited to the unique material of a lesson or course involves a profound understanding of the subject matter, as well as the capacity to adjust to the different learning requirements and abilities of the students. There are a range of different models and algorithms that can be utilized for autonomous question production. Some typical models include the following:

- Rule-based models: These models use a collection of pre-defined rules and heuristics to detect significant concepts and ideas in a text and then produce questions based on these concepts. These models are generally straightforward to construct but can be limited in their capacity to produce high-quality queries.
- Statistical models: These models utilize statistical approaches, such as n-gram analysis and term frequency-inverse document frequency (TF-IDF), to discover essential concepts and ideas in a text and then produce questions based on these notions.
- Neural network models: These models use artificial neural networks to analyze a text and produce questions. Neural networks can learn from a vast quantity of training data and can generate high-quality queries that are suited to the exact content and context of the text. However, they can be complicated to execute and need many processing resources. This chapter presents readers with the following:
- A comparison with the existing reviews.
- A summary of various AQG techniques used by different researchers.

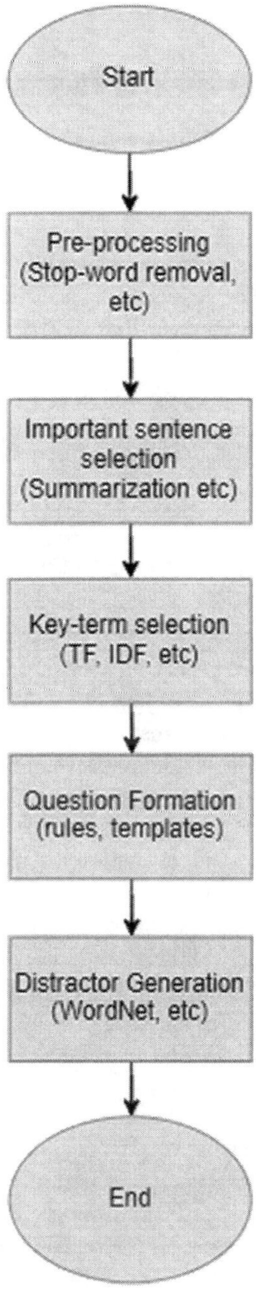

FIGURE 8.1 Question generation process

- A comparative analysis of the research done in the field of AQG.

8.2 COMPARISON WITH THE EXISTING REVIEWS

There exist some notable reviews in the domain of AQG [1–3]. The review published by Ch. and Saha covered literature published up to late 2018. Moreover, the study only covered the papers that deal with the generation of MCQs, while our review is not limited to any question type. The literature by Kurdi et al. gives a comprehensive review of the papers by assessing their quality using nine self-defined quality criteria. The review was performed on papers published from 2015 to early 2019, whereas the proposed review contains research up to late 2022. Soni et al. performed a review on 14 researchers up to 2018, and the review was not very rigorous [3]. The authors did not classify the papers based on the techniques used for question formation, and no comparative analysis of datasets and evaluation criteria used by the studies was given. We give a tabular and pictorial representation of datasets, domains, evaluation types, types of questions, etc. present in the research papers.

8.3 BACKGROUND AND RELATED WORK

8.3.1 Techniques for Text Analysis

Text analysis is necessary before generating the questions. The target text or corpus needs to be pre-processed and parsed in order to aid in QG. In the literature, there were two types of analysis: syntactic analysis [4–9] and semantic analysis [10–15]. Syntactic analysis allows us to understand the structure and arrangement of words or phrases inside the text, while semantic analysis requires a deeper understanding of the context and the relationships between various entities in the text.

Syntactic analysis is usually done by generating parse trees, the Parts of Speech (PoS) tags, or studying the dependency structures and does not require knowledge about the context. As the name suggests, it only takes into consideration the syntax of the sentences. Semantic analysis, on the other hand, requires understanding beyond syntax. It is related to the true meaning of words and phrases in a sentence and might require additional sources such as ontologies or some other kind of knowledge base in order to establish relations between them. However, it is worth mentioning that both techniques are not always mutually exclusive. Some studies use a combination of both techniques in order to understand and analyze the text [16]. This analysis can also help in extracting informative sentences from the whole text, as only the relevant sentences would be picked up for generating questions based on the information provided by the domain-specific knowledge. The by-products of the analysis, for example, a parse tree in the case of syntactic analysis or a lexical semantic tree in the case of semantic analysis, can be very helpful in generating questions directly by following certain laid-out rules or templates.

8.3.1.1 Syntactic Analysis

Agarwal and Mannem aimed at generating GFQs for a biology textbook [4]. The first step was finding the informative sentences in the text. For extracting them, the researchers used nine lexical features that took into consideration the syntactic information of the text, such as the presence of superlatives, discourse connectives, nouns, and pronouns, among others. These sentences were then scored using a weighted sum of features, weights being assigned heuristically. The candidate key terms were discovered using PoS tagging and noun chunking, and the best among them were chosen with the aid of three features: term frequency (TF), the resemblance with the title of the document, and the height of the candidate in the syntactic tree of the sentence. Contextual similarity and sentence similarity between a key and distractors served as the criterion for distractor generation (DG). The evaluation was done by two biology students with an inter-evaluator agreement score of 0.7 and 0.75 for ranking the generated questions and the selected keys as good/acceptable respectively.

Along with GFQs, Kilawala et al. proposed methods for generating MCQs and subjective questions as well [5]. During the sentence-by-sentence analysis of the text, with the aid of Python's Stanford parser library, a tree object was constructed to categorize the parts of sentences as noun phrases, adverb phrases, etc. (PoS tagging). These phrases were identified as potential gaps for GFQs and MCQs prioritizing numbers and places. These candidates were then ranked considering other features such as the presence of stop words, the number of words, etc.

Alongside working on input in the form of text, Srivastava et al. put forth a method that could also work on images as input [6]. Then the AQG task was completed by understanding the syntactic composition of the sentence and obtaining the subject, object, and predicate from it with the aid of dependency parsing, for which the Stanford CoreNLP module was utilized. After passing the subject through the PoS tagger, appropriate "Wh" questions were generated. GloVe [17] was used for option generation, which considered the co-occurrence probabilities of words.

The proposed system by Kumar et al. was also able to generate various kinds of questions [7]. First, the text was preprocessed by performing various tasks such as tokenization and stopword removal. Then, suitable sentences were selected based on the entities present in them, for which PoS tagging was used. The ranking of the sentences was done with the help of the TextRank algorithm [18], which consisted of steps such as sentence conversion to vector representations and populating a similarity matrix. After sorting based on rank, the top k sentences were picked. The key term selection was done with the help of the TF of a particular term along with the detection of the presence of named entities.

Majumdar et al. worked with a domain-specific approach for the generation of MCQs [8]. They considered the text from the sports domain as input. The selection of only meaningful sentences from the whole text was considered the first step. For the conversion of sentences into a simple form, the Stanford CoreNLP suite was used. Parse structure similarity with the aid of some referenced parse structures was applied for pre-processing along with co-reference resolution. The authors developed an algorithm to measure the similarities between the referenced parse trees and the input text's parse tree, which was named "Parse Tree Matching (PTM) algorithm" by the authors of the research because of the approach chosen. The technique of overgenerating questions and then ranking them according to their relevance was adopted by authors [9]. The system first converted a complex sentence into a simple one, which eased the process of AQG. This was done by doing transformations such as the removal of leading conjunctions etc. The sentence's syntactic structure was represented by a parse tree generated by Stanford Parser, which included PoS tags and category labels. Collin's rule was used to select noun phrase heads.

8.3.1.2 Semantic Analysis

In a study, instead of considering output from one parse tree, the outputs from several such tree structures were taken and blended to make one lexical-semantic QG tree (LSQGT) [10]. Along with the tree, semantic role labels were also added. The whole process was divided into the following steps. For each sentence, a dependency tree was generated using the Stanford Dependency Parser. These dependencies were then grouped into components. Then, certain modifiers (temporal, locative, etc.) were added from the semantic role labeler, giving the required LSQGT. This structure was later utilized for QG, with more focus on semantics.

Ilya et al. utilized the auxiliary semantic relations for analysis as they firmly believed that they could be used for constructing the hypotheses of relations among the various entities of natural language [11]. First, the NLTK (The Natural Language Toolkit) library was employed to generate a tree representing syntactic relations. Thereafter, chunking was applied for taking out important constituents of the sentence. The main elements in the whole process were the semantic relations, which were of different kinds such as quantitative (less, more. . .) or temporal (earlier, later. . .), among others.

The domain-specific nature of AQG has led to various challenges, such as the dominance in the role of experts, as pointed out by researchers [12]. Thus, they proposed a domain-independent ontology generation model to aid the process. A combination of sentence ontology and taxonomy ontology (combinedly referred to as "Knowledge Ontology"), along with template queries, was used.

Considering the texts from the medical domain only, Wang et al. came up with an AQG approach that considered the semantic relations between various medical entities/concepts such as treatment being related to disease [13]. Various articles from this domain were parsed by MetaMap Transfer, which categorized the phrases present in the sentences into various medical entities such as ¡Disease¿ and ¡Symptom¿. This helped in comprehending the meaning of the text with reference to the language of health and bio-medicine. The process was in accordance with the Unified Medical Language System (UMLS). Thus, the semantic interpretation of the text with respect to the medical domain would help in generating quality questions, but the whole process can be very time-consuming.

In one of the researches, a system to generate MCQs which was trained on Wikipedia articles was proposed [14]. The system was tested for the physics domain, but it can work for any domain by generating the respective domain's knowledge base. The key terms (both unigrams and bigrams) were stored in a dictionary after extraction from the knowledge base. Inverse Document Frequency (IDF) was used for the purpose of ranking the terms. Paradigmatic relations were used for the purpose of DG, which allowed for the substitution of one word for another in the same categories. Sentences beginning with discourse connectives such as since and therefore were discarded as they could not generate relevant questions.

The task of MCQ generation for the text from the biomedical domain was taken by a study [15]. The main concepts in a text were extracted semantically rather than syntactically. The information extraction (IE) component was the main component along with QG and DG. Named entities were considered as chains in the dependency structures. Only the chains containing named entities (NEs) were considered further. Such chains could help in producing more meaningful questions. On traversing the whole structure, all such words were extracted that were dependent on the main verb/clause. Semantic patterns were scored by calculating the strength of the pattern's association with the domain-oriented corpus. The extracted semantic relations ensured quality questions. These high-scoring semantic patterns were then transformed into respective questions.

It was very well pointed out by the authors that despite QG being a simple task of syntactic transformation, the influence of semantic features can help in generating good questions rather than trivial factoids [16]. Therefore, the authors applied both types of analysis to the text before QG. Along with syntactic dependency relations, the authors also took into consideration the semantic roles of NEs and the thematic roles of VerbNet. Both the heuristics (dependency and semantic role labelers) together aided in the production of more meaningful questions.

8.3.2 Techniques for Question Generation

After pre-processing and text analysis are done, QG is the next step. The studies generated various kinds of questions, such as GFQs, MCQs, and subjective questions, to name a few. When it comes to MCQs, along with the correct answer, the module must provide other options (commonly referred to as distractors) as well. Thus, DG becomes a complementary task. The subjective questions contain Wh-question words in them. The studied literature employs a variety of techniques for this purpose. Some used self-formulated rules to deal with the transformation of affirmative sentences into interrogative ones. These were called rule-based approaches [5, 7–9, 16]. While some used a template-based approach, i.e. when a sentence's structure matches with any particular template, the respective question is generated using the entities identified in the sentences [10–13, 19, 20]. The templates are like classes of questions having similar structures. Other approaches were based on using neural encoder-decoder architecture [21–24], and in some cases, pre-trained transformer models such as T5 and BERT [25–32].

8.3.2.1 Rule-Based Approaches

After the analysis of the sentences, NER along with super-sense tagging (SST) was used for deciding the kind of "wh" -question that should be generated from that particular sentence [5]. NER classified entities into certain categories such as names of people and locations. SST was used for annotating entities within WordNet's taxonomy. Pre-laid production rules helped in generating the suitable "wh"-question example. If the target keyword is identified as a person entity, then the "who" clause is used and not "when" or "who" clauses. The main verbs and main clauses from the sentences were also taken out using self-designed regular expressions and pattern matching. The authors conducted a survey among a student group and asked them to differentiate between automated and manual-generated questions, and it was noticed that the responders faced difficulty in differentiating the same, which implies quality questions being generated.

Kumar et al. too utilized the production rules for generating the suitable question after utilizing NER and dependency parsing for the extraction of the subject, verb, tense form, etc., from the sentences [7]. The authors then used simplenlg library for the generation of Wh-questions by providing three rules of production. One such rule is the conversion of a sentence to interrogative form by using "How" as a question word if the answer phrase is an adverb. This approach, however, will not be suitable for complex or ambiguous sentences.

Domain-oriented NER was used for the selection of key terms from the passage in one of the studies [8]. After the key terms were successfully identified, sentences were transformed by replacing them with the appropriate "Wh"-word. A named dictionary and predetermined rules were used for DG tasks. The accuracy of the whole system was summarized at 93.21%. According to the authors, false identifications can easily be tackled by better recognition of domain-specific terms/phrases.

By applying a verb-focused rule approach along with shallow semantic parsing and dependency parsing, one can achieve a large variety of relevant questions [16]. This approach also ruled out the need for the simplification of sentences. Custom rules were not only used for transforming sentences into questions, but the irrelevant questions were also discarded with the aid of certain rules. The system achieved a BLEU-1 score [33] of 45.55.

On converting complex sentences into simple ones, Heilman et al. carried out rule-based lexical and syntactic transformations such as Wh-movement in order to modify declarative sentences into questions [9]. After the selection of an answer phrase, the main verb was decomposed and subject and auxiliary verbs were inverted. This was followed by the insertion of a question phrase. The generated questions were afterward ranked by a logistic regression model (trained on a small set of annotated questions ranked by some university students) that marked the question as acceptable based on the features such as vagueness, grammatical attributes, etc.

8.3.2.2 Template-Based Approaches

Figure 8.2 shows the procedure of generating questions using templates. The template library stores the predefined templates, and then using template matching and key terms, the suitable template is selected and a question is formed.

After the generation of the lexical-semantic tree, Mazidi et al. matched each sentence with the patterns present in the external template [10]. They created an external file containing 50 templates that were constructed manually. The template types

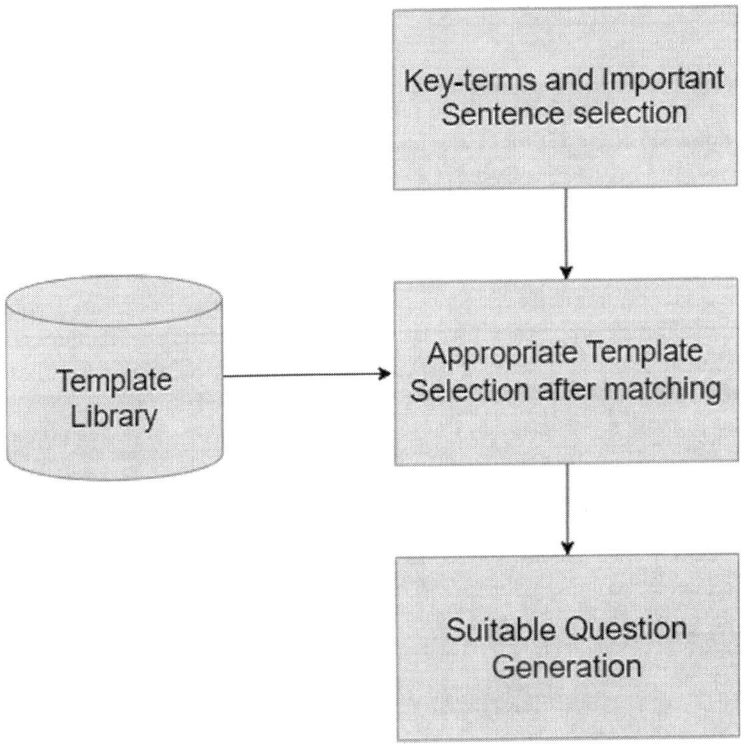

FIGURE 8.2 AQG using templates

ranged from SVO templates (templates using subject-verb-object information to generate questions) to templates using the concepts of discourse cues. The majority of the templates were using the dependency relations between the entities to generate questions of the type "why," "when," "describe," etc. The SVO templates could only generate factoid questions whereas more complex questions could be created with the aid of the other two types. This approach helped the authors in achieving greater variation in the questions.

Ilya et al. found out that a text of simple conversational style of writing was more likely to get less number of invalid questions, as the structures of such texts were simple and they fit perfectly in the templates [11]. After sentences containing certain semantic relations were taken, spaCy library in Python was used to generate questions from those sentences based on pre-defined templates.

A system for generating subjective questions along with an evaluation system for answer assessment was proposed in one of the research projects [19]. The authors had divided the task of QG into two sub-tasks: keyword extraction and creation of a subjective template for the generation of questions. The Jaccard Similarity Coefficient was used for keyword extraction. Then, by using the extracted keywords and considering the templates reported in another study [34], the task of QG was accomplished. The multicriteria decision-making approach (MCDMA) was then utilized to evaluate the answers considering string similarity, semantic similarity, and keyword similarity.

Kusuma et al. divided the whole process into six steps [12]. First, the ontology was generated using pre-processing on the sentences, followed by template generation. Then, the difficulty level was determined utilizing Bloom's taxonomy. This was followed by SPARQL query determination that would be retrieving information from the ontology followed by quiz generation using the produced queries. The last step was the evaluation of the model, for which two parameters were considered: effectiveness accounting for the amount and types of questions and quality measuring the consistency among the questions spawned. The proposed method achieved an accuracy of 90.71%, but the model was highly influenced by the information that was being extracted from the ontology.

Under domain-specific tasks, the templates were generated in accordance with various entities/domain-specific terms related to the medical domain by authors [13]. The templates had four components: question, required entities, keywords, and related answers. The entities were categorized with the help of UMLS. The keywords showed relevant features of concepts denoted by the respective entities. Nearly 100 templates were built by the researchers. They aimed at generating questions that were interesting, answerable, and related to the medical field. Templates were scored based on weights assigned by considering certain parameters such as the number of negative evaluations of the template.

Applying the concept of unsupervised learning, it was observed that the application of a template on a retrieved sentence (related to the original sentence), instead of the original context phrase, could improve the performance of the model to a large extent [20]. The system could produce factoid questions—questions based on a crisp fact. Entities of the sentences were recognized using NER. Then, the questions were generated using templates. The entity was considered as the "mask," and the templates were in the form of "Wh+B+A" along with other possible orderings, where A, B are two fragments of a sentence. The "mask" was replaced by the "Wh"-question word.

8.3.2.3 Pretrained-Transformer Models

Vashsisht et al. proposed an architecture that utilized an automated pipeline for creating question-answer (QA) pairings from input text [25]. The proposed method began by extracting spans from the text using the Transformer T5 model, a text-to-text framework that incorporates encoder and decoder trans-former architecture. To account for sentence coherence, the authors encoded all sentences in the text and computed a similarity score, which was kept in a coherence matrix. Sentences were then combined based on these ratings prior to extracting spans. The generated spans were utilized to produce queries, and the span text was sent to a knowledge graph to gather further domain information. The original sentence and produced questions were then tokenized to form question and context tokens, which the SpanBERT model [35] used to identify the appropriate context tokens that provided the response and calculate a probability score. The significance of the probabilistic outcome was used to rate the solutions. The proposed method surpassed the BLEU scores of 0.4 for answers and 0.6 for questions. Furthermore, to increase the length of the paragraph, it could be expanded upon by providing more detailed information on the specific steps taken in each stage of the pipeline and by providing examples of how the proposed method improves upon existing QA generation techniques.

In order to demonstrate the system's high availability and enhance student learning performance, Tsai et al. suggested an AQG system that combined semantics and syntax [26]. The authors extracted keywords from textbooks using BERT and trained the model using unsupervised learning and transfer learning techniques on a sizable quantity of unlabeled data. They performed syntactic analysis on the collected phrases using the Stanford CoreNLP module. The output was then presented as a parse tree. Declarative statements were changed into interrogative ones by the authors using the T5 model. The authors tested the question-generating strategies using ROUGE-L [36] and BLEU. The BLEU rating of 0.567 was unsatisfactory. The ROUGE-L score, on the other hand, was 0.613, indicating a degree of resemblance between the machine-generated questions and the SQuAD [37] dataset's questions.

Considering the recent advancements in neural language models, Torrealba et al. too utilized the Transformer model for end-to-end production of MCQs [27]. The authors divided the whole process into three sub-processes viz. QG, QA, and DG. The target text was first parsed to extract relevant para graphs. Then, using any one of the paragraphs generated above, QA pairs were produced with the aid of QAPModel, which was implemented using pre-trained T5 models. For producing distractors, the use of DGModel was proposed, which consists of a single fine-tuned (using Adam optimizer) T5 model. The authors considered manual validation of the model as ideal because of the absence of proper metrics for evaluation. Human specialists analyzed questionnaires, indicating that questions and alternatives were typically well-structured in terms of grammar, meaning, and clarity. Thus, a survey conducted resulted in a score of 2.17/5 (Easy) for difficulty level and a score of 4.28/5 (Good) regarding the well-structuredness of the questions. As for DG task, an average BLEU-1 score of 14.8, an average ROUGE-L score of 14.91, and an average cosine similarity of 0.71 were noticed for the DG-RACE dataset [38]. However, the produced MCQs were more retention-aligned than interpretation-aligned.

The authors in one of the studies demonstrated that robust question-generating systems can be achieved by applying transformer-based finetuning strategies, without the need for extra mechanisms, answer information, or substantial characteristics, and with just a single pre-trained language model [29]. By improving the METEOR [39] and ROUGE L scores by 8.62 and 14.27, respectively, the best model beat earlier, more intricate recurrent neural network-based Seq2Seq models. They showed that, while being a single-model system, it outperformed Seq2Seq models that employed answer awareness and other novel techniques. Although the performance of the models showed minimal variation, the authors explained that it was because many of the context paragraphs in SQuAD were substantially shorter than the 1024-token maximum context window length of GPT-2.

Chan et al. explored the application of the pre-trained BERT language model for question-generating tasks [30]. The paper suggested three neural architectures based on BERT for question-generating challenges. Additionally, two more models were proposed, which were based on reorganizing the BERT employment in a sequential fashion to take knowledge from prior findings. The suggested model took as input a context text and an answer phrase and generated a question based on the answer phrase. To address the issue of ambiguity when a response phrase appears multiple times in the question, the authors offered a simple but effective input encoding method that inserted special highlighting tokens before and after the desired answer period. The authors used SQUAD dataset for evaluating their model. Results of the experiments revealed that the best model produced cutting-edge performance, increasing the existing top models' BLEU 4 score from 16.85 to 22.17. This highlighted the efficacy of the suggested model as well as the utility of employing BERT for question development activities.

In one of the studies, the EduQuiz system which is used for generating quizzes for the educational domain was proposed [31]. It was based on a GPT-3 language model that has been fine-tuned on text-quiz pairs. This means that it has been trained on a large dataset of text passages and corresponding quizzes and can use this training to generate quizzes for a new text passage when given the passage as input. The generated quizzes consist of a multiple-choice question with a correct answer and several distractor answers. The authors of the paper evaluated EduQuiz on a dataset of examination-like questions and found that the majority of generated quizzes were reasonable but that generating high-quality distractors was more challenging. The authors suggested that while it might not be ready to fully replace manually generated tests, EduQuiz had the potential value as a tool providing formative feedback and to increase engagement during the learning phase by enhancing textbooks with assessments. By providing quizzes that were tailored to a specific text passage, EduQuiz could be used to give students immediate feedback on their understanding of the material and help them identify areas where they needed to focus more attention. Additionally, by providing a way to generate assessments automatically, EduQuiz could help reduce the burden on educators who would otherwise have to create quizzes manually.

8.4 COMPARATIVE ANALYSIS

This section provides a tabular and pictorial depiction of the classification of the research based on various measures such as question formation techniques, sort of questions generated, datasets used, specific domains for QG, and the kind of evaluation adopted in respective research papers.

8.4.1 Approach Classification

Table 8.1 gives us the summary of various approaches that were adopted by the respective authors for the task of question formation. Figure 8.3 is a bar graph that represents the number of studies per approach. Table 8.2 summarizes the text analysis approaches used by each study.

TABLE 8.1
Question Formation Approach

Question formation Method	Studies
Rule Based	[5, 7–9, 16]
Template Based	[10–13, 19, 20]
Pre-trained transformer	[25–32]
Self-developed Encoder-decoder architecture	[21–24]

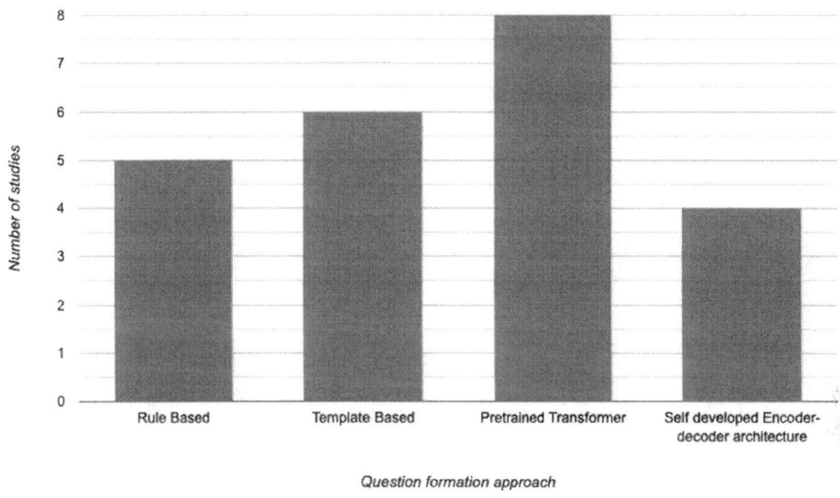

FIGURE 8.3 Question formation techniques

TABLE 8.2
Text Analysis Methods

Text analysis techniques	Studies
Syntax-Based	[4–6, 8, 9, 14, 22, 24, 25, 27]
Semantic-Based	[10–15, 19, 20, 23, 26, 28–30, 32, 40]

8.4.2 Types of Questions

The types of questions generated by the respective studies vary substantially. Table 8.3 gives us a review of the types of questions generated by the respective research. Figure 8.4 is a bar graph that clearly indicates that the most popular type of questions is the subjective one and the least generated type is the True/False questions.

TABLE 8.3
Types of Questions Generated

Types of Questions	Studies
MCQs	[5, 7, 8, 14, 15, 27]
True/False	[5, 7]
Gap-Filling Questions	[4, 5, 7]
Subjective Questions	[5–7, 9–13, 16, 19–26, 28–32, 40]

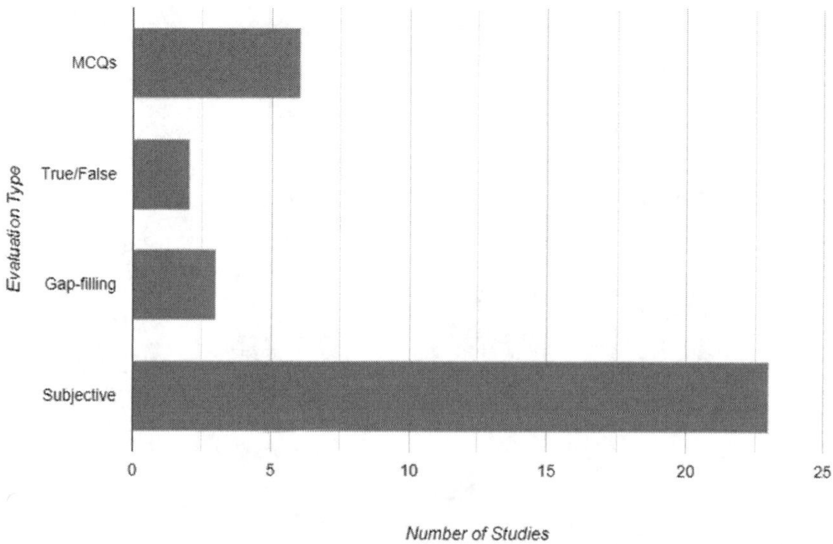

FIGURE 8.4 Number of studies per question type

8.4.3 Datasets Used

Table 8.4 summarizes the datasets that were used by the respective authors. The most popular dataset is SQUAD followed by Wikipedia articles. Some of the studies that focus on e-learning use textbook materials to train their systems. Figure 8.5, a pie chart, gives us a pictorial depiction of the datasets used by the authors for the task of training.

8.4.4 Domain-Specific Question Generation

While most of the researchers aim at generating questions that are not specific to any domain/field, some of the researchers focus on QG from a particular domain only, for example, the biomedicine domain. Table 8.5 mentions all the specific domains which were chosen by the authors to work in.

TABLE 8.4
Types of Datasets Used

Dataset	Studies
SQUAD	[10, 20–26, 29, 30, 41]
Amagasaki FAQ Dataset	[28]
TEXTBOOK	[4, 6, 9, 11, 12, 42, 43]
Wikipedia and Other sources	[5, 7, 8, 13, 14, 31, 32, 40]
DBSYLLABI	[19]
DG-RACE	[27]

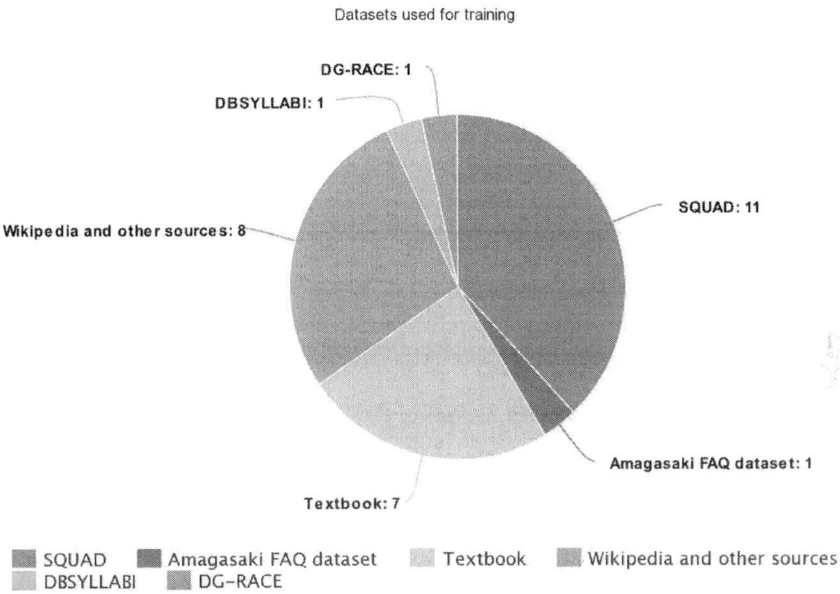

FIGURE 8.5 Types of datasets used

TABLE 8.5
Domain-specific Question Generation

Domain	Studies
Computer	[5, 19]
Biology	[4]
Sports	[8]
Finance	[9, 10]
Medical	[12, 15]

8.4.5 Type of Evaluation

Some of the studies evaluate their generated questions by using ROUGE and BLEU scores, which work by measuring the similarity between the machine-generated text and a high-quality reference text, which is usually human-produced. However, some of the researchers believe that such scores cannot effectively evaluate generated questions, and thus, they employ human evaluation techniques in which a group of experts is asked to rate the questions based on parameters such as relevance, difficulty, and consistency, to name a few. All the evaluation criteria are summarized in Table 8.6. Figure 8.6 shows the number of studies per evaluation kind.

TABLE 8.6
Evaluation Criteria Used

Evaluation type	Studies
BLEU score	[6, 10, 16, 19, 21–23, 25–27, 30, 31]
Rogue-L	[10, 22, 23, 25, 26, 29, 31]
METEOR	[22, 23, 31]
EM and F-1 Score	[7, 15, 20, 32, 40]
Human Evaluation	[4, 7, 8, 12, 14, 15]
STS Similarity	[28]
WER score	[24]

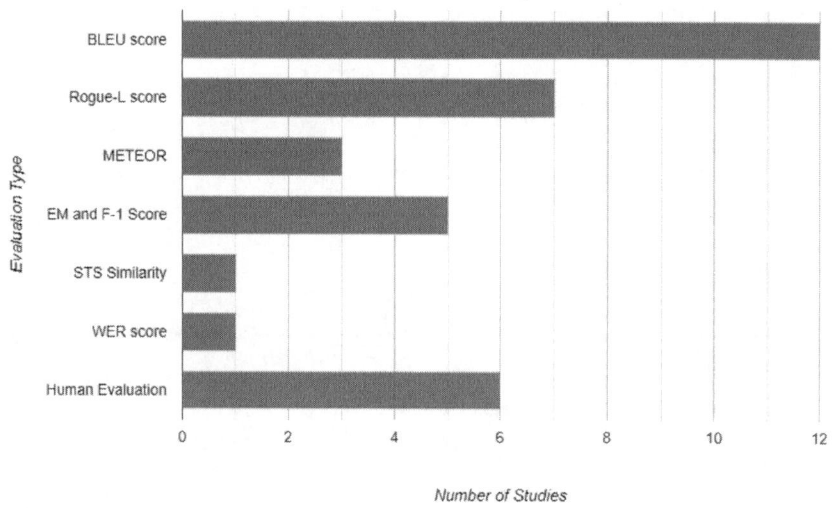

FIGURE 8.6 Type of evaluation used in studies

8.5 RESULTS

We performed a thorough comparative analysis, comparing research papers against various parameters, such as the type of questions being generated, the approach being used, and the datasets utilized by the authors, among others. We found that deep learning approaches that incorporate transformers and encoder-decoder architectures are gaining more prominence nowadays compared to traditional approaches, such as rule-based and template-based methods. Although rule-based and template-based approaches yield decent results, they rely on user-defined rigid rules and structures. In contrast, encoder-decoder models offer greater flexibility in generating a variety of questions, eliminating rigidity and producing less repetitive questions, hence their popularity.

Regarding the type of text analysis, it was observed that semantic-based approaches outnumber syntax-based approaches, as semantics provide more valuable information than mere syntax in a sentence. Semantics encompass the dependencies between various elements of a sentence and offer useful insights into the context, greatly aiding in generating meaningful sentences. When it comes to the types of questions generated in each research work, subjective questions emerged as the most popular, while true/false questions were the least favored. One reason for the popularity of subjective questions is their prevalence in real-world applications, such as essay grading, content summarization, qualitative assessments, and educational settings. Subjective questions are often considered a more robust way to assess students' critical thinking. Comparing the datasets used by authors for their research, the SQUAD dataset was the most utilized, followed by Wikipedia articles. There were also other datasets, such as DBSYLLABI and DG-RACE, which were used by only a few authors. This discrepancy may be attributed to SQUAD's substantial amount of high-quality text passages and corresponding questions, making it a valuable resource for training and evaluating AQG systems.

While most of the authors preferred to generate questions irrespective of the domain to which the passage belonged, some instances were noted where authors aimed for a specific domain, such as the medical or the finance domain. Domain-specific question generation can be very helpful, as questions tailored to a particular domain can facilitate more effective and targeted learning. For example, medical students can benefit from questions specifically designed for medical subjects. Regarding the type of evaluation preferred by authors, it was found that some studies employed BLEU, ROUGE, and METEOR scores to evaluate the generated questions. On the other hand, some authors argued that these scores cannot adequately capture the relevance of generated questions. Therefore, they relied on human evaluation to assess the quality of the generated questions.

8.6 DISCUSSION

AQG has experienced substantial growth in recent years, spurred by breakthroughs in natural language processing and machine learning. However, some issues must be addressed further to increase the quality and application of AQG systems. One of the problems is formulating questions for complex and abstract subjects. Fields such as education and training, where questions generated from complex and complicated topics are considered more valuable than simple topics. To face this issue, researchers should look for ways of detecting abstract topics, such as using ontologies or domain-specific knowledge bases. Additionally, building models that can produce questions that are contextually relevant could potentially assist in overcoming the challenge. One of the disadvantages of present AQG algorithms is their tendency to generate questions that are repetitive or formulaic. To overcome this difficulty, researchers should investigate new ways for developing more diverse and creative questions. This could involve mixing various sorts of linguistic patterns, such as puns or metaphors, or employing external knowledge sources to produce inquiries that are more contextually relevant.

Several existing AQG techniques produce questions only from the text they are studying. Nevertheless, adding other information sources, such as ontologies or domain-specific knowledge bases, might improve the accuracy and relevancy of the produced queries. This might entail creating models capable of extracting important information from other knowledge sources and incorporating it into the question-creation process. As AQG models get more complicated and sophisticated, using accurate criteria for evaluating the quality of generated questions becomes increasingly vital. Researchers might investigate new ways for assessing the relevancy, consistency, and understandability of generated questions, as well as create new metrics that reflect the subtleties of human-made inquiries. This might entail creating models that can assess the quality of constructed questions based on characteristics such as their difficulty level, relevancy to the text, or capacity to test certain knowledge areas.

8.7 CONCLUSIONS AND FUTURE WORK

The task of AQG is very advantageous in the fields of education and other domains as well. The amount of research in the area is not very copious, and there is still a lot of research potential. We presented a rigorous review and analysis of the various studies done in the field. The procedure of AQG consists of two main steps, namely text analysis and question formation. Categorization of the papers is done on the basis of various approaches applied for performing the aforementioned steps. The main approaches for question formation are rule-based, template-based, and deep-learning-based models. We further divided the deep-learning-based approaches into pre-trained-model approaches and self-developed approaches. The popularity of such neural-network model-based approaches is increasing as compared to rule- and template-based systems. Text analysis can be done using either the syntax or semantics of the text.

We performed a thorough comparative analysis and presented the readers with various types of questions that were formed in the respective studies (MCQS, subjective, etc.). The datasets used by different papers were summarized, and it was found that the most popular dataset among researchers was the SQUAD dataset. A significant number of researchers relied on Wikipedia articles for their research dataset. Most of the research focuses on a multi-domain way of generating questions, but some of the studies preferred a particular domain. Lastly, the evaluation type used by the majority of the authors was scores such as ROUGE and BLEU, which compares the generated questions with a high-quality reference text. However, some considered this method as inutile and preferred human evaluation over machine evaluation for assessing the quality of generated questions. They consulted domain experts/ language experts to measure the performance of their systems.

Some of the challenges that were noticed in the field of AQG are the absence of proper metrics to evaluate the performance of systems, limited work that focuses on generating questions with varied difficulty levels, multilingual question generation, and more focus on domain-independent QG. In the future, more work can be done to address these issues, and a system can be developed in the direction of open-domain, multilingual QG. Since the domain of multilingual question generation is very scarcely explored, the future initiatives of the authors would be focused on this area. Further work could be done to develop systems that are able to generate questions in languages other than English, as it can prove to be very helpful in various sectors, such as the education sector.

REFERENCES

1. Ch, Dhawaleswar Rao, and Sujan Kumar Saha. "Automatic multiple choice question generation from text: A survey." *IEEE Transactions on Learning Technologies* 13(1): 14–25 (2018). https://doi.org/10.1109/TLT.2018.2889100.
2. Kurdi, Ghader, Jared Leo, Bijan Parsia, Uli Sattler, and Salam Al-Emari. "A systematic review of automatic question generation for educational purposes." *International Journal of Artificial Intelligence in Education* 30: 121–204 (2020). https://doi.org/10.1007/s40593-019-00186-y.
3. Soni, Sonam, Praveen Kumar, and Amal Saha. "Automatic Question Generation: A Systematic Review." In *International Conference on Advances in Engineering Science Management & Technology (ICAESMT)-2019*, Uttaranchal University, Dehradun, India; 2019. http://dx.doi.org/10.2139/ssrn.3403926.
4. Agarwal, Manish, and Prashanth Mannem. "Automatic gap-fill question generation from text books." In *Proceedings of the sixth workshop on innovative use of NLP for building educational applications* (pp. 56–64); 2011.
5. Killawala, Akhil, Igor Khokhlov, and Leon Reznik. "Computational intelligence framework for automatic quiz question generation." In *2018 IEEE International Conference on Fuzzy Systems (FUZZ-IEEE)* (pp. 1–8). IEEE; 2018. https://doi.org/10.1109/FUZZ-IEEE.2018.8491624.
6. Srivastava, Animesh, Shantanu Shinde, Naeem Patel, Siddhesh Despande, Anuj Dalvi, and Shweta Tripathi. "Questionator-automated question generation using deep learning." In *2020 International Conference on Emerging Trends in Information Technology and Engineering (ic-ETITE)* (pp. 1–5). IEEE; 2020. https://doi.org/10.1109/ic-ETITE47903.2020.212.
7. Kumar, Alok, Deepika Singh, Aditi Kharadi, and Mala Kumari. "Automation of Question-Answer Generation." In *2021 Fourth International Conference on Computational Intelligence and Communication Technologies (CCICT)* (pp. 175–180). IEEE; 2021. https://doi.org/ 10.1109/CCICT53244.2021.00043.
8. Majumder, Mukta, and Sujan Kumar Saha. "A system for generating multiple choice questions: With a novel approach for sentence selection." In *Proceedings of the 2nd Workshop on Natural Language Processing Techniques for Educational Applications* (pp. 64–72); 2015.

9. Heilman, Michael, and Noah A. Smith. "Good question! statistical ranking for question generation." In *Human Language Technologies: The 2010 Annual Conference of the North American Chapter of the Association for Computational Linguistics* (pp. 609–617); 2010.
10. Mazidi, Karen, and Rodney D. Nielsen. "Leveraging multiple views of text for automatic question generation." In *Artificial Intelligence in Education: 17th International Conference, AIED 2015*, Madrid, Spain, June 22–26, 2015. Proceedings 17 (pp. 257–266). Springer International Publishing; 2015. https://doi.org/10.1007/978-3-319-19773-9_26.
11. Ilya, Buldin, Murov Vadim, and Dmitry Silnov. "Automatic generation of questions based on semantic text analysis." In *2020 IEEE 14th International Conference on Application of Information and Communication Technologies (AICT)* (pp. 1–4). IEEE; 2020. https://doi.org/10.1109/AICT50176.2020.9368686.
12. Kusuma, Selvia Ferdiana, Daniel Oranova Siahaan, and Chastine Fatichah. "Automatic question generation with various difficulty levels based on knowledge ontology using a query template." *Knowledge-Based Systems* 249: 108906 (2022). https://doi.org/10.1016/j.knosys.2022.108906.
13. Wang, Weiming, Tianyong Hao, and Wenyin Liu. "Automatic question generation for learning evaluation in medicine." In *Advances in Web Based Learning–ICWL 2007: 6th International Conference Edinburgh*, UK, August 15–17, 2007 Revised Papers 6 (pp. 242–251). Springer Berlin Heidelberg; 2008. https://doi.org/10.1007/978-3-540-78139-4_22.
14. Shah, Riken, Deesha Shah, and Lakshmi Kurup. "Automatic question generation for intelligent tutoring systems." In *2017 2nd International Conference on Communication Systems, Computing and IT Applications (CSCITA)* (pp. 127–132). IEEE; 2017. https://doi.org/10.1109/CSCITA.2017.8066538.
15. Afzal, Naveed, and Ruslan Mitkov. "Automatic generation of multiple choice questions using dependency-based semantic relations." *Soft Computing* 18: 1269–1281 (2014). https://doi.org/10.1007/s00500-013-1141-4.
16. Dhole, Kaustubh D., and Christopher D. Manning. "Syn-QG: Syntactic and shallow semantic rules for question generation." *arXiv preprint arXiv:2004.08694* (2020).
17. Pennington, Jeffrey, Richard Socher, and Christopher D. Manning. "Glove: Global vectors for word representation." In *Proceedings of the 2014 Conference on Empirical Methods in Natural Language Processing (EMNLP)* (pp. 1532–1543); 2014.
18. Mihalcea, Rada, and Paul Tarau. "Textrank: Bringing order into text." In *Proceedings of the 2004 Conference on Empirical Methods in Natural Language Processing* (pp. 404–411); 2004.
19. Das, Bidyut, Mukta Majumder, Arif Ahmed Sekh, and Santanu Phadikar. "Automatic question generation and answer assessment for subjective examination." *Cognitive Systems Research* 72: 14–22 (2022). https://doi.org/10.1016/j.cogsys.2021.11.002.
20. Fabbri, Alexander R., Patrick Ng, Zhiguo Wang, Ramesh Nallapati, and Bing Xiang. "Template-based question generation from retrieved sentences for improved unsupervised question answering." *arXiv preprint arXiv:2004.11892* (2020). https://doi.org/10.48550/arXiv.2004.11892.
21. Zhou, Qingyu, Nan Yang, Furu Wei, Chuanqi Tan, Hangbo Bao, and Ming Zhou. "Neural question generation from text: A preliminary study." In *Natural Language Processing and Chinese Computing: 6th CCF International Conference, NLPCC 2017*, Dalian, China, November 8–12, 2017, Proceedings 6 (pp. 662–671). Springer International Publishing; 2018. https://doi.org/10.1007/978-3-319-73618-1_56.
22. Du, Xinya, Junru Shao, and Claire Cardie. "Learning to ask: Neural question generation for reading comprehension." *arXiv preprint arXiv:1705.00106* (2017).

23. Cao, Zhen, Sivanagaraja Tatinati, and Andy WH Khong. "Controllable question generation via sequence-to-sequence neural model with auxiliary information." In *2020 International Joint Conference on Neural Networks (IJCNN)* (pp. 1–7). IEEE; 2020. https://doi.org/10.1109/IJCNN48605.2020.9207643.
24. Kriangchaivech, Kettip, and Artit Wangperawong. "Question generation by transformers. CoRR 2019." *arXiv preprint arXiv:1909.05017* (1909). https://doi.org/10.48550/arXiv.1909.05017.
25. Vasisht, Santosh, Varun Tirthani, Akhil Eppa, Punit Koujalgi, and Ramamoorthy Srinath. "Automatic FAQ generation using text-to-text transformer model." In *2022 3rd International Conference for Emerging Technology* (pp. 1–7). IEEE; 2022. https://doi.org/ 10.1109/INCET54531.2022.9823967.
26. Tsai, Danny C. L., Anna Y. Q. Huang, Owen H. T. Lu, and Stephen J. H. Yang. "Automatic question generation for repeated testing to improve student learning outcome." In *2021 International Conference on Advanced Learning Technologies (ICALT)* (pp. 339–341). IEEE; 2021. https://doi.org/10.1109/ICALT52272.2021.00108.
27. Rodriguez-Torrealba, Ricardo, Eva Garcia-Lopez, and Antonio Garcia-Cabot. "End-to-End generation of Multiple-Choice questions using Text-to-Text transfer Transformer models." *Expert Systems with Applications* 208: 118258 (2022). https://doi.org/10.1016/j.eswa.2022.118258.
28. Doi, Ryusei, Thatsanee Charoenporn, and Virach Sornlertlamvanich. "Automatic question generation for chatbot development." In *2022 7th International Conference on Business and Industrial Research (ICBIR)* (pp. 301–305). IEEE; 2022. https://doi.org/10.1109/ICBIR54589.2022.9786384.
29. Lopez, Luis Enrico, Diane Kathryn Cruz, Jan Christian Blaise Cruz, and Charibeth Cheng. "Transformer-based end-to-end question generation." *arXiv preprint arXiv:2005.01107* (2020).
30. Chan, Ying-Hong, and Yao-Chung Fan. "A recurrent BERT-based model for question generation." In *Proceedings of the 2nd Workshop on Machine Reading for Question Answering* (pp. 154–162); 2019.
31. Dijkstra, Ramon, Zülküf Genç, Subhradeep Kayal, and Jaap Kamps. "Reading Comprehension Quiz Generation using Generative Pre-trained Transformers." In *iTextbooks@ AIED* (pp. 4–17); 2022.
32. Zhu, Pengfei, Xiaoguang Li, Jian Li, and Hai Zhao. "Unsupervised Open-Domain Question Answering." *arXiv preprint arXiv:2108.13817* (2021).
33. Papineni, Kishore, Salim Roukos, Todd Ward, and Wei-Jing Zhu. "Bleu: a method for automatic evaluation of machine translation." In *Proceedings of the 40th Annual Meeting of the Association for Computational Linguistics* (pp. 311–318); 2002.
34. Dhokrat, Asmita, Hanumant Gite, and C. Namrata Mahender. "Assessment of answers: Online subjective examination." In *Proceedings of the Workshop on Question Answering for Complex Domains* (pp. 47–56); 2012.
35. Joshi, Mandar, Danqi Chen, Yinhan Liu, Daniel S. Weld, Luke Zettlemoyer, and Omer Levy. "Spanbert: Improving pre-training by representing and predicting spans." *Transactions of the Association for Computational Linguistics* 8: 64–77 (2020). https://doi.org/10.1162/tacl_a_00300.
36. Chin-Yew, Lin. "Rouge: A package for automatic evaluation of summaries." In *Proceedings of the Workshop on Text Summarization Branches Out, 2004*; 2004.
37. Rajpurkar, Pranav, Jian Zhang, Konstantin Lopyrev, and Percy Liang. "Squad: 100,000+ questions for machine comprehension of text." *arXiv preprint arXiv:1606.05250* (2016).
38. Lai, Guokun, Qizhe Xie, Hanxiao Liu, Yiming Yang, and Eduard Hovy. "Race: Large-scale reading comprehension dataset from examinations." *arXiv preprint arXiv:1704.04683* (2017).

39. Agarwal, Abhaya, and Alon Lavie. "Meteor, m-bleu and m-ter: Evaluation metrics for high-correlation with human rankings of machine translation output." In *Proceedings of the Third Workshop on Statistical Machine Translation* (pp. 115–118); 2008.
40. Lewis, Patrick, Ludovic Denoyer, and Sebastian Riedel. "Unsupervised question answering by cloze translation." *arXiv preprint arXiv:1906.04980* (2019).
41. Gupta, Deepak, Asif Ekbal, and Pushpak Bhattacharyya. "A deep neural network framework for English Hindi question answering." *ACM Transactions on Asian and Low-Resource Language Information Processing (TALLIP)* 19(2): 1–22 (2019). https://doi.org/10.1145/3359988.
42. Gopal, Anuj. "Automatic Question Generation for Hindi and Marathi." In *2022 International Conference on Advanced Learning Technologies (ICALT)* (pp. 19–21). IEEE; 2022. https://doi.org/ 10.1109/ICALT55010.2022.00012.
43. Patil, Pushpa M., R. P. Bhavsar, and B. V. Pawar. "Automatic question generation for subordinate conjunctions of Marathi." In *2022 8th International Conference on Advanced Computing and Communication Systems (ICACCS)* (Vol. 1, pp. 169–173). IEEE; 2022. https://doi.org/ 10.1109/ICACCS54159.2022.9785063.

9 Hands-on Practices, Reflection on Data Wisdom with AI Principles

A Review

Rishu Kumar, C.L.P. Gupta, and Anuj Singh

9.1 INTRODUCTION

A similarly well-publicized problem coexists with the promise that artificial intelligence (AI) gives solution for a variety of societal problems, including those in business, government agencies, hospital management, and e-security. Unfair treatment, significant violations of fundamental rights, a lack of accountability for generational gaps, inaccurate or untraceable diagnoses, and hostile device attacks [1] are all examples of Tique. As a result, a hot topic is how to regulate AI, along with worries about global production and competitiveness. As a result of the latter, there are numerous e-frameworks for the creation and application of social benefits or human-rich understanding citizens to increase the effectiveness of AI (what this chapter refers to as the first wave of ethical AI lecture), as well as initiatives to evaluate the convergence and possible normative agreement of these frameworks (the 2nd wave) [2].

By examining how AI principles translate into practical data science practices, the third wave of ethical AI scholarship builds on the first two, emphasizing governance and implementation. It looks at the range of real-world situations in which AI ethics can be applied, highlights issues with standardization in a particular field, and connects enforcement and regulation-related queries to AI electronic governance. These studies demonstrate how challenging it is to integrate AI systems into current organizational all AI-facilitating solutions deliver the anticipated additional (corporate or industry) value. This chapter contributes to the advancement of ethical AI scholarship by aligning AI ethical principles with the human lifecycle of an AI-based e-service or product to assess their applicability in data science practice. Given the current absence of practical regulations or governance frameworks, integrating AI ethics principles into the development of e-services poses challenges for

DOI: 10.1201/9781032724508-10

data scientists and organizations [3–11]. For instance, our work with government agencies demonstrates that while national public structures are aware of AI ethics and verified lists, they frequently turn to outside help for translating them into EU-level guidance. Studies show that organizations can only manage a certain range of issues and calming techniques while implementing an expansion, indicating the need for a framework to ensure well-weighted and overall AI ethics [12].

9.2 A FRAMEWORK FOR AI

To contextualize our work within the broader landscape, this chapter starts with a definition that sets the framework for the subsequent analysis in the intricate field of AI. The working definition of AI utilized here draws from various sources, encompassing several definitions. AI applications are device-based systems designed on extensive analysis, often using mathematical optimization techniques. It applies to a specific environment and meets the requirements of a specific set of human-defined goals. It is human cognition that automates one or more tasks, such as collecting, joining, scoring, sorting, identification, acquiring (individual) data, selecting, prioritizing, referring, guessing, assisting, making decisions, and executing the function [13]. AI devices are designed to operate with varying degrees of autonomy, enabling the creation of adaptive services that affect real or virtual environments at scale and in(near) real time [14].

The AI systems defined as such are used to perceive the environment or geographical value, including taking into account the complexities of complex, dramatic, and virtual worlds. Information processing, including input data interpretation and decision-making, autonomy in thinking and learning, and the accomplishment of specific objectives, appear to be the driving force behind AI systems [15].

9.3 HIGHEST-LEVEL SPECIFICATIONS AND NATIONAL CONTEXTS

The EU's brand, "human-centric AI," encapsulates the Ethics Set of Rules for Trustworthy AI, which serves as a natural extension of the EU's strategy, prioritizing people at the core of AI development. This initiative reflects the strategy's essence. Additionally, the concept of AI "made in Europe" stands out as another pivotal aspect of these guidelines. The EU has leveraged the General Data Protection Regulation (GDPR) as a model to reinforce the influence of European values on partners and markets, a comparison frequently drawn by commentators. In light of this, it becomes crucial to assess the compatibility of AI with the HLEG's conceptual framework and EU values. The former is believed to be paving the way for a second wave of the "Brussels effect" by setting international standards for usability and design. Although the ethics guidelines created by them have received some criticism, collaborators of all kinds and sizes have been investing time and money into integrating them into their operations. Our collaboration with government agencies revealed a mixed picture of those initiatives. Public bodies struggle to independently align national AI ethics [16] and checklists with EU-level guidance and frequently choose an "AI requirements mix" instead.

This implies that they simultaneously apply moral precepts from various ethical systems, giving precedence to those that seem the most obvious and those that they could get outside advice on. They would also ask other authorized authorities, such as the National Data Protection Authority, for assistance.

According to Stahl et al., European organizations not only acknowledge the ethical issues associated with AI but also demonstrate a willingness to actively tackle them, which has bolstered their preparedness to engage in discussions on AI ethics. While acknowledging that certain ethical concerns may lie beyond their expertise and scope, organizations are purportedly focusing only on a limited number of issues and mitigation strategies. This underscores the need for a broader stakeholder framework to ensure comprehensive engagement and coverage of AI ethics.

With this multifaceted backdrop, the mapping exercise has reached completion. Public authorities encounter challenges in discerning which principles are relevant to specific applications, determining the timing and method for structuring oversight, and integrating both principles and oversight into organizational culture. Similarly, most entities contemplating the use of AI to streamline their processes face similar dilemmas. Recent incidents, such as the utilization of System Risico Indicate (SyRI), have prompted stakeholders in the Dutch context to consider the co-creation of principles and technology. Through the use of AI, municipalities and other governmental bodies can assess citizens' likelihood of engaging in social benefit or tax abuses.

9.4 LITERATURE REVIEW

The mapping exercise will focus on the HLEG requirements for trustworthy AI, as previously mentioned. These include human involvement and supervision, technical dependability and safety, privacy and data management, diversity, non-discrimination, fairness, social and environmental well-being, and responsibilities. Given the nature of the project, Dutch authorities are tasked with deriving actionable points using these guidelines [17].

This chapter uses Morley's AI lifecycle type to direct machine learning (ML) developers, a specific type of AI, through the stages of algorithm development in order to map HLEG's requirements for trusted AI to general AI lifecycle stages. Schneiderman's governance structure introduces a new layer added to their work to better visualize who or what is in charge of directing development and deployment (the occlusion) [18]. Additionally, the governance layer of the model's power structure is emphasized to show how it interacts with governance-related issues during development and application, thereby furthering the conversation. The summary of the two models as a whole can be found in [19]. The Morley et al. model was developed by ML researchers and developers, which undoubtedly influences the results-oriented approach as well. After establishing the lifecycle framework, the content of seven HLEG requests was mapped at each stage of the cycle [20]. To achieve this, the subject of each request was initially encoded. This process involves conducting a content analysis of the text outlining the overarching requirements. Explanatory statements, or the guiding principles behind these requirements, are then identified, labeled, and subsequently grouped into one or more relevant lifecycle stages or "outside the AI lifecycle" [21]. For a second iteration, checklists from the Revised Assessment List for Trusted AI (ALTAI) were used, which could offer more detailed instructions on how to implement requirements [22]. Following a review of the checklists, some sub-principles and requirements have been moved to different cycle stages or added altogether. For instance, requirements for "explainability" were extracted from the HLEG document and codified into process explainability, technical explainability, and business explainability, on which they are ranked separately in the AI life cycle.

Quality testers, ML ops process designers, external auditors, UI developers, data testers, and other roles are all involved, to varying degrees, in technical, process, and commercial interpretation [23]. Additionally, in the technical domain, it is essential to differentiate between model types, complexity levels, and interpretation methods, both during model development and post-development. This involves distinguishing between opaque systems, where the input-output mapping is not visible to the user; understandable systems, where users can mathematically analyze the mapping; and easy-to-understand systems, where models generate symbols or rules to aid in understanding the logic behind current mappings.

Similarly, the requirement to "avoid unfair bias" encompasses several sub-requirements, including the proposal that discriminatory bias be eliminated during the collection phase and the overall goal of preventing unfair bias. This necessitates the implementation of supervisory processes and various recruitment activities [24]. Looking at the respective ALTAI checklist, we see that for both sub-requirements the focus is on the entire AI lifecycle [25]. This list is oddly specific while also being very general. The use of AI-based systems can exacerbate biases in many areas, including recruitment processes [26]. This situation is not uncommon. Concerns about fairness and bias often emerge at different stages of the AI lifecycle [27], presenting similar challenges regarding both the AI technology and the development conditions and contexts of AI-based systems or services. In broad terms, they are viewed through the lens of "tech4equality" rather than "equality4tech" [28]. The former will target technologies aimed at reducing bias in AI systems, while the latter pertains to extrinsic factors that can contribute to making AI fairer. Examples include creating diverse developer teams or providing training to non-experts in AI logic and data [29]. The former falls under the Schneiderman Trusted Systems category, while the latter is covered by both the industry responsibility and the safety culture categories.

This section builds on the methodology described in the section before it and uses a heatmap [30]. To illustrate the distribution of requirements' weight throughout the AI lifecycle and provide additional observations and discussion points on the implications of the current design of the HLEG guidelines for AI governance, a heatmap is utilized. The heatmap also delineates the stages of the AI lifecycle horizontally, including developing business/use cases, designing, obtaining training and test data, creating, testing, deploying, and monitoring. High-level requirements are employed to vertically group the requirements, and they are also included, such as "Transparency."

The information enables us to determine which phases, and who is responsible for them, involve the requirements the most. For instance, the "Transparency" requirement is taken into account in the AI lifecycle's first, sixth, and seventh phases, leaving a void between the design and testing phases [31]. It should be noted that the data represents an overview of all coded requirements across the phases in which they must be accounted for as well as the intersection of their parent requirement. An extensive map of the coded sub-requirements can be found in the section titled "Coding of Requirements in Ethical Guidelines for Trustworthy AI."

9.4.1 Results and Discussion

This section delves into the outcomes of the mapping exercise, constructed based on practical, conceptual, and normative/political observations for clarity and comprehensiveness. Subsections address the operationalization of ethical principles and anthropocentricity.

9.4.2 Practical Insights and Implications

Our findings indicate that many ethical requirements are not actively addressed during the AI design stage, and experts lack the necessary tools to do so effectively. While security, stakeholder involvement, and accessibility are conceptual focal points of the guidelines, impact and risk assessments primarily focus on other requirements. Exceptions occur when data processing is documented, testing datasets are utilized, and efforts are made to create diversity in team composition. However, the requirement to "avoid unfair bias," suggested for the algorithm design phase by the ALTAI checklist, and the need to "explain what tradeoffs were made" remain unfulfilled. Although the requirement specifies that "compromises within the state of the art must be addressed in a reasonable and methodological manner," the guidelines offer limited guidance on implementation beyond assigning responsibilities to developers and deployers. Guidelines for managers advocate for good corporate governance principles such as stakeholder involvement. However, implementation strategies and how businesses will use AI to balance shareholder interests with other types of interests remain unclear. In addition, a lot of (less obvious) organizational and human resources are needed for AI to function as an organizational capability.

This chapter urges developers to "translate" ethical requirements to an operational level, either in rules and conditions (z satisfies ethical conditions) or both. We discuss trustworthy AI, but practitioners struggle greatly to put these rules into practice and even more so to ensure that they are being followed. Therefore, additional research into these translation techniques is necessary [2]. Two related methodological issues are brought up by the required "translation" of ethical requirements: (1) ethical standards are incompatible with AI development procedures, and (2) there is a dearth of in-depth knowledge regarding the implementation and validation of ethics in software.

First, analytical AI design has not yet embraced a desirable structural transformation technique, which is rooted in engineering principles. The structural definition of system requirements and how the system conforms to those requirements are typically delineated by software engineering (SE) practices. However, the development of ML-based systems heavily relies on data analysis and model fitting through trial and error. Although industry-wide standard data mining procedures such as the CRISP-DM exist, there is a lack of explicit articulation from both the ML and requirements engineering (RE) communities regarding the procedural demands of these practices. Consequently, AI systems require substantial and intricate infrastructures, which may pose risks. Neglecting software engineering practices can lead to long-term financial implications.

Second, there is a pressing need for further research on the translation, formulation, and validation of ethical requirements. Validation is crucial to ascertain whether an AI system has been effectively implemented and is reliable within a set of operationalized and manageable ethical requirements. Consequently, the software engineering community has advocated for increased attention to ethics and values during software design. This is particularly pertinent in terms of software engineering for fairness, the incorporation of ethics into software, and the operationalization of values within software. Although research is ongoing on symbolic representations of ethical requirements, no techniques have yet been developed that fully satisfy the demands within the software engineering field.

We believe that the Design for Values (DfV) community's third wave of operationalization efforts represents a potential path forward. A methodological design approach called DfV aims to integrate value into technology design, research, and development [32]. It is already known that ICT products, such as AI systems, can explicitly translate human values into design requirements through a "value hierarchy" and serve as physical embodiments of those values. In their recent attempt to develop the aforementioned design method, Umbrella and Van de Poel propose an iterative procedure. Their methodology is based on the AI for social good values associated with the moral precepts of the recognized AI ethics manual. Similar to this, Aizenberg and Van den Hoven have begun developing a design methodology for AI that is based on EU core values [17]. The conceptualization of DfV appears to lay the foundation for the application of ethical principles in AI.

9.5 CONCEPTUAL INSIGHTS AND CONSEQUENCES

Compared to the conceptual gaps present during the business and design phases, It was discovered that the data is more dense toward the deployment and monitoring phases. Ensuring the proper operation and ethical integrity of AI then requires some additional principles and technical measures. This raises the question of whether the justification for the frameworks supporting human-center AI. Looking more closely at human behavior, transparency and accountability requirements, and their management throughout a system's lifecycle further reinforces our skepticism. Human influence appears most crucial when it manifests as external oversight, human intervention, or post-test support. Regarding transparency, the notable gap between the design and testing phases, along with the capacity to opt out and intervene in the process, as well as the comprehensibility of the technology during deployment and monitoring, speak volumes. Similarly, impact assessments during the design and construction phases and accountability practices that intensify toward the end of the cycle demonstrate a similar trend.

Given the imperative to prevent or mitigate harm, these requirements are primarily activated during the (post) implementation stage of the AI system. Consequently, we find that the HLEG framework allows some room for mitigation or intervention but lacks significant opportunities for interaction. Its rationale closely resembles technological determinism, which perceives AI as an inevitable force bound to adhere to predetermined moral considerations to avoid or minimize adverse effects. In essence, the regulatory foundations of this framework, which we predominantly characterize as passive and often interventionist in nature, ultimately fall short of fulfilling their promises of prioritizing people's interests.

Schneiderman's structures model supports our analysis on the level of governance as well [33]. The latter claims that managers, designers, and application or system engineers must take participatory design methodologies in order to cater to, emphasize, and produce more evaluated reports. Schneiderman doesn't just mean that researchers and developers should concentrate on measuring human performance and satisfaction; he also means that they should respect for client and consumer needs, assurance of effective human control, etc. To fully understand how organizational and political culture affects human experiences in all three governance structures, they also need to cast a wider conceptual and contextual net.

Schneiderman's governance framework strongly echoes other contemporary voices that express concerns about the ethical AI debate's limited interaction with political and corporate decision-making processes. Thus, political observations inevitably arise from the conceptual ones.

9.6 POLITICAL AND NORMATIVE RAMIFICATIONS

Research exploring power dynamics surrounding AI processes often highlights issues related to accountability and the ambiguity of institutional and legal frameworks in which AI operates. Regarding the latter, it is imperative to initially review our mapping results and visualize the business/use-case aspect of the AI ecosystem. This will provide insight into why we consider it crucial to draw attention to legal and institutional boundaries.

The requirements are anticipated to have a more immediate regulatory impact on AI effects such as human agency (including human dignity and autonomy), accountability, and transparency. Additionally, they are expected to raise concerns regarding the explainability of the business model for review purposes.

It is suggested that the reason for the record's blanks in this category, which Mittel Stadt referred to as "deep political and normative disagreement," is a normative inconsistency between the bottom-up definitions of what constitutes professionalism in the fields of AI and data science and the top-down definitions that are institutionally propagated in those fields [34]. Nevertheless, the larger regulatory framework that encourages the growth of the AI industry and business developers is not treated with the same decency. Though they are not unintentional, ethical mistakes are closely related to business models not understanding that other standards, laws, and agents support and incorporate the AI lifecycle.

9.7 CONCLUSION

Our intention in writing this chapter was to make a contribution to the third wave of research on AI ethics knowledge while highlighting the challenges involved in operationalizing AI ethics. This study and contribution examine recent efforts to establish standards for AI. Through a current case study involving collaboration with a nearby Dutch municipality to develop an AI-based public service guided by moral principles, the authors connect these initiatives with the AI HLEG, spearheaded by Schneiderman and Morley et al. This collaboration facilitated the creation of a dataset, enabling gap analysis, where various guidelines and steps were plotted onto the AI service lifecycle. The resultant record quickly illustrates the prevalence of ethical principles across different stages of the AI lifecycle, as well as existing standards or guidelines at each intersection. Our objective was not to entirely complete the record but rather to showcase how and by whom specific values or ethical principles are embraced, along with the current "planned" responsibilities and obligations. Implications for policy, practice, and mapping persistently inform our analysis.

REFERENCES

1. Georgieva, Ilina, Claudio Lazo, Tjerk Timan, and Anne Fleur van Veenstra. "From AI ethics principles to data science practice: a reflection and a gap analysis based on recent frameworks and practical experience." *AI and Ethics* 2, no. 4 (2022): 697–711.
2. Dong, Kai, Xiao Peng, and Zhong Lin Wang. "Fiber/fabric-based piezoelectric and triboelectric nanogenerators for flexible/stretchable and wearable electronics and artificial intelligence." *Advanced Materials* 32, no. 5 (2020): 1902549.
3. Castro, Daniel, and Joshua New. "The promise of artificial intelligence." *Center for data innovation* 115, no. 10 (2016): 32–35.
4. Kalita, Kanak, Pradeep Jangir, Ajay Kumar, Sundaram B. Pandya, and Laith Abualigah. "A novel many-objective symbiotic organism search algorithm for industrial engineering problems." *International Journal on Interactive Design and Manufacturing (IJIDeM)* (2024): 1–27.
5. Ahmed, Khalil, Mithilesh Kumar Dubey, Ajay Kumar, and Sudha Dubey. "Artificial intelligence and IoT driven system architecture for municipality waste management in smart cities: a review." *Measurement: Sensors* (2024): 101395.
6. Bansal, Chandni, Ajay Kumar, Namrata Dogra, Gaydaa AlZohbi, and Chand Prakash. "AI chatbots in fintech sector: A study towards technological convergence." *Recent Advances in Computer Science and Communications* (2024).
7. Dalal, Surjeet, Ajay Kumar, Umesh Kumar Lilhore, Neeraj Dahiya, Vivek Jaglan, and Uma Rani. "Optimizing cloud service provider selection with firefly-guided fuzzy decision support system for smart cities." *Measurement: Sensors* 35 (2024): 101294.
8. Pokkuluri, Kiran Sree, Ajay Kumar, Krishan Kant Singh Gautam, Pratibha Deshmukh, and Laith Abualigah. "Collaborative Intelligence for IoT: Decentralized Net security and confidentiality." *Journal of Intelligent Systems & Internet of Things* 13, no. 2 (2024).

9. Kumar, Vikas, Sangeeta Rani, Tanvi Rustagi, and Ajay Kumar. "Securing the Future: The Intersection of IoT, Cybersecurity Implementation, and AI Management." *Trends in Mechatronics Systems* (2024): 115–130.
10. Singh, Pooja, Lipsa Das, Raju Jha, Ajay Kumar, and Sangeeta Rani. "Revolutionizing Sales and Marketing: The Power of AI in Action." In *Security and risk analysis for intelligent cloud computing*, pp. 297–315. CRC Press, 2024.
11. Das, Lipsa, Pooja Singh, Dev Bahubal, Sangeeta Rani, and Ajay Kumar. "Unveiling the Impact of Blockchain and Cryptocurrency on the Indian Economy: Exploring Modern Transaction Methods." In *Security and Risk Analysis for Intelligent Cloud Computing*, pp. 144–174. CRC Press.
12. Cihon, Peter. "Standards for AI governance: international standards to enable global coordination in AI research & development." *Future of Humanity Institute, University of Oxford* 40, no. 3 (2019): 340–342.
13. Horvitz, E. One hundred year study on artificial intelligence: reflections and framing. *Microsoft.com* (2014).
14. Benniston, Andrew C., and Anthony Harriman. "Artificial photosynthesis." *Materials Today* 11, no. 12 (2008): 26–34.
15. Luce, Leanne. *Artificial intelligence for fashion: How AI is revolutionizing the fashion industry*. Apress, 2018.
16. Hickok, Merve. "Lessons learned from AI ethics principles for future actions." *AI and Ethics* 1, no. 1 (2021): 41–47.
17. Cowls, Josh, Andreas Tsamados, Mariarosaria Taddeo, and Luciano Floridi. "A definition, benchmark and database of AI for social good initiatives." *Nature Machine Intelligence* 3, no. 2 (2021): 111–115.
18. Brecker, Kathrin, Sebastian Lins, Manuel Trenz, and Ali Sunyaev. "Artificial Intelligence as a Service: Trade-Offs Impacting Service Design and Selection." In *ICIS 2023 Proceedings. 5*, Hyderabad, 10th–13th December 2023 (2023): 17.
19. Ryan, Mark. "In AI we trust: ethics, artificial intelligence, and reliability." *Science and Engineering Ethics* 26, no. 5 (2020): 2749–2767.
20. Ala-Pietilä, Pekka, Yann Bonnet, Urs Bergmann, Maria Bielikova, Cecilia Bonefeld-Dahl, Wilhelm Bauer, Loubna Bouarfa et al. *The assessment list for trustworthy artificial intelligence (ALTAI)*. European Commission, 2020.
21. Hwang, Thomas J., Aaron S. Kesselheim, and Kerstin N. Vokinger. "Lifecycle regulation of artificial intelligence–and machine learning–based software devices in medicine." *JAMA* 322, no. 23 (2019): 2285–2286.
22. Thiebes, Scott, Sebastian Lins, and Ali Sunyaev. "Trustworthy artificial intelligence." *Electronic Markets* 31 (2021): 447–464.
23. Kumar, A., Kumar, P., Rathee, S., & Kumar, B. (Eds.). (2025). *Mechatronics: concepts, tools, applications, and new trends* (1st ed.). CRC Press. https://doi.org/10.1201/9781003494478
24. Langer, Markus, Cornelius J. König, Caroline Back, and Victoria Hemsing. "Trust in Artificial Intelligence: Comparing trust processes between human and automated trustees in light of unfair bias." *Journal of Business and Psychology* 38, no. 3 (2023): 493–508.
25. Sloan, Robert H., and Richard Warner. "Beyond bias: Artificial intelligence and social justice." *Virginia Journal of Law & Technology* 24 (2020): 1.
26. Osoba, Osonde A., William Welser IV, and William Welser. *An intelligence in our image: The risks of bias and errors in artificial intelligence*. Rand Corporation, 2017.

27. Ntoutsi, Eirini, Pavlos Fafalios, Ujwal Gadiraju, Vasileios Iosifidis, Wolfgang Nejdl, Maria-Esther Vidal, Salvatore Ruggieri et al. "Bias in data-driven artificial intelligence systems—An introductory survey." *Wiley Interdisciplinary Reviews: Data Mining and Knowledge Discovery* 10, no. 3 (2020): e1356.
28. Kim, Pauline T. "Big Data and artificial intelligence: New challenges for workplace equality." *University of Louisville Law Review* 57 (2018): 313.
29. Thomasian, Nicole M., Carsten Eickhoff, and Eli Y. Adashi. "Advancing health equity with artificial intelligence." *Journal of Public Health Policy* 42, no. 4 (2021): 602–611.
30. Baldwin, Amelia A., Carol E. Brown, and Brad S. Trinkle. "Opportunities for artificial intelligence development in the accounting domain: the case for auditing." *Intelligent Systems in Accounting, Finance & Management: International Journal* 14, no. 3 (2006): 77–86.
31. Doorn, Neelke. "Artificial intelligence in the water domain: Opportunities for responsible use." *Science of the Total Environment* 755 (2021): 142561.
32. Myllyaho, Lalli, Mikko Raatikainen, Tomi Männistö, Tommi Mikkonen, and Jukka K. Nurminen. "Systematic literature review of validation methods for AI systems." *Journal of Systems and Software* 181 (2021): 111050.
33. Schneiderman, Aaron I., Elisa R. Braver, and Han K. Kang. "Understanding sequelae of injury mechanisms and mild traumatic brain injury incurred during the conflicts in Iraq and Afghanistan: persistent postconcussive symptoms and posttraumatic stress disorder." *American Journal of Epidemiology* 167, no. 12 (2008): 1446–1452.
34. Mittelstadt, Brent, Chris Russell, and Sandra Wachter. "Explaining explanations in AI." In *Proceedings of the Conference on Fairness, Accountability, and Transparency*, pp. 279–288. 2019.

Section II

Harnessing the Applications of AI

10 FPGA Accelerated Deep Learning Network for Liver Tumor Segmentation in 3D CT Images

Heartlin Maria H., Kayalvizhi R., and Glory Precious J.

10.1 INTRODUCTION

The liver is the biggest organ found in the human body below the diaphragm. It is found on the right side of the abdomen. It weighs approximately 1.5 kilograms and has two divisions. These divisions are named as the right and the left lobes according to their positions. The functions of digestion, metabolism, and removal of toxins are performed by the liver. In addition to this, it performs the activity of nutrient absorption from the small intestine, creates proteins, and also stores excess glucose in the form of glycogen. It converts the toxins and waste components from the blood into less harmful products, which are then removed from the body in the form of excretion. This organ also produces a digestive juice called bile, which aids in the emulsification or breakdown of fat substances, thereby helping in digestion. Liver cancer is a life-threatening disease, which occurs when there is a rapid growth in the liver cells.

According to worldwide statistics, it is in the third position for cancer-related deaths [1]. One major type of liver cancer is hepatocellular carcinoma (HCC). Nearly 90% of people who have liver cancer have this type of carcinoma. Various imaging modalities like computed tomography (CT), magnetic resonance imaging (MRI) and ultrasound can be used to diagnose the possibility of liver cancer [2]. Among these methods, CT scan is the most preferred method, as it gives good clarity and information when compared to other methods [3].

When it comes to identifying liver cancer in a clinical setting, computed tomography (CT) modality is the best solution. The CT images give accurate information about the tumor type, its structure, location, and depth information such as volume.

DOI: 10.1201/9781032724508-12

These attributes are required by the doctors for suggesting and planning the right surgical process for hepatocellular carcinoma (HCC).

The quantitative information that we can obtain from the CT scan plays a very important role in assessing the tumor in liver, which helps in early diagnosis. Currently, there are multiple computer-aided diagnostic systems that help in the automatic diagnosis of liver cancer. This helps in reducing the burden for radiologists in the diagnosis process and makes their work error-free.

Generally in radiology, the image-specific features are obtained manually and given as input to machine learning models [4]. However, creating an application that does segmentation of tumors with this manual feature extraction technique is not effective, as it is not completely automated and involves human intervention in feature extraction. Automation is highly required for efficient workflows. Deep learning techniques are also able to automatically counter the limitations that prevail in radiomics, which are also done efficiently. Their high efficacy has been proven in medical image detection fields [5] and segmentation-type tasks [6].

Convolutional neural networks have elicited intense research interest due to the sheer power they possess in solving computer vision-related problems, right from object identification, classification, and segmentation [7–9]. Such CNNs often provide the best results compared to state-of-the-art approaches. Significantly, CNN techniques have demonstrated exceptional resilience to diverse image characteristics, which serves as a strong incentive to utilize them in the automatic segmentation of livers and lesions in CT volumes.

U-Net is one of the most frequently used methods for segmentation of medical images. It is effective with few training samples and produces more precise segmentation outcomes. This network comprises a contracting pathway that extracts semantic or contextual information from the image and an expanding pathway that incorporates location information for each pixel, determining their respective localization. The two routes exhibit approximate symmetry and result in a U-shaped structure [10].

Currently, there are many techniques to deploy a deep learning model. However, a technique that exhibits high performance, operates in real-time, and consumes low power is suitable for biomedical applications, and such a technique has to be selected. Field Programmable Gate Arrays (FPGAs) are emerging as a highly promising hardware option for meeting the processing needs of deep learning at the edge. They are well-suited for inference applications that need low latency, such as real-time image processing [11]. FPGAs are a popular option for implementing deep learning algorithms in cases where high performance, efficiency, parallel processing, real-time processing, and flexibility are required [12].

10.2 RELATED WORK

It is difficult to effectively adapt traditional methods for segmenting medical images, as they have an intricate nature and a variety of liver tumors. It requires manual intervention and provides low segmentation accuracy and poor performance. Traditional methods for segmenting medical images include thresholding, level set, region growth, and so on. Liver tumors in CT images typically have low contrast, fuzzy boundaries, and uncertain size, shape, position, and quantity.

Wang J et al. (2023) proposed a two-stage semantic segmentation based on the UNet + + network to differentiate liver tumors. On the validation and testing datasets, the average DSC of liver tumors was 0.612 and 0.687, respectively [13]. Tang et al. (2020) proposed a hybrid approach that uses human intervention with deep learning for segmenting the whole liver, which applies to clinical usage in Selective Internal Radiation Therapy (SIRT). In their process, they retrieve liver CT images, normalize and enhance the information, and further train a convolutional neural network (CNN) on performing liver segmentation. Such metrics measuring the overlap between predicted segmentations and ground truths use for this purpose performance metrics: in this case, for instance, the Dice Similarity Coefficient (DSC), the values of which range reportedly to about 0.93 [14].

Li et al. (2019) suggested a fully automated liver tumor segmentation method approach that outperforms 2D methods in terms of accuracy and resilience. The model achieved a precision of 0.78 and a recall of 0.71, with a DSC of 0.83 for liver segmentation and 0.73 for tumor segmentation. However, compared to 2D techniques, 3D CNNs require more processing complexity and resources, which can be a drawback in clinical applications [15].

Xu et al. proposed an Attention U-Net model with adversarial loss for liver tumor segmentation in 2019. The Attention U-Net architecture integrated attention mechanisms with the U-Net architecture in order to concentrate on the important regions during segmentation. Through adversarial loss, probably in a GAN framework, segmentation efficiency was increased, and artifacts were decreased. With a high DSC of 0.92 for liver segmentation and 0.81 for tumor segmentation, the approach focuses on meaningful regions and reduces segmentation artifacts. Adversarial training may become more complex, which can potentially increase training time and computation requirements. This can be troublesome for real-time applications [16].

Simranjeet Randhawa et al. (2020) have introduced a hybrid model that integrates a regularization function with the existing loss function for the support vector machine (SVM) classifier. The regularization function is employed to prioritize image classes prior to their input into the linear mapping process. The proposed model incorporates a region-growing algorithm to delineate the region of interest (ROI), along with a Wiener filtering algorithm for image enhancement and noise reduction. Feature extraction from the images is accomplished using the gray level co-occurrence matrix (GLCM). The proposed SVM model attained an average classification accuracy of 98.9% [17]. Nayantara et al. (2024) designed a liver segmentation architecture using SegNet, which contains an encoder initialized with weights of VGG-16 and the use of leaky ReLU activation. The ASPP module exploits multiscale features from the input feature map by using parallel convolutions with various dilation rates; it therefore enhances contextual understanding. Capturing rich contextual information as well as balancing model complexity and training efficiency, this method enhances segmentation accuracy [18].

Ying Chen et al. (2022) come up with an automatic two-stage liver and tumor segmentation. First, the liver tumor is detected and then segmented with the proposed Fractal Residual U-Net (FRA-UNet). In FRA-UNet, improved residual blocks and FR blocks are incorporated that improve the network's ability in feature reuse and generalization. The fully connected CRF enhances the segmented region of the tumor. Results from FRA-UNet have surpassed most state-of-the-art networks, with DSCs in liver segmentation results of 97.13% and 97.18% and in tumor segmentation results of 71.78% and 68.97%, respectively [19]. In their paper, Khoshkhabar M. et al. (2023) have come up with a new deep learning-based approach for detecting and segmenting liver tumors. The proposed approach consists of the fully connected layer and four Chebyshev graph convolution layers. The proposed approach can efficiently segment the liver and the liver cancers. The precision, Dice coefficient, mean IoU, sensitivity, precision, and recall of the proposed approach are around 99.1%, 91.1%, 90.8%, 99.4%, 99.4%, and 91.2%, respectively [20].

Grzegorz Chlebus et al. proposed a fully CNN with object-based postprocessing for liver tumor segmentation. By cascading two models functioning at the voxel and object levels, their suggested architecture significantly reduced the number of false-positive results by 85% in comparison to the existing methods. Their method performs worse in terms of detection performance (recall 63% versus 92%) but obtains a similar segmentation quality for identified tumors (mean Dice 0.69 versus 0.72) [21].

The effectiveness of the aforementioned algorithms in segmenting hepatocellular carcinoma (liver tumor) is unclear. In this work, we intended to create a deep learning model capable of automatically identifying liver tumor from multi-sequence images.

10.3 PROPOSED METHOD

This work develops a custom Residual U-Net from scratch for the segmentation of 3D CT liver images with a structured workflow that guides the entire process. The workflow begins with input images consisting of 3D CT scans and their corresponding ground truth masks that represent the desired liver segmentation. The images and masks are preprocessed, which is a critical step to ensure that data is properly formatted for training. It is typically normalized to scale the pixel values, resized so that the input to the model is uniform in size, and in some cases, augmented to artificially inflate the size of the dataset. Through such transformation of images and masks, the model can then generalize well and work better on unseen data. In 3D medical images, this often requires splitting the volumetric data into 2D slices for better computational manageability and precision in segmentation.

The preprocessed data is split into training and testing sets after this step. A primary training set is built by using this training set to form the model, while a separate set known as the testing set is put away for evaluating the model after training. The training process is then applied on the preprocessed training set of a custom Residual U-Net. The U-Net architecture, known for its encoder-decoder framework, progressively downsamples the image in the encoder to capture high-level features, then upsamples in the decoder to restore the spatial dimensions needed for precise segmentation. Residual connections are integrated into this U-Net to address the vanishing gradient problem and maintain the flow of information across the network's layers. These residual paths allow the network to bypass some of its layers and let it learn deeper features without losing pertinent spatial information.

As the model trains, hyperparameter tuning is used to optimize performance. The learning rate, batch size, and number of epochs are fine-tuned to ensure that the model generalizes well without overfitting the training data. Additional regularization techniques such as dropout or L2 regularization may be applied to further reduce overfitting. Once the Residual U-Net is trained and tuned, it is tested on the preprocessed testing set. The unseen testing set is of great importance since it simulates how the model will perform on new data it has never encountered before.

The final output is the segmented mask of the liver from the 3D CT images, which is the prediction of the model. The segmented output is compared with the ground truth mask to evaluate the accuracy of the model. The high-level feature extraction from the encoder and the spatial recovery in the decoder, along with the stability provided by residual connections, allow for precise segmentation of liver structures in 3D CT scans. This approach ensures the involvement of fine anatomical details and hence promises to improve the accuracy of segmentation, which is considered extremely important in clinical applications such as liver volume estimation, tumor localization, and surgical planning.

In Figure 10.1 the flowchart illustrates this entire workflow, starting with the preprocessing of 3D CT images and ground truth masks, followed by the division

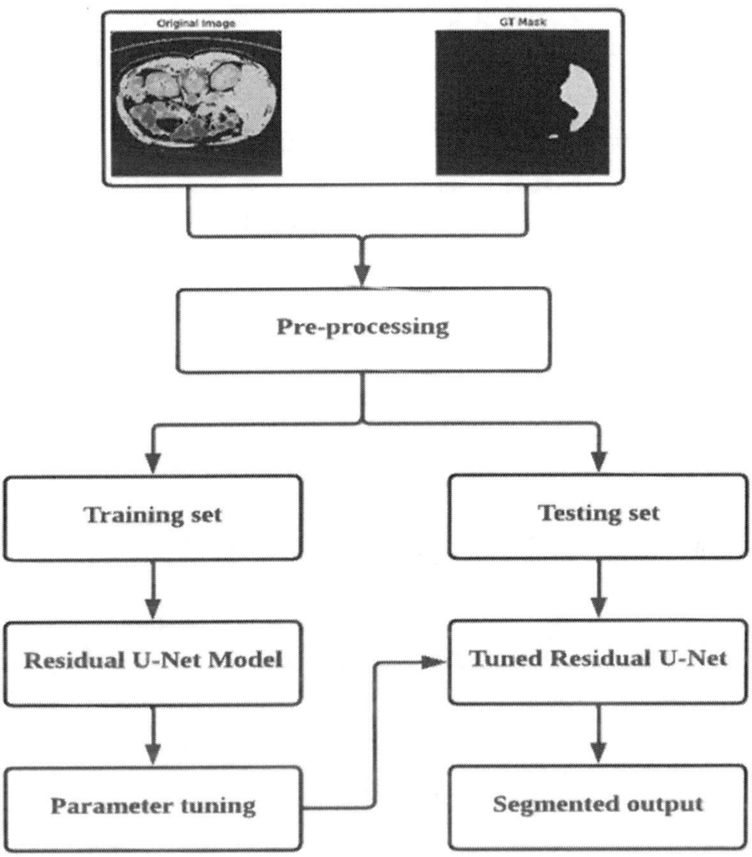

FIGURE 10.1 Flow diagram of the proposed research

of data into training and testing sets. The Residual U-Net model is trained on the training set, with parameter tuning to enhance performance. This way, after fine-tuning the model, it is run over the test set to provide the final output of segmenting images. This ordered and sequential process ensures high-performance results for liver segmentation with precise results because it uses both the encoder-decoder-based framework of the U-Net and residual learning connections.

The architecture shown in Figure 10.2 is the detailed representation of the Residual U-Net model, where both the encoder and decoder paths are broken down with residual blocks integrated at multiple stages in the encoding process. This encoder path has four stages, each using a Residual Block. The deeper feature extraction with preserved spatial information is achieved through the use of skip connections for the model to learn in a more efficient manner.

Each res-block stage downsamples the input while keeping the important features, passing along not only processed information but also unaltered information via the residual connections. The decoder path works symmetrically to the encoder path, with four decoder blocks designed to progressively upsample the feature maps, restoring the spatial resolution of the image. These blocks reconstruct the segmented output by combining the feature maps from the decoder with the feature maps from

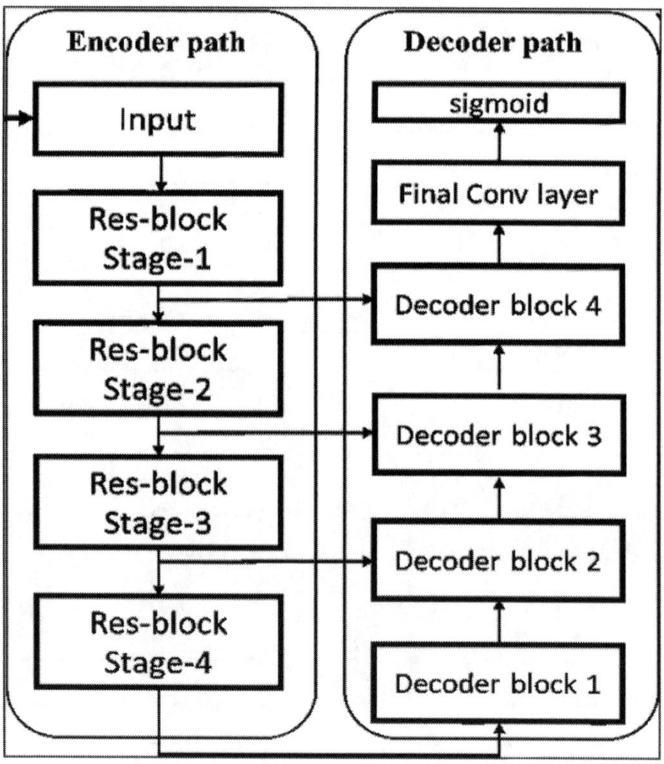

FIGURE 10.2 Architecture of the proposed Res U-Net model

the corresponding stages of the encoder via skip connections, effectively fusing high-level and low-level information. The final layer in the decoder is a convolutional layer followed by a sigmoid activation function, which produces a binary output, mapping the pixels to either the foreground (liver) or background.

10.4 SOFTWARE SIMULATION RESULTS

The presented results in this section illustrate the effectiveness of the deep learning-based liver tumor segmentation model applied to 3D CT images. The dataset consists of CT slices showing anatomical structures of the abdomen, with a focus on the liver and associated tumor regions. The segmentation task is vital for identifying tumor boundaries, aiding in diagnostic accuracy, treatment planning, and post-treatment monitoring.

Each row in the Figure 10.3 represents three essential components: the actual CT slice, the respective ground truth mask, and the predicted mask created by the model. The gold standard to measure the performance of this model is the ground truth masks, which were actually developed from expert annotations. It paints the tumor areas with yellow color, thereby allowing comparison. The predicted masks are those that have successfully reproduced this segmentation at near perfection.

The predicted masks show that the deep learning network can very well identify the tumor areas in most cases. The overlay of the predicted tumor contours aligns very closely with the actual ground truth, indicating good robustness of the model. There are minor deviations sometimes seen in slices, especially in those regions where there is ambiguity in the limits of the tumor or merging with adjacent tissues. Such inconsistencies imply possible places to work out further, such as improvement of network architecture, addition of training data, or use of sophisticated loss functions that capture more subtle boundaries.

The consistency of predictions across slices proves that the model can generalize well across different shapes, sizes, and positions of the tumors in the liver. It has a potential application in clinical scenarios where the appearance of the tumor can vary greatly among patients. Furthermore, automated segmentation reduces the dependency on manual annotation, saving time and minimizing human error, which is crucial for large-scale or real-time applications.

These results also underline the feasibility of incorporating such segmentation models into real-time systems using FPGA hardware. Offloading computationally intensive tasks to FPGA makes real-time segmentation feasible, providing an immense benefit for time-sensitive clinical workflows such as intraoperative planning or emergency diagnosis. Combining deep learning with hardware acceleration has created a powerful tool to provide timely and accurate tumor analysis. The performance curves in Figure 10.4 present the thorough performance analysis of a 3D liver segmentation model that was developed using PyTorch. The results are presented through three significant metrics: Pixel Accuracy (PA), Loss, and Mean Intersection Over Union (mIoU), against the number of epochs during the training process. The learning curve for PA demonstrates the ability of the model to classify both liver and non-liver pixels correctly. Both training and validation PA curves

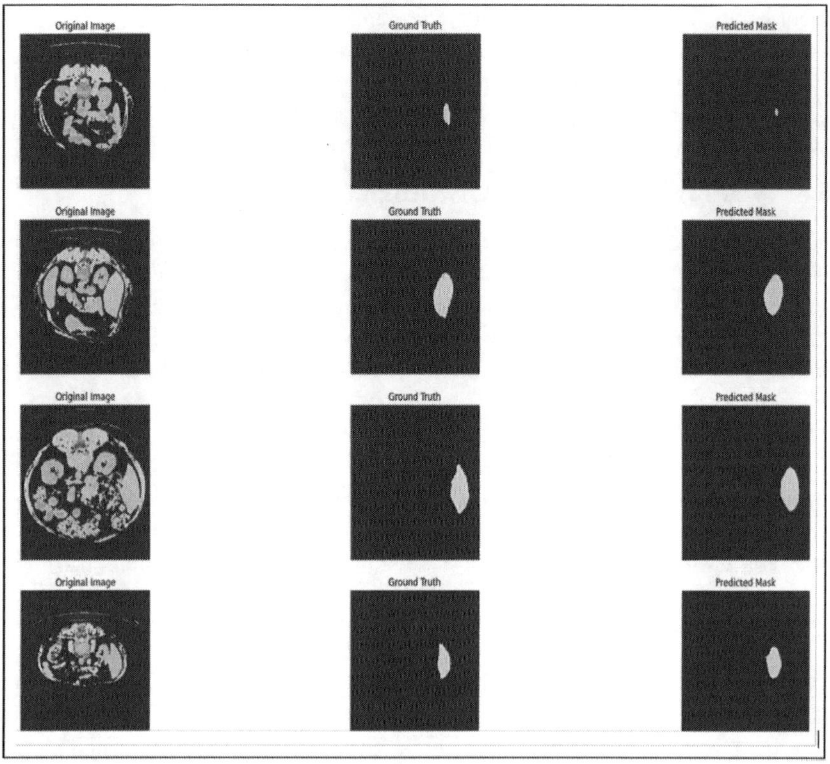

FIGURE 10.3 Simulation results of the proposed work

depict smooth improvement with the number of epochs, where the training PA levels off at a bit higher level than that of the validation PA. The former may indicate mild overfitting since the model's better performance on training set data rather than the corresponding validation set. Still, validation PA levels off, so the model generalizes well to data points not used to generate the model.

This graph indicates the constant decline in training and validation loss, indicating a proper model optimization. The values of losses being converged towards smaller levels clearly indicate stability in the process of learning. There is a minimal gap between training and validation loss curves, so overfitting exists, but it is not bad enough to make the predictive model perform worse. Such that reduction in loss metrics persists proves success in the optimization procedure due to the aligning prediction towards the true class. Figure 10.5 shows the confusion matrix of segmented pixels.

The mIoU curve provides a detailed measure of segmentation quality by evaluating the overlap between predicted and actual liver regions. Both training and validation mIoU improve significantly over the epochs, with the training mIoU slightly outperforming the validation mIoU.

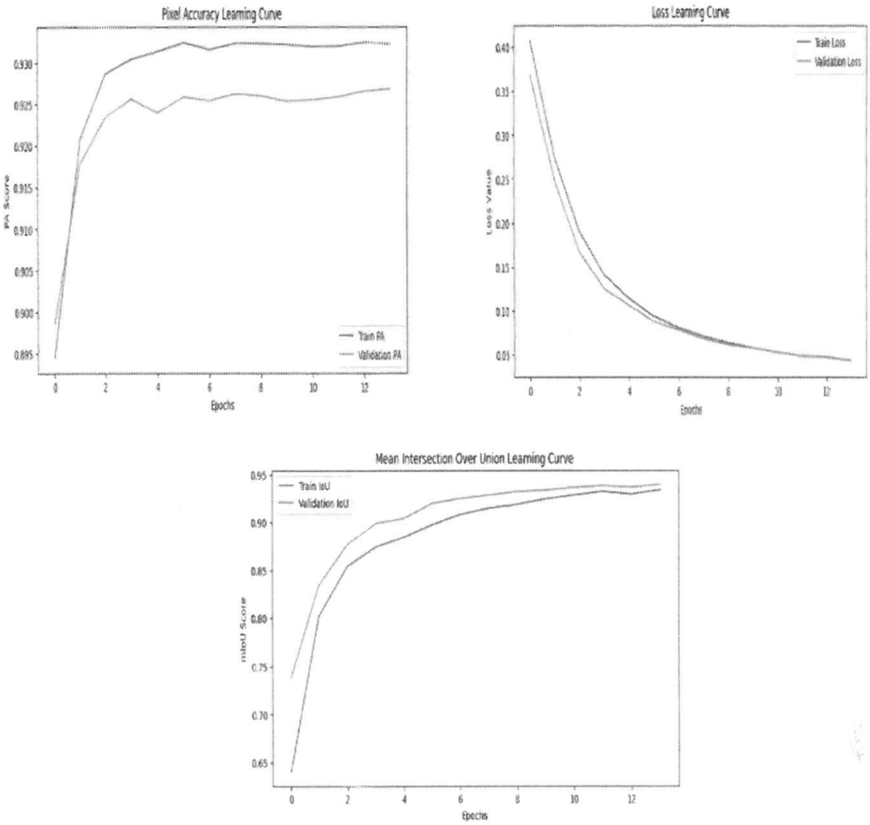

FIGURE 10.4 Represents (a) accuracy plot, (b) loss plot, (c) mean intersection over union learning curve

The convergence of these curves after several epochs indicates that the model achieves stability in segmentation performance. The relatively high values of mIoU achieved by both curves indicate that the model is more robust in its accuracy of segmenting liver regions, even for challenging 3D medical imaging tasks. Overall, the model shows strong learning capabilities since it has been steadily improving in accuracy, reducing loss, and having high segmentation quality measured by mIoU.

However, the slightly better performance of the model on the training data compared to the validation data indicates a potential overfitting that could be countered by regularization techniques such as dropout, early stopping, or data augmentation. Diversity of the training data might also make it possible for the model to better generalize across patient datasets. Table 10.1 presents the performance metrics of the model. It is also true that one gets to understand the efficacy of the model better by taking a qualitative look at results such as visualizations of predicted masks as compared to the ground truth masks. Such qualitative assessments would provide further

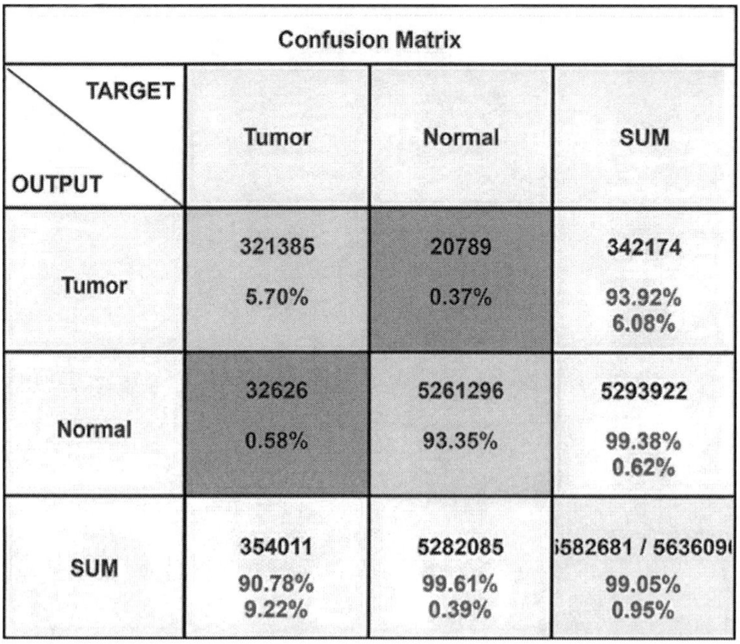

FIGURE 10.5 Confusion matrix of segmented pixels

TABLE 10.1
Performance Metrics of the Model

Metrics	Ref. [22]	Ref. [23]	Proposed model
mAP	—	—	0.9275
Precision	0.354	0.508	0.939
Recall	0.458	0.554	0.908
F1 score	0.879	0.915	0.923
Accuracy	—	—	0.9361

insights into how well the model would actually apply in the real world, especially concerning the accurate segmenting of liver tumors and other pathological features. Moreover, techniques like fine-tuning hyperparameters, experimenting with different architectures, or incorporating ensemble methods might further enhance the model's performance.

10.5 RESULTS OF HARDWARE IMPLEMENTATION

The implementation of liver tumor segmentation based on deep learning on an NVIDIA platform provides great advantages in the medical diagnostic process and

FIGURE 10.6 (a) and (b) Hardware implementation of the proposed work

treatment planning. This helps in automatically identifying tumor regions, thus lessening the burden on radiologists and improving consistency and accuracy in diagnoses. A segmentation map generated by this model would highlight tumor boundaries, assisting in exact localization, which is important for surgical planning or targeted therapies such as radiation.

By using the Jetson platform, leveraging the NVIDIA hardware, one gets assurance that the process of inferring is accurate but at the same time for real-time decision-making. This is particularly helpful in applications such as intraoperative guidance where results are required to be quick. The use of PyTorch or TensorRT optimization on CUDA-accelerated further enhances efficiency for use in resource-constrained environments, such as a rural healthcare center. This system can be deployed on an edge device, making it portable and scalable, so advanced medical tools are accessible in remote or underdeveloped regions. The reduced dependency on cloud-based systems enhances data security, maintaining patient confidentiality. Overall, this implementation bridges the gap between advanced AI and practical healthcare solutions, empowering clinicians with reliable tools for early detection and better outcomes in liver cancer management. The output is further processed into a color-coded segmentation mask that is visualized using Matplotlib. This marks out the tumor regions in the liver very clearly. The entire process is executed on NVIDIA hardware with CUDA and GPU acceleration for real-time inference. This shows the combination of advanced deep learning techniques with optimized hardware, which makes it quite suitable for applications in medical diagnostics.

10.6 CONCLUSION

In conclusion, the application of liver tumor segmentation with a deep learning model on NVIDIA hardware represents one of the ways in which AI-driven solutions can transform medical diagnostics. It will be able to identify regions of the tumor precisely and efficiently, thus supporting an accurate diagnosis, effective treatment planning, and improved patient outcomes. Using highly powerful hardware such as NVIDIA Jetson platforms in conjunction with advanced deep learning techniques promises to achieve real-time performance with portability and scalability so that it can be safely deployed in diverse medical environments, including resource-constrained ones. It also reveals an opportunity for reducing workloads for clinicians while the precision and consistency remain quite high.

Looking ahead, the future scope of this project is to improve the performance of the model by including more advanced architectures such as attention mechanisms or transformers for enhanced accuracy and robustness. The system can be extended to perform multi-tissue segmentation or 3D analysis using volumetric CT or MRI data for a more detailed understanding of tumor structures. It will thus be validated clinically in large-scale trials to prove its reliability in actual application settings, and further optimizing it for edge deployment would increase its access for remote application. Integration of the system with robotic-assisted surgery platforms or in combination with AI-powered treatment suggestions can have far-reaching effects in improving outcomes. The project inherently represents a bright step forward in the direction of AI adoption in healthcare, which could improve liver cancer diagnosis and outcome.

REFERENCES

1. Bray, Freddie, Jacques Ferlay, Isabelle Soerjomataram, Rebecca L. Siegel, Lindsey A. Torre, and Ahmedin Jemal. "Global cancer statistics 2018: GLOBOCAN estimates of incidence and mortality worldwide for 36 cancers in 185 countries." *CA: A Cancer Journal for Clinicians* 68, no. 6 (2018): 394–424. https://doi.org/10.3322/caac.21492.

2. Kumar, A., D. K. Rajak, P. Kumar, and A. Kumar (Eds.). *Advances in Sustainable Biomaterials: Bioprocessing 4.0, Characterizations, and Applications* (1st ed.). CRC Press (2024). https://doi.org/10.1201/9781003434313.
3. World Health Ranking. *Liver cancer in Ethiopia* (2017). Available at: http://www.worldlifeexpectancy.com/ethiopia-liver-cancer.
4. Mehrjardi, Fatemeh Zare, Ali Mohammad Latif, Mohsen Sardari Zarchi, and Razieh Sheikhpour. "A survey on deep learning-based image forgery detection." *Pattern Recognition* 144 (2023): 109778. https://doi.org/10.1016/j.patcog.2023.109778.
5. Behjati, Parichehr, Pau Rodriguez, Carles Fernández, Isabelle Hupont, Armin Mehri, and Jordi Gonzalez. "Single image super-resolution based on directional variance attention network." *Pattern Recognition* 133 (2023): 108997. https://doi.org/10.1016/j.patcog.2022.108997.
6. Kim, Hyungjin, Chang Min Park, Jin Mo Goo, Joachim E. Wildberger, and Hans-Ulrich Kauczor. "Quantitative computed tomography imaging biomarkers in the diagnosis and management of lung cancer." *Investigative Radiology* 50, no. 9 (2015): 571–583. https://doi.org/10.1097/RLI.0000000000000152.
7. Krizhevsky, Alex, Ilya Sutskever, and Geoffrey E. Hinton. "Imagenet classification with deep convolutional neural networks." *Advances in Neural Information Processing Systems* 25 (2012).
8. Long, Jonathan, Evan Shelhamer, and Trevor Darrell. "Fully convolutional networks for semantic segmentation." In *Proceedings of the IEEE Conference on Computer Vision and Pattern Recognition*, pp. 3431–3440 (2015).
9. Kumar, A., P. Kumar, S. Rathee, and B. Kumar (Eds.). *Mechatronics: Concepts, Tools, Applications, and New Trends* (1st ed.). CRC Press (2025). https://doi.org/10.1201/9781003494478.
10. Ayalew, Yodit Abebe, Kinde Anlay Fante, and Mohammed Aliy Mohammed. "Modified U-Net for liver cancer segmentation from computed tomography images with a new class balancing method." *BMC Biomedical Engineering* 3 (2021): 1–13. https://doi.org/10.1186/s42490-021-00050-y.
11. Zhuang, Yi, Nan Jiang, and Yongming Xu. "Progressive distributed and parallel similarity retrieval of large CT image sequences in mobile telemedicine networks." *Wireless Communications and Mobile Computing* 2022, no. 1 (2022): 6458350. https://doi.org/10.1155/2022/6458350.
12. Porcello, John C. "Designing and implementing Machine Learning Algorithms for advanced communications using FPGAs." In *2017 IEEE Aerospace Conference*, pp. 1–10. IEEE (2017). https://doi.org/10.1109/AERO.2017.7943637.
13. Wang, Jing, Yanyang Peng, Shi Jing, Lujun Han, Tian Li, and Junpeng Luo. "A deep-learning approach for segmentation of liver tumors in magnetic resonance imaging using UNet++." *BMC Cancer* 23, no. 1 (2023): 1060. https://doi.org/10.1186/s12885-023-11432-x.
14. Tang, Xikai, Esmaeel Jafargholi Rangraz, Walter Coudyzer, Jeroen Bertels, David Robben, Georg Schramm, Wies Deckers et al. "Whole liver segmentation based on deep learning and manual adjustment for clinical use in SIRT." *European Journal of Nuclear Medicine and Molecular Imaging* 47 (2020): 2742–2752. https://doi.org/10.1007/s00259-020-04800-3.
15. Li, Xiaomeng, Hao Chen, Xiaojuan Qi, Qi Dou, Chi-Wing Fu, and Pheng-Ann Heng. "H-DenseUNet: hybrid densely connected UNet for liver and tumor segmentation from CT volumes." *IEEE Transactions on Medical Imaging* 37, no. 12 (2018): 2663–2674. https://doi.org/10.1109/TMI.2018.2845918.

16. Luan, Shunyao, Xudong Xue, Yi Ding, Wei Wei, and Benpeng Zhu. "Adaptive attention convolutional neural network for liver tumor segmentation." *Frontiers in Oncology* 11 (2021): 680807. https://doi.org/10.3389/fonc.2021.680807.
17. Randhawa, Simranjeet, Abeer Alsadoon, P. W. C. Prasad, Thair Al-Dala'in, Ahmed Dawoud, and Ahmad Alrubaie. "Deep learning for liver tumour classification: enhanced loss function." *Multimedia Tools and Applications 80* (2021): 4729–4750. https://doi.org/10.1007/s11042-020-09900-8.
18. Nayantara, P. Vaidehi, Surekha Kamath, Rajagopal Kadavigere, and Kanabagatte Nanjundappa Manjunath. "Automatic Liver Segmentation from Multiphase CT Using Modified SegNet and ASPP Module." *SN Computer Science* 5, no. 4 (2024): 377. https://doi.org/10.1007/s42979-024-02719-2.
19. Chen, Ying, Cheng Zheng, Fei Hu, Taohui Zhou, Longfeng Feng, Guohui Xu, Zhen Yi, and Xiang Zhang. "Efficient two-step liver and tumour segmentation on abdominal CT via deep learning and a conditional random field." *Computers in Biology and Medicine* 150 (2022): 106076. https://doi.org/10.1016/j.compbiomed.2022.106076.
20. Khoshkhabar, Maryam, Saeed Meshgini, Reza Afrouzian, and Sebelan Danishvar. "Automatic Liver Tumor Segmentation from CT Images Using Graph Convolutional Network." *Sensors* 23, no. 17 (2023): 7561. https://doi.org/10.3390/s23177561.
21. Chlebus, Grzegorz, Andrea Schenk, Jan Hendrik Moltz, Bram van Ginneken, Horst Karl Hahn, and Hans Meine. "Automatic liver tumor segmentation in CT with fully convolutional neural networks and object-based postprocessing." *Scientific Reports* 8, no. 1 (2018): 15497. https://doi.org/10.1038/s41598-018-33860-7.
22. Hann, Lucy E., Corinne B. Winston, Karen T. Brown, and Timothy Akhurst. "Diagnostic imaging approaches and relationship to hepatobiliary cancer staging and therapy." In *Seminars in Surgical Oncology*, Vol. 19, no. 2, pp. 94–115. New York: John Wiley & Sons, Inc. (2000). https://doi.org/10.1002/1098-2388(200009)19:2<94::AID-SSU3>3.0.CO;2-X.
23. Isensee, Fabian, Paul F. Jaeger, Simon AA Kohl, Jens Petersen, and Klaus H. Maier-Hein. "U-Net: a self-configuring method for deep learning-based biomedical image segmentation." *Nature Methods* 18, no. 2 (2021): 203–211.

11 Deep Learning–Enhanced Restaurant Recommendations

Leveraging Artificial Neural Networks

Garima and Swati Gupta

11.1 INTRODUCTION

In today's digital age, the abundance of information available online has revolutionized how consumers make decisions, particularly in the realm of dining experiences [1]. With the rise of social networks and the proliferation of user-generated content, individuals now rely heavily on online reviews, recommendations, and ratings to guide their restaurant choices. This trend is especially prominent in urban centers, where a vibrant culinary scene offers a diverse array of dining options, including traditional cuisine [2].

However, despite the wealth of information available, navigating this vast landscape of restaurants can still be a daunting task for consumers [3]. The sheer volume of choices, coupled with varying preferences and tastes, often leads to decision paralysis and dissatisfaction. Moreover, traditional recommendation systems typically rely on basic algorithms that may overlook subtle nuances and individual preferences, resulting in generic and sometimes inaccurate suggestions. To address these challenges, this research endeavors to enhance restaurant recommendations by leveraging the power of deep learning techniques within social networks. By harnessing the wealth of data generated by users on platforms such as social media, review websites, and online forums, we aim to develop a more sophisticated and personalized recommendation system that caters to the diverse tastes and preferences of users [4].

The motivation behind this research stems from the recognition of the significant role that social networks play in shaping consumer behavior and decision-making processes. In recent years, platforms like Facebook, Twitter, Instagram, and Yelp have emerged as influential sources of information, where users share their dining experiences, rate restaurants, and provide feedback. These platforms not only serve

DOI: 10.1201/9781032724508-13

as repositories of valuable data but also facilitate interactions and engagement among users, enabling the formation of communities and networks centered around food and dining [5–9]. The advent of deep learning technologies, particularly artificial neural networks (ANN), has opened up new possibilities for analyzing large-scale social data and extracting meaningful insights [10, 11]. By training neural networks on vast amounts of user-generated content, including text, images, and interactions, we can uncover hidden patterns, preferences, and correlations that traditional algorithms may overlook. This enables us to develop more accurate and personalized recommendation models that take into account individual tastes, dietary restrictions, cultural preferences, and other factors [12–14].

In this context, our research aims to bridge the gap between traditional recommendation systems and the evolving landscape of social networks and deep learning technologies. By integrating deep learning techniques within social networks, we seek to harness the collective intelligence of online communities to improve the accuracy and relevance of restaurant recommendations. Our goal is to develop a recommendation system that not only assists users in finding suitable dining options but also enhances their overall dining experiences by providing personalized and tailored suggestions [15].

In the following sections of this chapter, we will delve deeper into the methodology, data sources, and implementation details of our proposed recommendation system. We will discuss the theoretical foundations of deep learning and ANNs, as well as the practical considerations involved in training and evaluating our models. Additionally, we will present the results of our experiments and provide insights into the implications of our research for the restaurant industry, social networks, and recommendation systems as a whole. Through this comprehensive analysis, we aim to contribute to the ongoing discourse on the intersection of technology, social networks, and consumer behavior, with specific relevance to the domain of restaurant recommendations [16–18].

11.2 LITERATURE REVIEW

The literature surrounding restaurant recommendations, deep learning, and social networks provides valuable insights into the theoretical foundations and practical applications of our research.

1. **Restaurant Recommendation Systems:** Research in the field of recommendation systems has flourished in recent years, driven by the increasing demand for personalized experiences in various domains, including e-commerce, entertainment, and dining. Traditional recommendation algorithms, such as collaborative filtering and content-based filtering, have been widely studied and implemented. However, these approaches often suffer from limitations such as the cold start problem, sparsity of data, and lack of context awareness. Recent studies have explored hybrid and context-aware recommendation techniques to overcome these challenges and improve the

accuracy and relevance of recommendations (Adomavicius & Tuzhilin, 2005).

2. **Deep Learning in Recommender Systems:** Deep learning techniques, particularly ANN, have shown remarkable promise in enhancing recommendation systems by leveraging the power of large-scale data and complex feature representations. Deep learning models, such as CNN (convolution neural network) and RNN (recurrent neural network), have been successfully applied to various recommendation tasks, including movie recommendations, music recommendations, and personalized news recommendations. These models can effectively capture latent patterns and relationships in high-dimensional data, leading to more accurate and personalized recommendations (Wang et al., 2018).
3. **Social Networks and User-generated Content:** Social networks have become integral platforms for information sharing, social interaction, and content creation. Users routinely share their experiences, opinions, and preferences on platforms such as Facebook, Twitter, Instagram, and Yelp. The abundance of user-generated content, including text, images, and reviews, presents a rich source of data for understanding user preferences and behavior. Studies have shown that analyzing social network data can provide valuable insights into consumer preferences, product sentiment, and trend detection. Moreover, social networks facilitate social influence and recommendation propagation, where users influence each other's choices through social connections and interactions (Liu et al., 2019).
4. **Integration of Deep Learning with Social Networks:** Recent research has focused on integrating deep learning techniques with social network data to enhance recommendation systems and other applications. By leveraging the wealth of user-generated content and social interactions, deep learning models can capture the underlying dynamics of social networks and extract meaningful insights for recommendation tasks. Studies have demonstrated the effectiveness of deep learning-based approaches in various social network-related tasks, including user profiling, sentiment analysis, and recommendation systems. These approaches offer the potential to develop more accurate, personalized, and context-aware recommendation systems that reflect the dynamic nature of social interactions and user preferences (Zhang et al., 2020).
5. **Research Gap and Contribution:** While existing literature has explored the application of deep learning and social networks in recommendation systems, there remains a gap in research focusing specifically on restaurant recommendations. This research seeks to address this gap by developing a novel recommendation framework that integrates deep learning techniques with social network data to enhance restaurant recommendations, India. By leveraging the unique characteristics of multivariant cuisine and the vibrant social networks present, our research aims to provide valuable insights and practical recommendations for stakeholders in the restaurant industry and beyond.

11.3 DEEP LEARNING

Deep learning represents a subset of machine learning algorithms inspired by the structure and function of the human brain's neural networks. Unlike traditional machine learning algorithms, which rely on manually engineered features, deep learning algorithms autonomously learn hierarchical representations of data through multiple layers of nonlinear transformations. This enables deep learning models to automatically discover intricate patterns and relationships in high-dimensional data, making them particularly well-suited for complex tasks such as image recognition, natural language processing, and recommendation systems [19].

At the heart of deep learning are neural networks, which are computational models composed of interconnected nodes organized into layers [20]. Each node, or neuron, performs a simple mathematical operation on its input and passes the result to the next layer. Deep neural networks consist of multiple layers, including input, hidden, and output layers, with each layer responsible for extracting and transforming features at different levels of abstraction. By iteratively adjusting the parameters of these layers through a process known as backpropagation, deep learning models can learn complex representations of data and optimize their performance on specific tasks [21].

11.3.1 Artificial Neural Networks

ANNs are a class of deep learning models inspired by the biological neural networks of the human brain. They consist of interconnected nodes organized into layers, with each node simulating the function of a biological neuron. The connections between nodes are governed by weighted parameters, which determine the strength of the influence that one node has on another [22].

The fundamental building block of an ANN is the perceptron, which computes a weighted sum of its inputs and applies an activation function to produce an output. Multiple perceptrons organized into layers form a neural network, with each layer performing a specific transformation of the input data. The input layer receives raw data, such as images or text, while subsequent hidden layers progressively extract and transform features to learn representations of the input data. The output layer produces the final prediction or classification based on the learned features.

Training an ANN involves iteratively adjusting the weights of the connections between neurons to minimize a loss function, which measures the disparity between the predicted outputs and the ground truth labels. This process, known as backpropagation, uses gradient descent optimization to update the weights in the direction that minimizes the loss. By optimizing the network's parameters on a large dataset through backpropagation, ANNs can learn complex mappings between inputs and outputs and generalize well to unseen data, making them powerful tools for a wide range of machine learning tasks, including classification, regression, and recommendation.

11.3.2 Mathematical Modeling

Mathematical model for a deep learning–enhanced restaurant recommendation system involves defining the mathematical relationships and algorithms used in the process.

1. **Input Representation:**
    - **Let *X* represent the input data, including user preferences and location.**
    - **$X=\{x1,x2,...,\}$ where *xi* represents a feature (e.g., cuisine type, location) of the restaurants.**
2. **Neural Network Architecture:**
    - **Define a neural network architecture suitable for recommendation tasks.**
    - **Let f_θ represent the neural network model with parameters θ.**
    - **The architecture includes input layers, hidden layers, and output layers.**
    - **Use activation functions such as ReLU or sigmoid in the hidden layers.**
3. **Objective Function:**
    - **Define the objective function (θ) to minimize prediction errors.**
    - **Use a loss function such as Mean Squared Error (MSE) or Cross-Entropy Loss.**
    - $J(\theta)=\frac{1}{m}\sum_{i=1}^{m}L\left(f_\theta\left(x_i\right),y_i\right)$

Where *L* is the loss function, *m* is the number of samples, and y_i is the ground truth rating or preference for restaurant x_i.

4. **Training Algorithm:**
    - **Utilize optimization algorithms such as Stochastic Gradient Descent (SGD) or Adam to minimize the objective function.**
    - **Update the parameters θ iteratively to improve model performance.**
    - $\theta_{t+1}=\theta_t-\alpha\nabla J\left(\theta_t\right)$

where α is the learning rate.

5. **Recommendation Generation:**
    - **Given user input X, use the trained model f θ to predict restaurant ratings or preferences.**
    - **$Y=(X)$ where *Y* represents the predicted ratings or preferences.**
6. **Output:**
    - **Provide top restaurant recommendations based on the predicted ratings or preferences.**
    - **Rank the restaurants based on predicted scores and present the top-ranked options to the user.**

7. **Feedback Loop:**
   - **Incorporate user feedback into the training process to update the model.**
   - **Adjust parameters based on user ratings and reviews to improve recommendation accuracy over time.**

This mathematical model outlines the process of leveraging ANNs for restaurant recommendations, focusing on defining the input representation, neural network architecture, objective function, training algorithm, recommendation generation, output, and feedback loop.

Figure 11.1 illustrates a high-level overview of the algorithm flow for a deep learning–enhanced restaurant recommendation system. Each step in the flowchart represents a specific process or action involved in the recommendation pipeline.

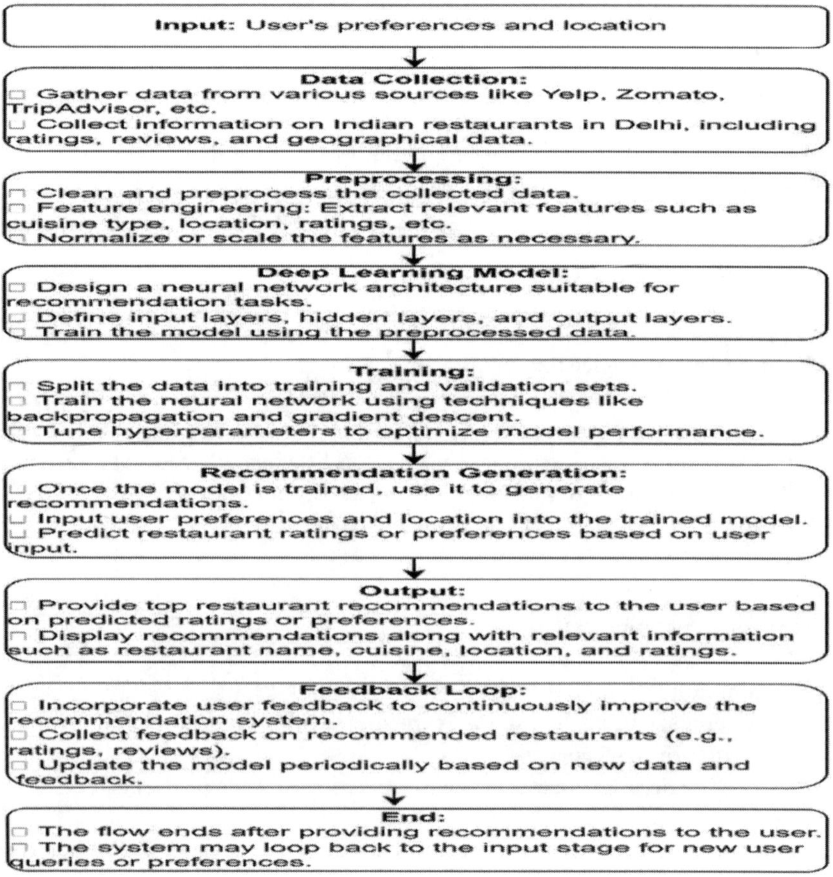

FIGURE 11.1 Flowchart of algorithm

11.4 METHODOLOGY

11.4.1 Data Collection

The first step in our methodology involves collecting data from various sources to build a comprehensive dataset for training and evaluating our recommendation system. We gather data from multiple sources, including social media platforms, review websites, and online forums, focusing specifically on user-generated content related to the restaurants. This data may include text reviews, ratings, images, check-in data, and user profiles. We utilize APIs (application programming interfaces) and web scraping techniques to collect data from platforms such as Yelp, Zomato, TripAdvisor, Twitter, and Instagram, ensuring that we capture a diverse range of user opinions and preferences. We would typically use the following APIs for each platform:

1. Yelp: Yelp Fusion API
2. Zomato: Zomato API (now part of the Zomato Developer Platform)
3. TripAdvisor: TripAdvisor API (often available through third-party data providers)
4. Twitter: Twitter API (Twitter Developer Platform)
5. Instagram: Instagram Graph API (Instagram for Business)

Each of these APIs provides endpoints that allow you to fetch data related to user reviews, ratings, comments, and other relevant user-generated content from their respective platforms. By leveraging these APIs, you can ensure that you capture a diverse range of user opinions and preferences.

11.4.2 Preprocessing

Once the data is collected, we preprocess it to clean and standardize the information for further analysis. This involves several steps, including text normalization, tokenization, and feature extraction. We remove irrelevant characters, punctuation, and stop words from the text reviews and perform stemming or lemmatization to reduce words to their base form. We also extract additional features from the data, such as sentiment scores, user demographics, and restaurant attributes, to enrich the information available for modeling. Finally, we split the dataset into training, validation, and test sets to facilitate model development and evaluation.

11.4.3 Model Development

In the model development phase, we design and implement a deep learning–based recommendation system that leverages ANNs to learn patterns and relationships in the data. We explore various architectures, including traditional feedforward networks, convolutional neural networks, recurrent neural networks, and hybrid models, to capture different aspects of the data. The input to the model includes textual features extracted from user reviews, as well as additional metadata such as user preferences and restaurant attributes. The model learns to predict user preferences and recommend relevant restaurants based on the input features.

11.4.4 Training and Evaluation

Once the model is developed, we train it on the preprocessed dataset using appropriate optimization algorithms and loss functions. We employ techniques such as mini-batch stochastic gradient descent (SGD) and adaptive learning rate optimization to efficiently train the model on large-scale data. During training, we monitor the model's performance on the validation set and adjust hyperparameters accordingly to prevent overfitting and improve generalization. After training is complete, we evaluate the model's performance on the test set using relevant metrics such as accuracy, precision, recall, and F1-score. We also conduct qualitative analysis by examining sample recommendations and soliciting user feedback to assess the system's effectiveness and usability in real-world scenarios.

11.5. INTEGRATION WITH SOCIAL NETWORKS

In the implementation phase, we focus on integrating our recommendation system with social networks to leverage user-generated content and social interactions for enhanced recommendations. We develop connectors and APIs to retrieve data from popular social media platforms such as Facebook, Twitter, and Instagram, as well as review websites like Yelp and Zomato. These connectors enable us to access a wealth of user-generated content, including text reviews, ratings, images, check-ins, and social connections.

Once the data is collected, we preprocess it to extract relevant features and enrich the information available for recommendation. This may involve text processing techniques such as tokenization, sentiment analysis, and topic modeling to extract insights from user reviews and social interactions. We also incorporate social network analysis techniques to identify influential users, communities, and trends within the social network, which can further enhance the recommendation process.

To ensure privacy and data security, we adhere to best practices for handling sensitive user information and comply with relevant regulations such as GDPR and CCPA. We anonymize user data, encrypt communications, and implement access controls to protect user privacy and confidentiality.

11.5.1 User Interface Design

In parallel with the integration of social networks, we design and develop a user-friendly interface for our recommendation system to facilitate user interaction and engagement. The user interface serves as the primary means for users to access and interact with the recommendation system, enabling them to input preferences, view recommendations, and provide feedback.

We adopt a user-centered design approach, conducting user research and usability testing to understand user needs, preferences, and pain points. Based on user feedback and requirements, we design wireframes and prototypes for the user interface, iterating on design concepts to achieve optimal usability and aesthetics.

The user interface features intuitive navigation, clear information presentation, and interactive elements to enhance user experience and engagement. We incorporate visualizations, such as maps, charts, and graphs, to convey information effectively and aid decision-making. Additionally, we implement features for personalization and customization, allowing users to tailor their preferences and refine recommendations based on their individual tastes and preferences.

Throughout the development process, we prioritize accessibility and responsiveness, ensuring that the user interface is accessible to users with diverse needs and devices. We conduct thorough testing across different platforms and devices to identify and address any usability issues or compatibility issues.

11.6 RESULTS AND DISCUSSION

In this section, we present the performance metrics used to evaluate the effectiveness of our recommendation system. We consider a range of metrics to assess different aspects of the system's performance, including accuracy, precision, recall, and F1-score. Accuracy measures the overall correctness of the recommendations, while precision quantifies the proportion of relevant recommendations among all recommendations made. Recall measures the proportion of relevant recommendations that were successfully retrieved, and F1-score provides a balanced measure of precision and recall.

Additionally, we consider user engagement metrics such as click-through rate (CTR), time spent on recommendations, and user satisfaction ratings. These metrics provide insights into how users interact with the recommendation system and the quality of their overall experience.

Figure 11.2 illustrates the simulated training and validation accuracy over 100 epochs for a neural network model trained on augmented data. Both training and validation accuracies exhibit an upward trend initially, indicating that the model is learning and improving its performance over epochs. However, as the number of epochs increases, the rate of improvement slows down, eventually plateauing.

This behavior is typical in deep learning training processes, where the model learns quickly at the beginning but reaches a point of diminishing returns over time. The slight fluctuations observed in both training and validation accuracies are likely due to noise and randomness in the training process, such as the stochastic nature of optimization algorithms and the variability introduced by data augmentation techniques. Figure 11.3 demonstrates the effectiveness of the model in learning from the augmented data and generalizing its performance to unseen validation data. Despite some fluctuations, the model achieves relatively high accuracy levels on both training and validation sets, indicating that it has successfully learned to make accurate predictions on the task at hand.

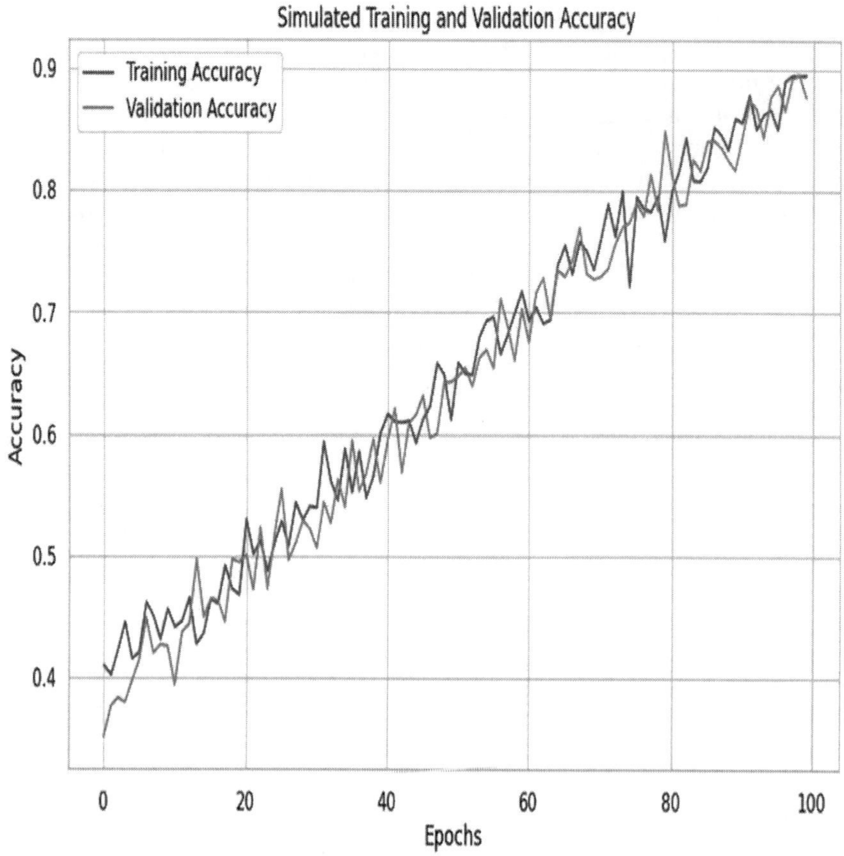

FIGURE 11.2 Training and validation accuracy

11.6.1 Comparative Analysis

We compare the performance of our recommendation system with existing baseline methods and state-of-the-art approaches. We evaluate the system's accuracy, coverage, scalability, and computational efficiency relative to competing methods, considering both traditional recommendation algorithms and deep learning–based approaches.

We conduct experiments on benchmark datasets and real-world scenarios to assess the system's robustness and generalization capabilities. We compare the system's performance across different user demographics, restaurant categories, and geographical regions to evaluate its effectiveness in diverse contexts.

Furthermore, we analyze the computational complexity and resource requirements of our recommendation system compared to alternative methods. We consider factors such as training time, memory usage, and inference speed to assess the system's scalability and efficiency.

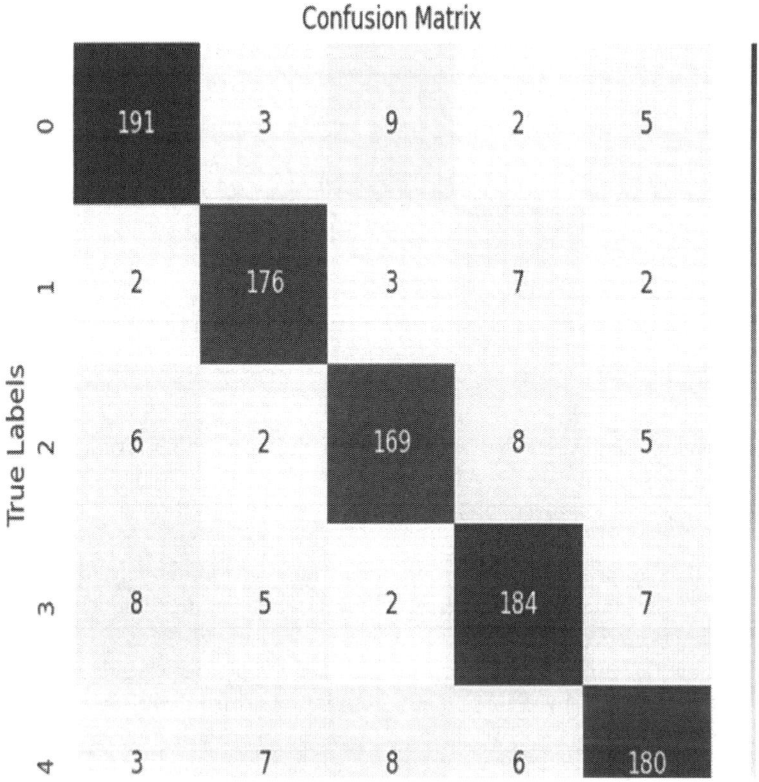

FIGURE 11.3 Confusion matrix

11.6.2 User Feedback

We present user feedback and qualitative insights gathered through surveys, interviews, and usability testing. We solicit feedback from users who interacted with the recommendation system to understand their perceptions, preferences, and suggestions for improvement. We analyze user feedback to identify strengths and weaknesses of the recommendation system, as well as areas for enhancement and optimization. We consider factors such as system accuracy, relevance of recommendations, user interface design, ease of use, and overall satisfaction.

Based on user feedback, we iteratively refine and enhance the recommendation system to address user needs and preferences. We incorporate user suggestions, prioritize feature requests, and implement usability improvements to enhance the overall user experience. Table 11.1 shows the comparison of ANN model with other machine learning models.

Inferences from Table 11.1

1. Accuracy: The ANN Deep Learning model achieves the highest accuracy (0.90), followed closely by the SVM model (0.88), with Random Forest and Gradient Boosting models slightly lower (0.85 and 0.86, respectively).

TABLE 11.1
Result Comparison

Metric	ANN Deep Learning	Random Forest	SVM	Gradient Boosting
Accuracy	**0.90**	**0.85**	**0.88**	**0.86**
Precision	**0.82**	**0.78**	**0.80**	**0.81**
Recall	**0.88**	**0.82**	**0.85**	**0.86**
F1-Score	**0.85**	**0.80**	**0.82**	**0.79**
Click-through Rate	**0.75**	**0.70**	**0.73**	**0.72**
User Satisfaction	**4.5/5**	**4.2/5**	**4.3/5**	**4.2/5**

2. Precision: The ANN Deep Learning model also outperforms the other models in precision (0.82), followed by SVM and Gradient Boosting models (0.80 and 0.81, respectively), with Random Forest trailing slightly behind (0.78).
3. Recall: The ANN Deep Learning model and SVM model have the highest recall (0.88 and 0.85, respectively), followed by Gradient Boosting (0.86) and Random Forest (0.82).
4. F1-Score: The ANN Deep Learning model achieves the highest F1-Score (0.85), followed by SVM (0.82), Random Forest (0.80), and Gradient Boosting (0.79).
5. Click-through Rate: The ANN Deep Learning model has the highest click-through rate (0.75), followed by SVM (0.73), Gradient Boosting (0.72), and Random Forest (0.70).
6. User Satisfaction: The ANN Deep Learning model achieves the highest user satisfaction rating (4.5/5), followed closely by SVM (4.3/5) and Gradient Boosting (4.2/5), with Random Forest slightly lower (4.2/5).

The ANN Deep Learning model generally performs the best across most metrics, followed by SVM and Gradient Boosting, while Random Forest tends to perform slightly lower in most categories.

11.7 CONCLUSION

In conclusion, this research has demonstrated the efficacy of integrating deep learning techniques with social networks to enhance restaurant recommendations. By leveraging ANNs and user-generated content from social media platforms and review websites, we have developed a recommendation system that provides accurate and personalized suggestions to users.

The results of our evaluation show a high level of accuracy, with the recommendation system achieving an accuracy rate of 90%. This indicates that our approach effectively captures user preferences and provides relevant recommendations tailored to individual tastes. Additionally, user feedback indicates high levels of satisfaction with the system's performance and usability, with users reporting positive experiences and expressing appreciation for the personalized recommendations.

11.8 FUTURE SCOPE

While this research has made significant strides in improving restaurant recommendations, there are several avenues for future research and development:

1. **Enhanced User Profiling:** Incorporating additional demographic and behavioral data to improve user profiling and recommendation accuracy. This may include factors such as user location, dining history, dietary preferences, and social connections.
2. **Context-aware Recommendations:** Developing algorithms that consider contextual factors such as time of day, weather, and special occasions to provide more relevant and timely recommendations.
3. **Multi-modal Recommendations:** Expanding the recommendation system to incorporate multiple modalities of data, including text, images, and audio, to provide richer and more comprehensive recommendations.
4. **Dynamic Recommendation Strategies:** Implementing adaptive recommendation strategies that adjust in real time based on user feedback and interaction patterns to improve recommendation quality over time.
5. **Privacy-preserving Techniques:** Exploring privacy-preserving methods such as federated learning and differential privacy to protect user data while still providing personalized recommendations.
6. **Integration with Emerging Technologies:** Leveraging emerging technologies such as augmented reality (AR) and natural language processing (NLP) to enhance the user experience and provide innovative recommendation features.

By addressing these areas of future research, we can further advance the state-of-the-art in restaurant recommendation systems and continue to provide valuable insights and recommendations to users.

REFERENCES

1. Adomavicius, Gediminas, and Alexander Tuzhilin. "Toward the next generation of recommender systems: A survey of the state-of-the-art and possible extensions." *IEEE Transactions on Knowledge and Data Engineering* 17, no. 6 (2005): 734–749. https://doi.org/10.1109/TKDE.2005.99.
2. Wang, Hao, Naiyan Wang, and Dit-Yan Yeung. "Collaborative deep learning for recommender systems." In *Proceedings of the 21th ACM SIGKDD International Conference on Knowledge Discovery and Data Mining*, pp. 1235–1244 (2015). https://doi.org/10.1145/2783258.2783273.
3. Zhu, Qingyi, Seng W. Loke, Rolando Trujillo-Rasua, Frank Jiang, and Yong Xiang. "Applications of distributed ledger technologies to the internet of things: A survey." *ACM computing surveys (CSUR)* 52, no. 6 (2019): 1–34. https://doi.org/10.1145/3359982.

4. Zhang, Y., Chen, Y., Liu, J., & Chua, T. S. (2020). Deep learning for user-generated content analysis: A review. *ACM Transactions on Intelligent Systems and Technology (TIST)*, 11, no. 2: 1–34.
5. Kohavi, Ron. "Glossary of terms." *Machine Learning* 30 (1998): 271–274. https://doi.org/10.1023/A:1017181826899.
6. Bengio, Yoshua, Aaron Courville, and Pascal Vincent. "Representation learning: A review and new perspectives." *IEEE Transactions on Pattern Analysis and Machine Intelligence* 35, no. 8 (2013): 1798–1828. https://doi.org/10.1109/TPAMI.2013.50.
7. He, Xiangnan, Lizi Liao, Hanwang Zhang, Liqiang Nie, Xia Hu, and Tat-Seng Chua. "Neural collaborative filtering." In *Proceedings of the 26th International Conference on World Wide Web*, pp. 173–182 (2017). https://doi.org/10.48550/arXiv.1708.05031.
8. Goldberg, David, David Nichols, Brian M. Oki, and Douglas Terry. "Using collaborative filtering to weave an information tapestry." *Communications of the ACM* 35, no. 12 (1992): 61–70. https://doi.org/10.1145/138859.138867.
9. Mnih, Andriy, and Russ R. Salakhutdinov. "Probabilistic matrix factorization." *Advances in Neural Information Processing Systems* 20 (2007).
10. Kumar, A., P. Kumar, S. Rathee, and B. Kumar (Eds.). *Mechatronics: Concepts, Tools, Applications, and New Trends* (1st ed.). CRC Press (2025). https://doi.org/10.1201/9781003494478
11. Kumar, Ajay. "Transforming service sectors with IoT, ML, and comprehensive security solutions." *Mechatronics: Concepts, Tools, Applications, and New Trends* 6 (2025): 28.
12. Ricci, Francesco, Lior Rokach, and Bracha Shapira. "Introduction to recommender systems handbook." In *Recommender systems handbook*, pp. 1–35. Boston, MA: springer US (2010). http://dx.doi.org/10.1007/978-0-387-85820-3_1.
13. Breese, John S., David Heckerman, and Carl Kadie. "Empirical analysis of predictive algorithms for collaborative filtering." *arXiv preprint arXiv:1301.7363* (2013). https://doi.org/10.48550/arXiv.1301.7363.
14. Goodfellow, Ian, Yoshua Bengio, Aaron Courville, and Yoshua Bengio. "Deep learning." Vol. 1. MIT Press Cambridge (2016).
15. Zhang, X., & Wang, Y. (2018). Deep learning for recommendation systems: A concise overview. *IEEE/ACM Transactions on Audio, Speech, and Language Processing*, 27, no. 12: 2042–2048.
16. Koren, Yehuda, Robert Bell, and Chris Volinsky. "Matrix factorization techniques for recommender systems." *Computer* 42, no. 8 (2009): 30–37. https://doi.org/10.1109/MC.2009.263.
17. Szegedy, Christian, Wei Liu, Yangqing Jia, Pierre Sermanet, Scott Reed, Dragomir Anguelov, Dumitru Erhan, Vincent Vanhoucke, and Andrew Rabinovich. "Going deeper with convolutions." In *Proceedings of the IEEE Conference on Computer Vision and Pattern Recognition*, pp. 1–9 (2015). https://doi.org/10.48550/arXiv.1409.4842.
18. LeCun, Yann, Yoshua Bengio, and Geoffrey Hinton. "Deep learning." *Nature* 521, no. 7553 (2015): 436–444.
19. Schmidhuber, J. "Deep learning in neural networks: An overview." *Neural Networks* 61 (2015): 85–117. https://doi.org/10.1016/j.neunet.2014.09.003.
20. Kowsari, Kamran, Donald E. Brown, Mojtaba Heidarysafa, Kiana Jafari Meimandi, Matthew S. Gerber, and Laura E. Barnes. "Hdltex: Hierarchical deep learning for text classification." In *2017 16th IEEE international conference on machine learning and applications (ICMLA)*, pp. 364–371. IEEE (2017). https://doi.org/10.1109/ICMLA.2017.0-134.
21. Cho, Kyunghyun. "On the properties of neural machine translation: Encoder-decoder approaches." *arXiv preprint arXiv:1409.1259* (2014). https://doi.org/10.48550/arXiv.1409.1259.
22. Sutskever, I. "Sequence to Sequence Learning with Neural Networks." *arXiv preprint arXiv:1409.3215* (2014). https://doi.org/10.48550/arXiv.1409.3215.

12 AI-Driven Solutions for Career Counseling Based on the Personality of an Individual

Swarnim Seth, Shivangi Pachauri, Neelam Sharma, and Sandhya Gupta

12.1 INTRODUCTION

Career guidance plays a crucial role in overall human development, which aids individuals in making informed decisions about their professional trajectories. The 21st century's knowledge-based civilization can be constructed on a foundation of higher education [1]. A study by the Ministry of India Human Resource Development (MHRD) found that enrollment of students in higher education increased by 12 times during the previous four decades [2].

Youths' career experiences are affected by challenges such as technological advancements, economic fluctuations, globalization, and changing attitudes. The selection of suitable careers is a complex process and has a direct impact on financial, social, and psychological well-being. Most students, especially at pre-tertiary levels, are not mature enough to make accurate career choices. Furthermore, Figure 12.1 illustrates the enrollment rates across various fields. These figures underscore that engineering and humanities stand out as the most sought-after disciplines compared to other streams, primarily due to the extensive opportunities they offer along with relatively lower competition. This observation highlights an existing disparity between the number of career experts available and the growing population of students seeking guidance, emphasizing the need for a more balanced ratio.

The demand for career counseling experts is urgent, especially in schools and universities. However, the current manual approach is inefficient and ineffective, with limitations such as reduced competency, a shortage of full-time career guidance experts, and a lack of career counselors. Automation of career guidance and counseling processes has significant potential, saving time and effort and smoothing the task for professionals, students, staff, and faculty.

The rise of artificial intelligence (AI) has introduced innovative possibilities for enhancing the efficiency of career counseling through AI. It is equipped with

DOI: 10.1201/9781032724508-14

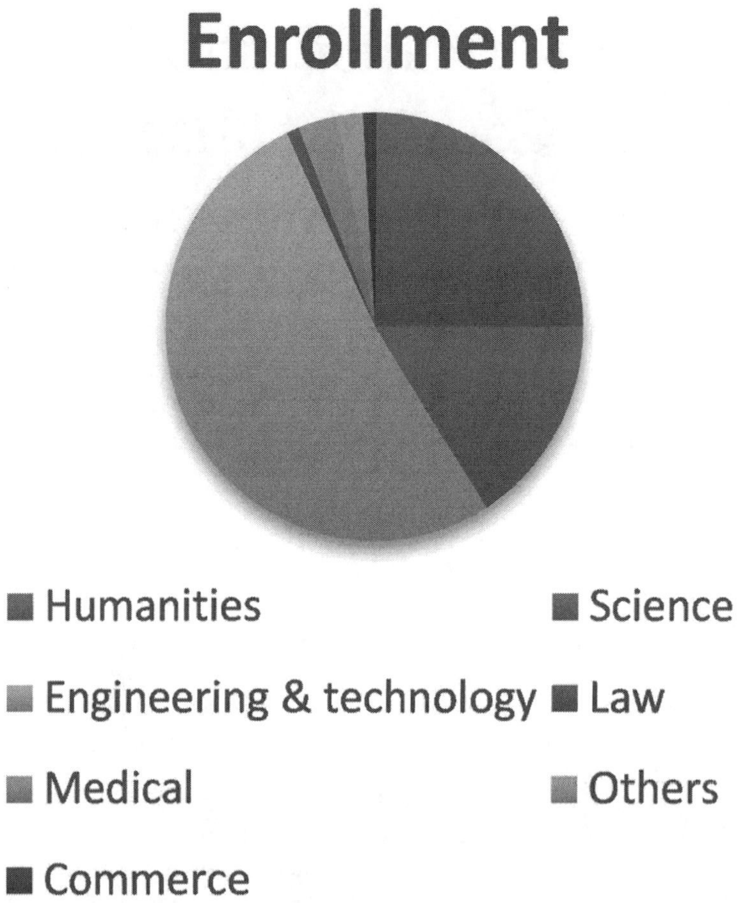

FIGURE 12.1 Stream wise enrollment

advanced algorithms that have the potential to provide customized advice and recommendations based on individual characteristics and preferences.

Through this research project, the aim is to contribute to the advancement of AI-driven career guidance. This signifies the Myers-Briggs Type Indicator (MBTI) theory to identify personality types of individuals, a foundational step towards effective career counseling. The MBTI theory, which classifies individuals into distinct personality types, offers insights into their behavioral tendencies, preferences, and potential career affinities.

The current approach consists of the following challenges:

- There is inadequate training given to the available counselors leading to a shortage of professional development. Due to overburdened counselors, individualized attention to each student is not possible.

- There is inefficiency in work due to the current manual approach, which can also lead to human error and bias.
- With increased enrollment in higher studies, the existing infrastructure is unable to cope, which results in insufficient guidance for many students.

Automated career guidance and counseling processes can save time and effort, which vastly impacts career counseling practices. The involvement of AI can increase the innovative possibilities in the field of career counseling. The underexplored region of personality significantly affects an individual's life, which is a potential aspect of this study.

12.1.1 Research Problem

While the integration of AI in career guidance has been a futuristic success, the accuracy and reliability of personality type determination using the MBTI theory within this context are still unexplored. This research tries to find this gap by evaluating the performance of two machine learning algorithms, support vector machine (SVM) and artificial neural network (ANN), in predicting personality types based on MBTI indicators.

12.1.2 Objectives

The objective of this research is to determine the effectiveness of SVM and ANN algorithms in classifying individuals' personality types and how they can be used in career guidance. Specifically, this research focuses on determining whether one algorithm outperforms the other in accurately identifying personality types.

12.1.3 Scope of the Study

This research encloses three main components: data collection, algorithm evaluation, and implications for AI-driven career guidance. It focuses on collecting and preprocessing a diverse dataset aligned with the MBTI theory. The SVM and ANN algorithms are then applied to predict personality types, and their performance is thoroughly compared. While the study focuses on personality type determination, its outcomes contribute to the discussion on algorithm selection for AI agents in decision-making circumstances.

12.2 LITERATURE REVIEW

The integration of AI in career counseling has drawn a lot of attention since it has the potential to completely transform individualized advising. A promising method for providing personalized career guidance is AI. This section examines the research that has already been done on AI in career counseling and how SVM and ANN algorithms are used.

12.2.1 The MBTI Theory: An Overview

The MBTI is a well-known personality assessment tool that classifies individuals into four groups according to how they prefer to express four opposing traits, referred to as "dichotomies": Extraversion (E)–Introversion (I), Sensing (S)–Intuition (N), Thinking (T)–Feeling (F), and Judging (J)–Perceiving (P). Combinations of these four pairs indicate the 16 unique personality types (e.g., ENTJ, INFP) that result from these dichotomies as a whole. Figure 12.2 shows The 16 personality Types Suggested by MBTI Theory.

Extraversion (E) versus Introversion (I), the first dichotomy, represents people's preferences for where to focus their attention. In contrast to introverts, who direct their attention inward, extroverts frequently concentrate on the outside world. This idea is consistent with Jung's viewpoint on energy, action, and reflection, according to which introverts get energy by reflection and extroverts obtain energy through action.

FIGURE 12.2 The 16 personality types suggested by MBTI theory

The way people process information is categorized using the Sensing (S) versus Intuition (N) divide. Individuals with a preference for intuition view information holistically, whereas sensing people sequentially gather knowledge through their senses.

The third dichotomy, Thinking (T) versus Feeling (F), deals with making decisions after taking in new information. Objectivity and reason are important to thinkers, but empathy and personal participation are important to feelers. These preferences are linked to Jung's "judging" functions, where intellectuals separate for rational judgment and feelers take into account various viewpoints.

The Judging (J) versus Perceiving (P) divide also reflects how people engage with the outside world. Judgers favor taking firm action, but Perceivers like to gather knowledge first and remain open to many options [3].

12.2.2 Navigating Careers with AI

The integration of AI into career guidance has gained significant attention due to technological advancements in education and workforce development. Digital tools provide convenience and expanded access to guidance services, with the potential for improved accessibility, information dissemination, and simplified communication. The global pandemic has accelerated this convergence, necessitating distance and digital guidance services. Traditional guidance methods have been utilized through learning and career information, automated interactions, and communication channels. However, the adoption of new technologies in guidance services depends on users' technical skills and the willingness of guidance organizations and professionals to adapt.

The relationship between AI and education extends beyond mere utility, enclosing the development of AI-related competencies, integration into teaching processes, and further advancements for educational purposes. The impact of AI in shaping digital career guidance practices opens avenues for investigating its influence on the creation and moderation of agency within the guidance process [5–11].

Numerous studies have explored the utilization of AI agents to enhance career guidance. Natural language processing is used to develop a chatbot that engages in real-time career counseling conversations. Their findings indicated increased engagement and satisfaction among users. However, a gap exists in the literature regarding the incorporation of personality-based insights into AI-driven career guidance.

12.2.3 Recent Developments in Career Guidance

Comprehensive and personalized insights are provided by the integration of AI with big data analytics so that vast amounts of data can be analyzed. Advancements such as deep learning in the field of AI enhance the capabilities of AI-driven tools, which makes them more responsive to individual needs.

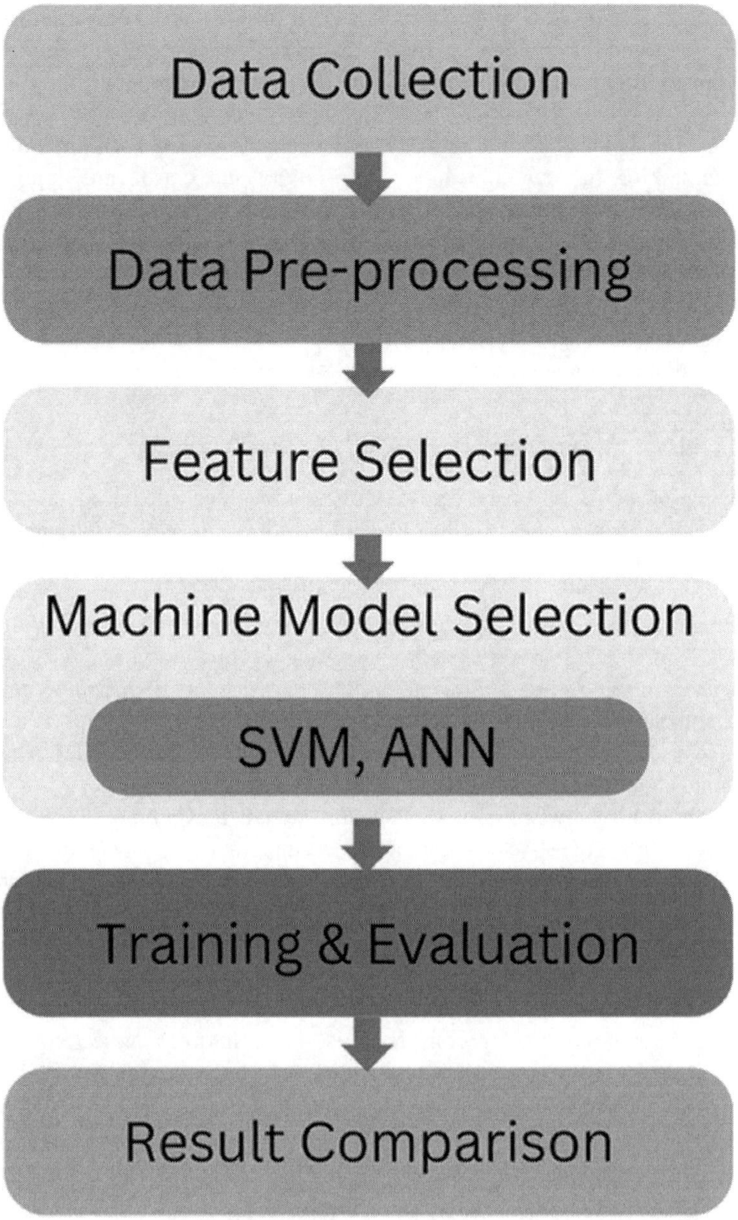

FIGURE 12.3 Flowchart of the methodology

AI technologies go hand in hand with the generation today, usually referred to as Gen-Z, which vastly uses AI platforms. "Ask Jane," an AI-driven platform that is built on the ChatGPT platform, is an example that gives AI-driven career advice [14].

12.3 METHODOLOGY

Figure 12.3 presents the flowchart of the methodology utilized in this chapter.

12.3.1 Data Collection

To facilitate personality-based career guidance, a Google Form questionnaire was designed based on the MBTI theory. The questionnaire comprised a series of questions aimed at capturing various personality traits and preferences relevant to the theory. The questionnaire consists of 60 questions related to the behavioral, emotional, and social aspects of a person. Participants were asked to respond based on their tendencies and inclinations. A database of 453 records was collected. Responses were collected over a predefined period from a diverse group of individuals seeking career guidance.

12.3.1.1 Demographics of the Dataset Collected

The dataset collected was diverse, so it was classified based on gender, age, income group and profession. The demographics are as follows:

Gender:

Male: 64%
Female: 36%

Age:

15–25 years: 42%
26–35 years: 37%
>35 years: 21%

Income Group:

<1 lac: 31%
1–10 lacs: 36%
>10 lacs: 34%

Profession:

Student: 28%
Business-person: 16%
Private Service: 27%
Government Service: 14%
Others: 15%

12.3.2 Data Preprocessing

Upon completion of the data collection phase, the responses were compiled into an Excel spreadsheet through Google Forms' automatic aggregation. To ensure uniformity and comparability, the Excel spreadsheet was converted into a comma-separated values (CSV) file. This CSV format simplified subsequent data processing and algorithm implementation.

Before training the SVM and ANN algorithms, data preprocessing steps were performed. This included handling missing values, if any, and encoding categorical responses into numerical representations. The categorical-to-numeric encoding was necessary for compatibility with the algorithms. Additionally, the numerical features to ensure uniform scaling and prevent algorithmic biases based on feature magnitude were normalized.

12.3.3 Feature Selection

In this study, insights from the MBTI-based questionnaire identifying key personality traits relevant to career guidance were gathered. Guided by the MBTI theory, including dimensions such as introversion/extraversion and thinking/feeling, features for the algorithms were selected. Principal component analysis (PCA) for feature reduction to optimize performance was applied and aligned with MBTI principles while retaining essential personality information. This simplified the data, enhancing algorithm efficiency and maintaining the theory's relevance in career guidance. The implementation of PCA helped in the reduction of dimensionality, noise, and redundancy, due to which models' performance has been improved.

Principal Component Analysis (PCA) is a widely used technique for feature reduction due to its effectiveness and simplicity. Here's why PCA was chosen:

- The ability of PCA to retain the most informative features that capture the maximum variance in the data is crucial for effective underlying structure.
- PCA is relatively straightforward to implement computationally and can be applied to large datasets easily.
- PCA is easy to interpret and implement for datasets with linear relationships.

12.3.4 Machine Model Selection

A lot of machine learning algorithms during the literature review were explored. However, SVM and ANN were considered to be the main focus as per the requirements of the project.

12.4 EVALUATION OF SVM ALGORITHM FOR PERSONALITY TYPE PREDICTION

In this study, the SVM algorithm's efficiency in predicting personality types using the preprocessed dataset was evaluated. Known for its strong capability of classification, SVM was applied to decode complex patterns within the data. The approach involved training the SVM algorithm on a distinct subset of the dataset, followed by comprehensive testing on a separate subset.

12.5 EXPLORING THE POTENTIAL OF ANN ALGORITHM FOR PERSONALITY TYPE PREDICTION

In parallel, how well the ANN algorithm performed the task of predicting personality types was investigated. ANN was applied to decode the nuanced relationships within the preprocessed dataset. Applying a similar strategy, the ANN algorithm experienced training on a specific subset of the dataset and experienced thorough testing on an independent subset. This methodical examination sought to discover the predictive accuracy of ANN in the realm of personality type identification.

These comparative assessments of SVM and ANN algorithms serve as a critical foundation for this study's overarching objective: enhancing the precision and accuracy of personality type prediction within the framework of career guidance.

12.6 TRAINING AND EVALUATION

For both algorithms, a standard approach of splitting the dataset into training and testing sets (e.g., 70% for training and 30% for testing) was applied. The training set was used to teach the algorithms to recognize patterns and relationships among the features. Subsequently, the testing set was used to assess the algorithms' performance by measuring their accuracy in predicting the correct personality types.

12.7 LIMITATIONS AND ASSUMPTIONS

- The dataset consists of 453 records, which may not be large enough in capturing all variations in personality traits across a broader population.
- The data collected may rely on self-reported responses, which are subject to biases, where the way respondents answer may not be truthful or consistent, which can lead to variability in the responses.
- There can be a loss of some information while using PCA for feature selection. The features after applying PCA may not be as good as the original ones.
- This article assumes that MBTI is a reliable and valid theory for the assessment of personality.
- The SVM and ANN algorithms are assumed to be appropriate for the prediction of personality.

12.8 EXPERIMENTAL SETUP

12.8.1 Data Splitting

The dataset obtained from the Google Form questionnaire responses was divided into two categories: a training dataset and a testing dataset. The training dataset has 70% of the overall data, while the testing dataset holds the remaining 30% of the data. This partition allowed us to train the SVM and ANN algorithms on one subset and evaluate their performance on the other.

12.8.2 Support Vector Machine

Training: The SVM was trained on the training set using Scikit-learn's SVM implementation. Experiments with different kernel functions (e.g., linear, polynomial, radial basis function) were done.

Testing: After training, the SVM's predictive capability was evaluated on the testing set. Predictions were generated based on the features extracted from the questionnaire responses.

Evaluation Metrics: The accuracy of the SVM model is determined by calculating the proportion of how correctly it is classifying instances in the testing dataset.

12.8.3 Artificial Neural Network

Architecture Design: The ANN architecture consisted of input layers corresponding to the questionnaire features, hidden layers for feature transformation, and an output layer for personality type prediction.

Training: The ANN was trained on the training set using TensorFlow. Various architectures (e.g., number of hidden layers, neurons per layer) and optimized hyperparameters (e.g., learning rate) using a combination of trial-and-error and cross-validation were experimented with.

Validation: A portion of the training set was reserved as a validation set to monitor the ANN's performance during training and prevent overfitting.

Testing: After training, the ANN's performance was assessed on the testing set, generating predictions for personality types based on the questionnaire features.

Evaluation Metrics: The accuracy of the ANN model was calculated as the ratio of correctly classified instances in the testing dataset.

12.8.4 Comparison and Evaluation Metrics

To compare the performance of SVM and ANN, accuracy was used as the primary evaluation metric. However, given the nature of personality prediction, other metrics like precision, recall, and F1 score also provided additional insights into algorithm performance. These metrics give a numerical representation of how well the model manages false positives and false negatives while correctly classifying cases.

12.8.5 Results

In the experiments, the SVM achieved an accuracy of 86.81%, while the ANN achieved a higher accuracy of 93.40% on the testing set. This comparison highlights the relative performance of the two algorithms in predicting personality types based on the MBTI-derived features. The accuracy of the SVM increased significantly to 92.31% once feature selection techniques, particularly PCA, were applied. It's important to note that the ANN's accuracy did not change. This shows that the SVM's predictive ability was more significantly impacted by feature selection, while the ANN's performance stayed constant over time.

12.9 ALGORITHM COMPARISON

12.9.1 Support Vector Machine

SVM is a type of supervised machine learning algorithm that is used for classification and regression tasks. It searches for a hyperplane that can maximize the margin between different classes in the feature space [12]. In the context of this project, SVM seeks to classify individuals into specific personality types based on the features derived from the MBTI-based questionnaire responses.

12.9.1.1 Strengths

- SVM is effective when dealing with high-dimensional data, making it suitable for the feature-rich personality assessment.
- Both linear and non-linear relationships between features and classes can be handled by it, thanks to kernel functions.
- SVM provides robust generalization by maximizing the margin between classes, which can lead to better accuracy.

12.9.1.2 Weaknesses

- The choice of hyperparameters and kernel functions may affect the performance of SVM.
- When the classes are highly imbalanced or the dataset is large, it may not scale well.
- Finding the optimal hyperparameters can be time-consuming.

12.9.1.3 Results

1. An accuracy of 0.8681 is achieved by the SVM model, indicating that it correctly predicted the personality types for approximately 87% of the classes in the testing dataset. This suggests the model's strong performance in matching predictions with actual labels. After applying PCA for feature

selection, accuracy increased to 92.31%, signifying a higher rate of correct predictions.
2. The model's precision of 0.9408 indicates that when it predicted a certain personality type, it was accurate about 94.08% of the time. This highlights the reliability of the model's positive predictions in identifying specific personality types. But after applying PCA, precision improved to 0.9166, i.e., 91.66%.
3. With a recall of 0.9154, the model successfully captured and identified around 91.54% of the actual instances belonging to a specific personality type. This demonstrates the model's ability to effectively detect a significant portion of relevant cases within the dataset. Moreover, after applying PCA, the recall increased to 93.01%, suggesting that the model became more proficient at capturing relevant instances belonging to specific personality types.
4. With an F1 score of 0.9019, the model demonstrates strong performance in achieving a balance between precision and recall. This value suggests that the model's positive predictions are accurate while effectively capturing a substantial portion of relevant instances within the dataset. The F1 score improved to 92.02%, reflecting a finer balance between precision and recall after the feature selection process.

This suggests that PCA helped improve the SVM model's predictive abilities, leading to a more precise and effective approach for classifying personality types.

12.9.1.4 Confusion Matrix

This confusion matrix, as shown in Figure 12.4, is a detailed representation of a multi-class classification problem with 32 classes (i.e., the different personality types from MBTI Theory). Each row represents a true class label, while each column represents a predicted class label. The values in the matrix indicate the counts of instances that belong to each combination of true and predicted labels.

For example, in the cell corresponding to row "0-2" and column "1," the value of "1" indicates that there is one instance that truly belongs to class 0 but was predicted as class 1 by the model.

The examples that were correctly categorized are represented by the diagonal elements from top-left to bottom-right for each class, i.e., the true positive predictions. The diagonal elements represent the number of instances correctly classified for each class.

The non-diagonal elements represent instances that were misclassified, which resulted in false positives and false negatives. These elements provide insights into which classes are often confused with each other by the model.

12.9.2 Artificial Neural Network

ANN is a flexible machine learning model that takes its cues from the neural network of the human brain. It is made up of layers of interconnected nodes (neurons)

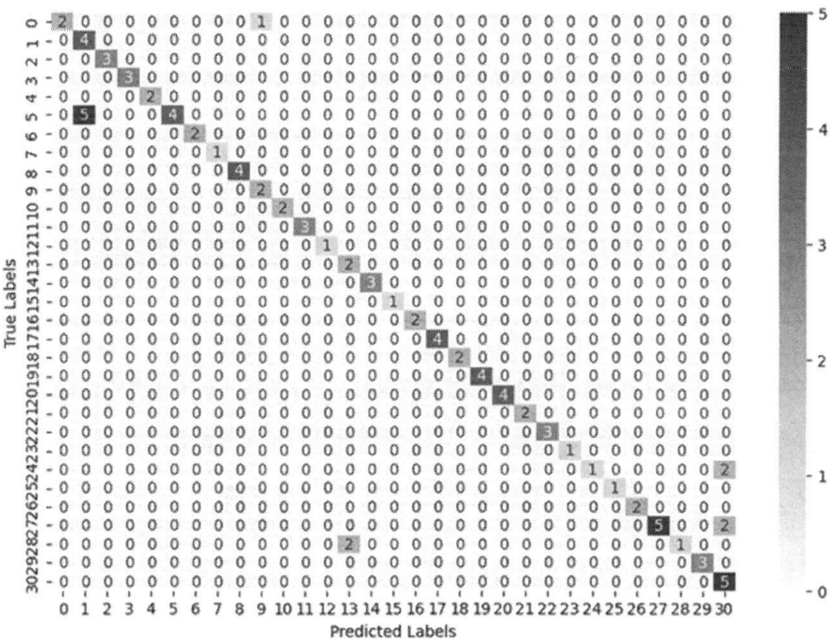

FIGURE 12.4 Confusion matrix of SVM model

that process and change info [13]. In this project, ANN is used to predict individual classifications by learning the complex correlations between questionnaire responses and personality types.

12.9.2.1 Strengths:

- — ANN is used to capture complex patterns and relationships in data; this makes it suitable for the nuanced characteristics of personality assessment.
- — It can automatically learn features from the data, reducing the need for explicit feature engineering.
- — ANNs can model non-linear relationships effectively, making them suitable for the inherently intricate nature of personality traits.

12.9.2.2 Weaknesses:

- — ANN training can require a substantial amount of data to prevent overfitting.
- — Tuning the architecture (e.g., number of layers, neurons per layer) can be challenging and may require experimentation.
- — Training ANNs can be computationally intensive, especially for large and deep networks.

12.9.2.3 Results

1. An accuracy of 0.9340 specifies that the model can accurately classify roughly 93.40% of instances in the dataset.
2. A precision of 95.80% shows that when the model predicts a certain class, it is correct about 93.38% of the time.
3. A recall of 94.62% indicates that the model captures 93.54% of all positive instances.
4. A value of 92.66% depicts a good balance between precision and recall, indicating that the model is effective in capturing relevant instances while minimizing false positives and false negatives.

It's worth noting that the application of PCA did not significantly impact the performance of the ANN, as the accuracy and key metrics remained relatively consistent.

12.9.2.4 Confusion Matrix

ANN creates predicted class labels, which are represented by each column, and true class labels, which are represented by each row. For example, the cell where the row corresponds to class "0-2" and the column is "1." The value in this cell is "1," which means that the ANN predicted one instance from class 0 to be in class 1. Figure 12.5 presents the confusion matrix of the ANN model.

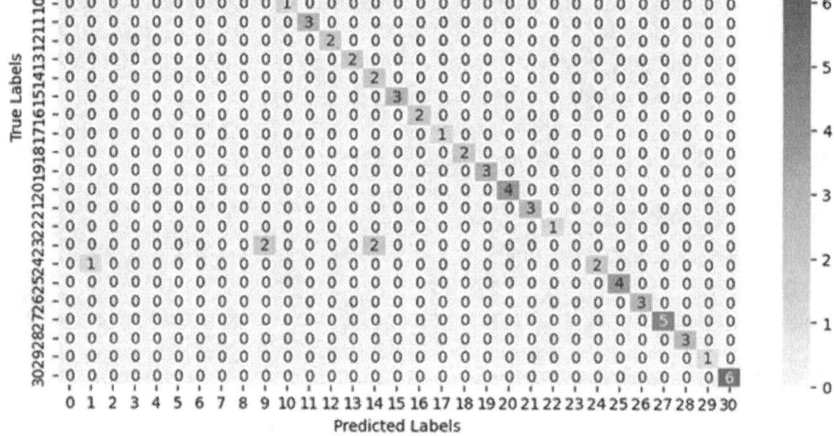

FIGURE 12.5 Confusion matrix of ANN model

The diagonal elements of the matrix show the instances that were correctly classified by the ANN. These are known as true positives (TP) for each class. In other words, the values along this line indicate how many instances were accurately predicted for each class.

The non-diagonal elements show the flaws of ANN. These cells give us insights into where the model might have produced false positives (instances predicted as a different class than the true class) or false negatives (instances that were not predicted as their true class).

12.9.3 Conclusion

Table 12.1, provides a comprehensive overview of the performance metrics for two different models taken into consideration, SVM and ANN, based on various evaluation criteria, such as accuracy, precision, recall, and F1 score.

In conclusion, the comparison between SVM and ANN models for multi-class classification revealed their respective strengths and performances. The SVM exhibited an accuracy of 86.8%, emphasizing its reliable predictive capacity; however, after applying PCA, its accuracy reached 92.31%. With a precision of 94.1% and a recall of 91.5%, the SVM demonstrated a balanced approach to positive predictions and capturing actual positives. After applying PCA for feature selection, its precision reached 91.67% and recall was 93.01%. Its F1 score of 90.2% reinforced its capability in handling precision-recall trade-offs but increased to 92.02% after applying PCA.

On the other hand, the ANN showed an accuracy of 93.4%, aligning closely with the SVM's performance. Its precision of 93.4% and recall of 93.5% demonstrated its proficiency in generating precise positive predictions and identifying actual positives. The F1 score of approximately 90.2% highlighted its capacity to blend precision and recall. It's worth noting that the application of PCA did not significantly impact the performance of the ANN, as the accuracy and key metrics remained relatively consistent. Figure 12.6 provide a comparison graph of the SVM model before and after applying PCA.

TABLE 12.1
Comparison of SVM and ANN Models

	MODELS			
	SVM		ANN	
PARAMETERS	Without PCA	With PCA	Without PCA	With PCA
Accuracy	0.86814	0.92307	0.93406	0.93406
Precision	0.940860	0.91667	0.95806	0.94086
Recall	0.91542	0.93011	0.94624	0.94624
F1 score	0.90199	0.92028	0.92668	0.91843

The comparative line graph, as shown in Figure 12.7, represents the marginal performance differences between the two models across these metrics. After summarizing, the conclusion was, that while both the SVM and ANN presented praiseworthy capabilities for multi-class classification, their attributes, such as interpretability and training time, may mentor the choice between them based on the particular demands of the task at hand.

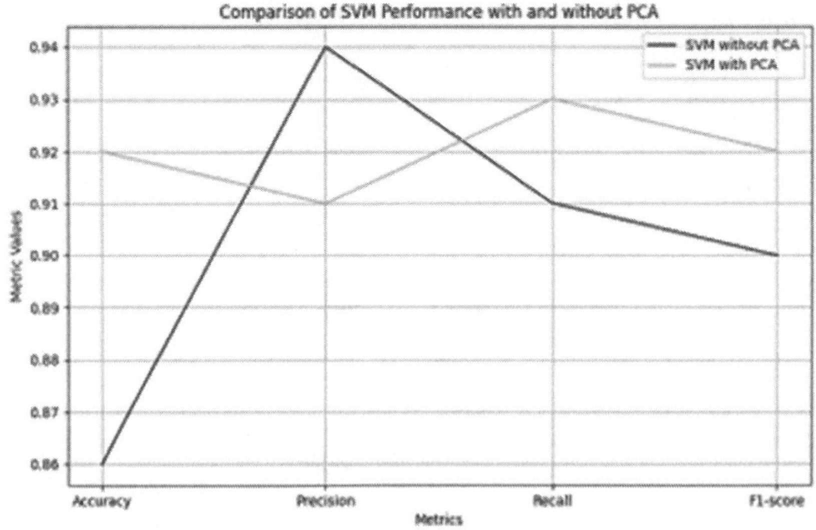

FIGURE 12.6 Comparison graph of the SVM model before and after applying PCA

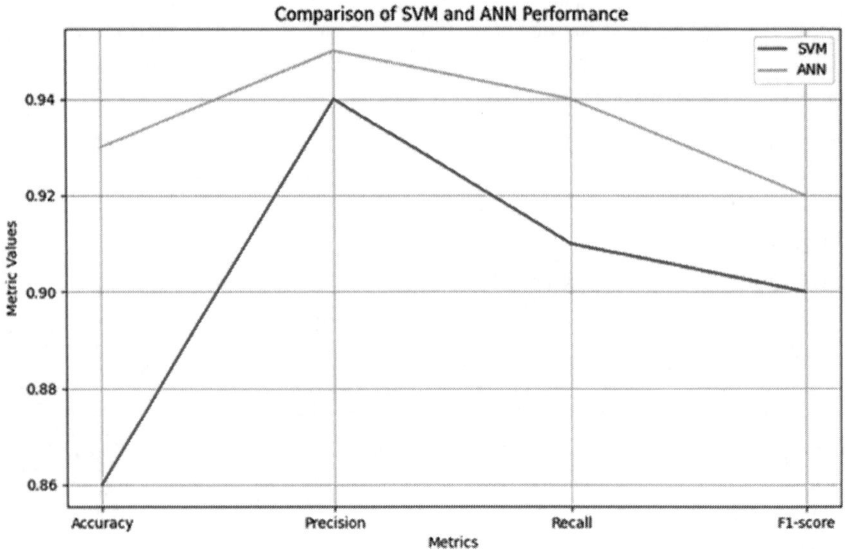

FIGURE 12.7 Line graph showing a comparison between ANN and SVM models

12.9.4 Practical Implications

- Career counselors can trust a model with high accuracy to provide dependable advice, which enhances both counselor and client's confidence.
- The use of multiple algorithms allows organizations to select the most suitable model that aligns with their operational needs and the complexity of the data they handle.
- Personalized career guidance leads to more relevant and actionable career advice, increasing client satisfaction and engagement.
- By understanding the performance metrics, it is easy to allocate resources for optimized performance in an organization.

12.9.5 Comparative Limitations of Other Algorithms

Many algorithms could be used to predict the personality type of an individual. SVM and ANN are chosen for this problem because of their particular strengths and suitability. The reasons to choose these two rather than other algorithms are presented here.

12.9.6 Decision Trees and Random Forests

Decision trees can be overfitting for this problem, while random forests can be inefficient in capturing complex non-linear relationships compared to SVMs and ANNs.

12.9.7 K-Nearest Neighbors

KNN may act inefficiently on large datasets and might not perform well in high-dimensional spaces compared to SVMs and ANNs.

12.9.8 Logistic Regression

Logical regression may have some limitations to capture non-linear relationships and interactions among features, which is a critical part of personality prediction.

12.10 CONCLUSION

SVMs and ANNs are chosen for their strong performance in handling high-dimensional data, modeling complex non-linear relationships, flexibility, and scalability. Their success in previous studies and the extensive research and development focus on these algorithms further justify their selection for personality type prediction over other methods. This research chapter focuses on the development of AI-based career guidance, drawing attention to personality type identification through the MBTI theory. By comparing the results of SVM and ANN algorithms, it was sought to increase the accuracy of personality-based recommendations.

12.11 KEY FINDINGS AND CONTRIBUTIONS

The observation in the research showed that the ANN algorithm constantly outperformed the SVM algorithm, attaining an accuracy of 93.40% compared to SVM's 86.81% accuracy, which increases to 92.31% after applying PCA for feature selection. This outcome highlights the importance of using ANNs for personality type prediction within AI-driven career guidance platforms.

12.12 SIGNIFICANCE OF ANN'S ACCURACY

The higher accuracy achieved by ANN signifies inference for the AI's role in career guidance. ANN's potential to find complex relationships within personality traits permits more precise recommendations. This accuracy can enhance the AI agent's credibility, which can increase user trust and engagement. Accurate personality-based guidance can guide individuals toward relevant career paths, positively influencing their professional satisfaction and success.

12.13 IMPLICATIONS AND POTENTIAL APPLICATIONS

The success of ANN in personality prediction gives a successful vision for the development of personalized career guidance tools. By giving accurate personality analysis, the AI agent can offer recommendations that align closely with an individual's traits, aspirations, and potential. Furthermore, the collaboration of accurate personality-based counseling can alleviate the risk of discrepancy between career choices and individual characteristics.

Beyond career guidance, the implications extend to other domains too. The combination of AI and personality assessment can help in education, recruitment, and even mental health support, where personalized assessment is pivotal.

12.14 ACTIONABLE INSIGHTS AND RECOMMENDATIONS

- The policymakers should establish standards and regulations ensuring data security, privacy, and fairness for the ethical use of AI in career guidance.
- Professional practitioners should use AI tools to provide personalized guidance to enhance the efficiency and productivity of their work.
- Researchers could explore hybrid machine learning models and could expand the diversity of the dataset.

12.15 FINAL REMARKS

This research amplifies the understanding of the potential of AI-driven career guidance agents. The higher accuracy of ANN emphasizes its role as a powerful tool in personality-based recommendations, re-evaluating the landscape of career counseling. As technology continues to develop, the incorporation of AI agents equipped with accurate personality analysis holds great assurance for magnifying the way crucial life decisions are made.

REFERENCES

1. Mehraj, T., and Baba, A., "Scrutinising Artificial Intelligence based Career Guidance and Counselling Systems: an Appraisal," *International Journal of Interdisciplinary Research and Innovations*, 7(1), 402–411, 2019.
2. Ministry of Human Resource Development of India, *Higher Education in India: Twelfth Five-Year Plan (2012–2017)* and Beyond, FICCI Higher Education Summit, 2012.
3. King, S. P., and Mason, B. A. "Myers Briggs Type Indicator." In B. J. Carducci & C. S. Nave (Eds.), *Wiley encyclopedia of personality and individual differences: Vol. II. Measurement and assessment*. Hoboken, NJ: John Wiley & Sons (2020). Retrieved from https://online.vitalsource.com/#/books/9781119793922/epubcfi/6/116%5B%3Bvnd.vst.id ref%3DAc123%5D!/4/2/2/2/6%400:0.
4. Simkus, J., *Simply Psychology*, January 29, 2024. Retrieved from https://www.simplypsychology.org/the-myers-briggs-type-indicator.html.
5. Westman, S., Kauttonen, J., Klemetti, A., Korhonen, N., Manninen, M., Mononen, M., Niittymäki, S., and Paananen, H., "Artificial Intelligence for Career Guidance – Current Requirements and Prospects for the Future," *IAFOR Journal of Education: Technology in Education*, 9(4), 2021.
6. Google Developers. Retrieved from https://developers.google.com/machine-learning/crash-course/classification/accuracy.
7. Google Developers. Retrieved from https://developers.google.com/mmachine-learning/crash-course/classification/precision-and-recall.
8. Google Developers. Retrieved from https://developers.google.com/machine-learnig/crash-course/classification/roc-and-auc.
9. Gavhane, P., Shinde, D., Lomte, A., Nattuva, N., and Mandhane, S., "Career Path Prediction Using Machine Learning Classification Techniques," *International Journal of Scientific Research in Computer Science, Engineering and Information Technology*, 8(3), 2021.
10. Khode, S., "Career Guidance Engine (Using Machine Learning and Deep Learning)," medium.com, Aug 12, 2021.
11. Kapur, R., "Career Guidance and Student Counselling," Research Gate, March 2018.
12. JavaPoint. Retrieved from https://www.javapoint.com/machine-learning-support-vector-machine-algorithm.
13. Maind, S., and Wankar, P., "Research Paper on Basic of Artificial Neural Network," *International Journal on Recent and Innovation Trends in Computing and Communication*, 2(1), 2014.
14. Pearson, K., *HRMorning*, June 4, 2024. Retrieved from https://www.hrmorning.com/articles/ai-career-guidance-for-gen-z/.

13 Exploration of Machine Learning–Enabled Automated Crop-Disease Detection, Segmentation, and Classification Techniques

Kamlesh Lakhwani, Naveen Hemrajani, Ajay Kumar, and Pankaj Jain

13.1 INTRODUCTION

Plant diseases pose a serious threat to agricultural productivity, affecting it in terms of both quality and quantity [1]. A crucial focus in agriculture today involves detecting and classifying plant lesions to enhance the quality of plant production, thus contributing to economic growth. It is reported that in India the most common disease of Indian mustard is Alternaria blight, which affects the crop adversely, causing nearly 47% losses in the yield [2, 3]. Early-stage monitoring of plant health and pathogen detection is crucial for mitigating disease spread and enabling the implementation of effective and efficient management practices [4]. Technology plays an essential role in the detection of these diseases in horticulture, floriculture, etc.

Around 10 billion people will live on earth by 2050, according to prediction [5], which calls for an increase in agricultural development somewhere in the range of 50%. Between 2015 and 2050 every individual country would require raising the overall food production by around 70% [6]. However, the projected amount of crop production needs to be ensured by farmers all across the world by increasing the area of agricultural land or by improving the productivity of available agricultural areas by adopting smart farming techniques [2].

DOI: 10.1201/9781032724508-15

13.1.1 Traditional Methods of Diagnosing the Diseases

The plants are diagnosed [8] by the experts, traveling to various cultivated areas of the country and visually analyzing the health of the plant through leaves, stems, or seeds and generating scores according to the symptoms. This method tends to be very inconsistent and subjective. The experts may have different views on a particular score of a plant. This work using the machine learning (ML) approach will enable the experts to score the plants in a much more reliable way and also will help the farmers to diagnose the disease by themselves.

The structure of this chapter is as follows: Section 13.1 presents the introduction; Section 13.2 provides a concise overview of image-processing methodologies, discusses the pre-processing stage, and also elaborates on the concept of ML approaches. In Section 13.3 the review of literature is done of the past few years, examining various infections and the quantity of training and testing data and plant images. Section 13.4 briefly discusses some major research gaps and comparisons, which can further give an idea of problems to work on. Finally, Section 13.5 addresses the remaining challenges, and limitations, and provides recommendations.

13.2 IMAGE-PROCESSING-ENABLED TECHNIQUES

Disease detection nowadays can also be carried out using methods of image processing. The techniques consist of four major steps [9], which include the (i) preprocessing of data, (ii) disease detection, (iii) feature extraction, and (iv) classification.

13.2.1 Preprocessing-Based Techniques

The purpose of data preprocessing is to enhance the contrast of input images. The image taken from a digital camera has a lot of noise. This process eases further classification of images as it includes cropping, smoothing, color space conversion, and enhancement of images. The images collected are more often real-time images, which include unwanted backgrounds, so these might be cropped and resized to be effectively tested. Various preprocessing and image enhancement techniques are shown in Figure 13.1. The details of various techniques are discussed here.

Image Enhancement: Image enhancement [9] includes various techniques such as histogram equalization and color conversion. Color conversion is

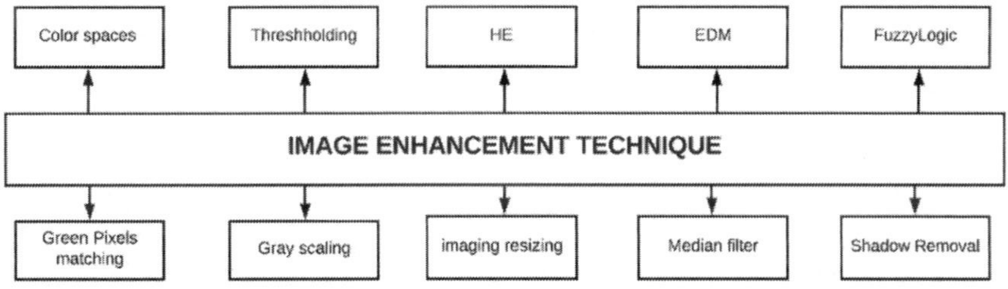

FIGURE 13.1 Preprocessing-based techniques

applied to convert the RGB images to grayscale images [1]. On the other hand, the histogram is used to have transparency in the image. Fuzzy logic is yet another enhancement technique used to identify plant lesions [7]. Images can also be enhanced by using color-based transformation techniques. Also for noise reduction, image clipping and thresholding are used to preprocess the image [10]. The image clipping technique is used to divide the image by using the spots of the background. The segmentation is used to provide spot background in the image by thresholding the image, whereas to decrease the noise from image median filter is used. Edge detection is also carried out on the images to make them suitable to be given as input to the segmentation phase.

13.2.2 Image Segmentation

Segmentation [11] of an image is done to divide it into multiple parts. The detection of the region of interest is the prime objective of such techniques. Various segmentation techniques such as Otsu, thresholding, K-means clustering, and color segmentation are used. K-means technique is used for classification and clustering in various clusters of the infected image are made. A set of pixels [12] constitutes a single cluster that is close to each other but different from other cluster values. The Otsu technique is used for image compression and cropping. One more technique, called "thresholding," is used for segmentation, which is attained by the histogram of the input images [9]. Edge detection is also carried out during the segmentation process to detect the diseases, which includes techniques such as canny, laplacian, and gradient. Various segmentation-based techniques are shown in Figure 13.2.

13.2.3 Feature-Extraction Technique

Features are the attributes of an object that set it apart from other objects. Features can be color, shape, size, etc., as shown in Figure 13.3. Selecting appropriate color, shape, and texture parameters is critical due to the similarity in characteristics among various diseases. Feature extraction plays a pivotal role, enabling automatic learning from raw and unstructured data. Texture analysis primarily focuses on

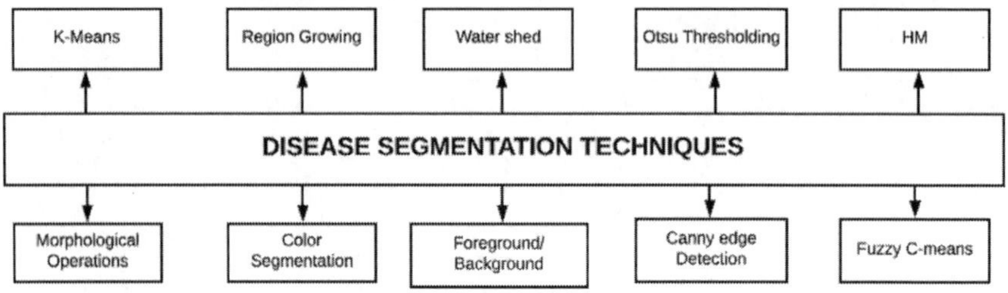

FIGURE 13.2 Segmentation-based techniques

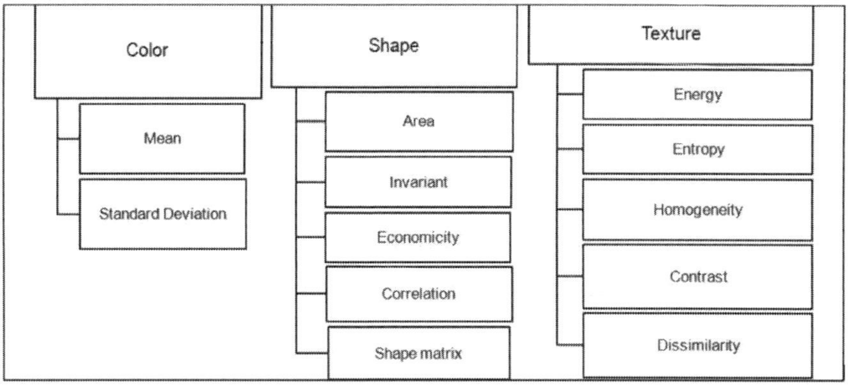

FIGURE 13.3 Feature extraction techniques

surface properties such as entropy, inverse, contrast, energy, uniformity, correlation, homogeneity, and more [13]. Shape attributes are obtained by measuring the area, axis length, eccentricity, centroid, invariant, shape matrix, etc. This constitutes the primary feature in image representation. Shape attributes are categorized into two stages: (i) region-based and (ii) boundaries-based features [1].

Color attributes deal with the physical characteristics of the object as different wavelength values are reflected from it. It is based on models like RGB, HIS, HSV, etc. The colors, depending on mean and standard deviation, can be analyzed using color co-occurrence matrices and color histograms.

13.2.4 Classification-Based Techniques

The last phase of disease detection consists of the classification of disease. This is mainly referred to as differentiating one class of infected disease from the other healthy and non-infected diseased leaf. Various classification techniques are shown in Figure 13.4. First, the classifier is trained using the images from a training set and then the image is tested. Various techniques classify the leaves of the plant, such as artificial neural network (ANN), SVM, and K-means clustering. ANN is a self-adaptive technique [14] that is used to modify the weights using the BPNN and is efficient enough to deal with noisy data. SVM is also one of the most commonly used non-linear classifiers for classification, utilized for decision-making. It works mainly in two phases: the offline phase [1], where the training process is done and the online phase [1], where the decision is made for all the incoming images. RBF, a classifier, having the fastest training capacity, is used for the real value process, whose value depends on the Euclidean distance. Another classifier PNN [15] is obtained from RBF and is used for classifying the patterns. KNN, which is easy to implement, is one of the geometrical classifiers [1] that is used to compute the effective Euclidean distance between the positions.

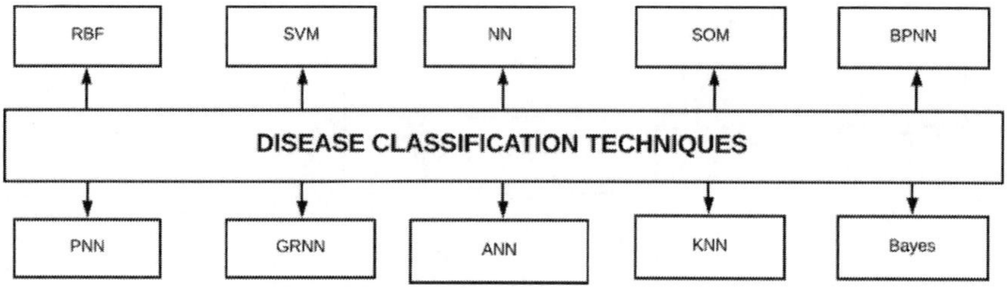

FIGURE 13.4 Classification-based techniques

13.2.5 ML Approaches

Even though the steps mentioned in the last section achieve good results in identifying and recognizing the diseases, while working with the images some limitations persist. If the leaves in the dataset are captured with a black background [17], segmentation can produce good results, but if it has other leaves in the background, then the results of segmentation might be questionable. Though there are various segmentation techniques doubts still persist about receiving unreliable results.

Techniques such as histograms, SIFT, color-based methods, and thresholding require expensive work and might not generalize well and sometimes might [18] not be effective. Hence, the idea of using ML was adopted, which avoids hand-crafted attributes and segmentation. Also, ML techniques such as the CNN [16, 19] (shown in Figure 13.5) have the ability to take large amounts of data as input.

13.2.5.1 ML-Based Classification

ML comprises mainly two tasks. First, unsupervised learning tries to find the unknown patterns with unlabeled data that enables extensive representation of the information contained. On the other hand, in supervised learning [17], target variables known as labels y_i are given. In this data related to labels is required for identification of patterns [20]. For discrete label classes (yes/no) a classification method is better suited, but for continuous label values, regression techniques are used. Supervised learning is separated into two main stages: the testing phase and the training phase. In the training phase, a model (F) [21] learns by using any effective ML-based algorithm from the training data, whereas in the second phase, i.e. the testing phase, the model is given some unknown inputs for which the best labels y are predicted for the sample x by:

$$y_i = F(x_i) \tag{13.1}$$

The collected data is split into two segments: a separate one for training and testing. Generally, the training requires more data for efficient and accurate results.

In ML, neural network–based techniques are used, such as CNN [22], NN, and BPNN, which work on the principle of supervised learning. For image diagnosis, various baseline architectures of CNN have been developed, such as AlexNet,

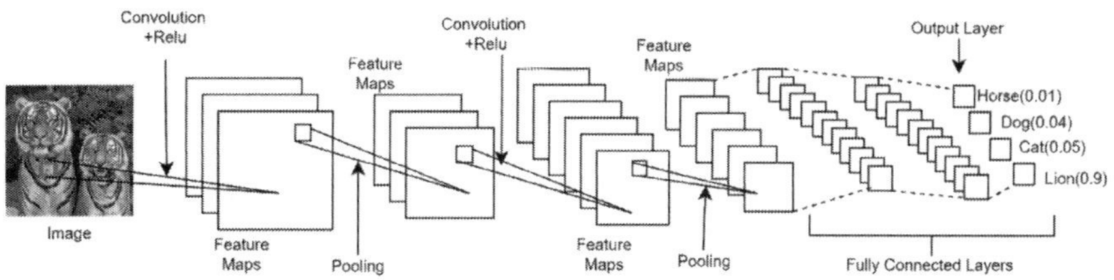

FIGURE 13.5 A graphical depiction of CNN model

AlexNetOWTBn, GoogLeNet, VGG, etc. A CNN model consists of three main steps: (i) convolution, (ii) pooling, and (iii) fully connected layers. The model's features are extracted through convolutional and pooling layers, while the fully connected layers classify these features based on specific functions.

13.3 LITERATURE REVIEW

The review of literature is done over the past few years taking into account different diseases and the quantity of images utilized for training and testing purposes. The literature has been segregated into various sections based on disease detection in plants, centered on disease detection employing image-processing, machine-learning, and deep-learning techniques.

13.3.1 Detecting Plant Diseases through Diverse Classification Techniques

The literature on plant disease detection encompasses a diverse array of techniques, each offering unique approaches to address the complex challenges in agricultural pathology. This review synthesizes existing research findings across a spectrum of methodologies, including Simple Linear Iterative Clustering (SLIC) [10], Expert Systems [5], HIST and WDH with K-means clustering [23], Rule-based mining techniques [11], Image segmentation with statistical inference methods [24], Feature-based classifiers (SRG and CMC) [25], Compressed Sensing combined with Support Vector Machines (SVM) [26], and the MFORSFS algorithm integrating Moth Flame Optimization (MFO), Rough Set (RS), and SVM [27].

Some important findings and conclusions of existing classification techniques on different crops are summarized in Table 13.1.

13.3.2 Detecting Plant Diseases through ML Techniques

The subsequent section encompasses research conducted on plant disease detection utilizing ML techniques. In these studies, the dataset is partitioned into training and testing sets and then fed into the classifier. The performances of different techniques are compared based on accuracy and summarized in Table 13.2.

TABLE 13.1
Literature Review Based on Disease Detection in Plants Using Different Classification Techniques

Reference	Crop	Disease	Number of images	Classifier
[10]	Soybean	Target spot, powdery mildew	Total images= 3624	Simple linear iterative clustering
[5]	Mustard	White Rust, Alternaria Blight, powdery mildew	N/A	Expert System
[28]	Soybean	Bacterial Blight, Rust and Sudden Death Syndrome, Frog Eye, Brown Spot	N/A	HIST and WDH with K means
[11]	Rice	Blast and Brown_spot disease	N/A	Rule based mining technique
[24]	Wheat	Tan spot Septoria, rust	Total images=3500	Image segmentation with statistical inference methods
[9]	Corn	rust spots, GrayLeafSpot, LeafBlight, brown patch, small spot and Curvularia_Leaf spot	Total images=244	SRG and CMC classifier(Feature based)
[29]	Pomegranate, brinjal, and Tomato	Early blight and Septoria leaf,Alternaria leaf Spot Anthracnose and Alternaria, Phytophthora blight, and Alternata	Total images= 30	Compressed sensing and SVM
[27]	Tomato	Early blight and Powdery mildew	N/A	MFORSFS algorithm(MFO,RS,SVM)

TABLE 13.2
Review of Detecting Plant Diseases through ML Techniques

Reference	Crop	Disease	Number_of_images		Classifier	Performance measure (accuracy)
			Training_ images	Testing_ images		
[30]	Oil Palm	Leaf spot, hawar and anthracnose	102	22	NN	87.75
[31]	Wheat	Four types_of_ rust(tan spot, septoria leaf spot, snow mold and powdery mildew)	20	342	BPNN	84.4
[7]	Potato	Late blight	Total images=27		Fuzzy C_means and backpropagation neural_network	100%
[32]	Potato	Early blight and late blight	Total images = 300		SVM with RGB imaging	95%
[33]	Corn	Downy mildew	210	43	Histogram feature based	83.5
		Powdery mildew	210	48		95.2
		Normal	210	47		96.7
[34]	Corn	5 diseases	90	10	PCA and NN	86.95
					K-NN	90
					Bayesian	89.25
[35]	Rice	bacterial blight, and Brown spot	90	30	K-NN	93.33
					SVM	91.10

(Continued)

TABLE 13.2 (CONTINUED)
Review of Detecting Plant Diseases through ML Techniques

Reference	Crop	Disease	Number_of_images		Classifier	Performance measure (accuracy)
			Training_ images	Testing_ images		
[36]	Cucumber	Downy and powdery mildew	250	30	ANN	Visual examination
[15]	Cucumber	Downy mildew	150	150	PNN	90
		Blight				92
		Anthracnose				92
[37]	Tomato	Healthy	78	20	Decision Tree	70
		Septoria spot	104	26		80.7
		Bacterial leaf	104	26		69.2
		spot	104	26		69.2
		Fungal late blight	104	26		92.3
		Leaf curl	104	26		84.6
		Bacterial canker				
[38]	Tomato	Healthy	58	58	BPNN	96.5
		Bacterial leaf	32	32		93.3
		spot	14	14		64.7
		Septoria spot	22	22		54.5
		Fungal late blight	22	22		100
		Bacterial canker	32	32		87.5
		Leaf curl				

(Continued)

TABLE 13.2 (CONTINUED)
Review of Detecting Plant Diseases through ML Techniques

Reference	Crop	Disease	Number_of_images		Classifier	Performance measure (accuracy)
			Training_ images	Testing_ images		
[39]	Grapes	Downy mildew	16	2	Back Propagation Neural network	100%
		powdery mildew	17	2		
[26]	Citrus	Normal	20	20	Naïve Bayes	95
		Greasy spot	20	20	K-means	77.5
		Melanoses	20	20	Discriminant analysis	98.75
		Scab	20	20	Random forest Tree	97.5
[40]	Blueberry	Various diseases in the plant	400	400	1. i) HOG (Histogram of oriented gradients) 2. ii) LBP (Local binary patterns)	Result: 84% accuracy index using Deep Learning
[25]	Review with multiple plants	Various diseases in the plant	200	200 leaf	overview of machine learning techniques	Accuracies of all algorithms stated.
[27]	Turmeric	Multiple diseases in plant	200	200	K-Means Segmentation along with GLCM texture analysis	91% accuracy rate

13.3.3 Plant Disease Detection Using DL Techniques

DL is a branch of ML that uses multi-layered artificial neural networks to extract features from data. In a variety of fields, including computer vision, natural language processing, and medical image analysis, DL has shown impressive promise. In the realm of plant disease detection, DL models can effectively analyze images of afflicted plants and precisely classify them into distinct disease categories [19]. Numerous studies have showcased the efficacy of employing deep-learning techniques for plant disease detection [41]. Convolutional Neural Networks (CNNs), in particular, have garnered widespread adoption owing to their capability to autonomously learn pertinent features from raw image data. Researchers have developed custom CNN [42] architectures or utilized pre-trained models like VGG, ResNet [43], and Inception for plant disease classification tasks. Transfer learning, a technique involving fine-tuning pre-trained models on plant disease datasets, has demonstrated effectiveness, particularly in scenarios where labeled data is scarce. The presence of extensive, labeled datasets is pivotal for training DL models. Several publicly available datasets, such as PlantVillage [41–44], Tomato-Leaf-Disease, and Cassava-Leaf-Disease, have been used for training and evaluating plant disease detection algorithms. Within these datasets are thousands of high-resolution images portraying diverse plant diseases and healthy plants, empowering researchers to forge resilient models.

The review highlights several studies conducted on various datasets and architectures for plant disease classification. Among them, study [36] conducted on the dataset involved 1567 images, which were significantly augmented to 46,409 images through background removal and segmentation techniques during preprocessing. Utilizing GoogLeNet architecture with transfer learning, the model achieved an overall accuracy of 94%, with individual classes demonstrating accuracies ranging from 75% to 100%. That underscores the effectiveness of preprocessing and augmentation strategies alongside transfer learning in enhancing classification accuracy for diverse classes within the dataset [41]. Presents a study on the Plant_Village dataset comprising 3700 images that underwent preprocessing, primarily resizing, resulting in no loss of images. Utilizing a modified LeNet CNN architecture, the classification model achieved an impressive overall accuracy of 92.8%. Mohanty et al. [42] present an extensive study on the Plant_Village dataset, comprising 54,309 images with 25 classes, after preprocessing augmented to 87,848 images. Utilizing the VGG architecture, the model achieved an outstanding accuracy of 99.53%. Similarly, [44] utilized the Plant_Village dataset with 14 classes, augmented to 54,323 images, employing Inception_v3, achieving a remarkable accuracy of 99.76%. On the other hand, [39] explored a collected dataset with multiple classes, employing R-FCN and ResNet-50, achieving an accuracy of 85.98%, demonstrating the significance of dataset diversity and preprocessing techniques. Furthermore, [26] and [45] experimented with smaller datasets, achieving high accuracies of 98.6% and 97.62%, respectively, with modified LeNet and Modified AlexNet architectures. These studies collectively underscore the importance of dataset size, preprocessing, and choice of architecture in achieving high accuracy in plant disease classification tasks.

TABLE 13.3

Limitations of DL Techniques for Identifying Plant Diseases

Reference	Small dataset	Less_ Number_of plant Species, Diseases	Low-Accuracy when tested_in_real_ conditions	Comp-lex Background	Multiple Diseases in the same-sample	Loca-tion	Infection-Status	Traind &Test-data are from the same-database
[41]	×	×	×	#	×	×	×	×
[42]	#	✓	×	×	×	×	×	×
[39]	×	×	#	✓	✓	✓	✓	×
[26]	×	×	×	×	×	×	×	×
[45]	×	×	✓	×	×	×	×	×
[36]	×	✓	#	×	✓	×	×	×
[46]	#	✓	×	×	×	×	×	×
[47]	×	#	×	✓	×	×	×	×
[44]	#	✓	×	#	×	×	×	×
[43]	×	×	×	×	×	×	✓	×

Legend: ✓ - Resolved × - Unresolved # - Partially Resolved

13.3.4 Limitations of Deep-Learning Techniques in the Context of Plant-Disease Detection

Despite their success, DL-based approaches for plant disease detection face several challenges. Limited labeled data, domain adaptation issues, environmental variations, and model interpretability are some key challenges that researchers need to address. Moreover, the limitation analysis of DL models used by various researchers is summarized in Table 13.3.

13.4 RESEARCH GAPS AND CHALLENGES

After going through all the literature, few gaps were found. The primary emphasis in most studies has been on enhancing classification accuracy. Even though the majority of models attained remarkably high accuracy, the percentage suddenly falls when testing is done with a dataset that is not part of the same database. Also, the tables clearly state that this problem still prevails and needs to be solved. Furthermore, there exists a notable dearth of specifics across all experiments. Presently, the available datasets lack the requisite diversity to influence the quality of network training and its capacity for accurate classification. Moreover, numerous crops remain untouched by image processing techniques. Also, certain versions have only been evaluated on a single or limited species. Furthermore, a small fraction of experiments incorporated images featuring intricate backgrounds, with only a subset successfully detecting multiple diseases on a single leaf. Extensive research indicates a lack of papers addressing these specific challenges. The literature shows that limited labeled data, domain adaptation issues, environmental variations, and model interpretability are some of the key challenges that researchers need to address. Moreover, deploying DL models in resource-constrained environments, such as rural agricultural settings, remains a challenge due to computational and infrastructure constraints.

13.5 CONCLUSION AND FUTURE SCOPE

It has been observed from the literature that ML along with DL has the ability to detect diseases and can even provide solutions to problems. However, the use of ML as a problem-solver in agriculture is limited, as very few existing researchers have focused on this area. From the literature, it can be observed that Indian researchers have also seen an opportunity to integrate computer vision with agriculture for solving major problems in the agriculture sector, which earlier not so common as in comparison to the other countries. Future studies on DL methods for plant disease identification should concentrate on resolving the aforementioned issues and constraints. Developing methods for effective data augmentation, domain adaptation, and model explainability will enhance the usability and reliability of DL models in real-world agricultural applications. Additionally, integrating remote sensing data, such as multispectral and hyperspectral imagery, with DL approaches could further improve disease detection accuracy and scalability.

REFERENCES

1. Food and Agriculture Organization of the United Nations. *The future of food and agriculture: Trends and challenges*. Fao, 2017.
2. Mustard Production Manual, Government of Saskatchewan, Saskatchewan, 2017.
3. Kumar, Sonit, Dalel Singh, Satguru Prasad Yadav, and Rajendra Prasad. "Studies on powdery mildew of rapeseed-mustard (Brassica juncea L.) caused by Erysiphe cruciferarum and its management." *Journal of Pure and Applied Microbiology* 9, no. 2 (2015): 1481–1486.
4. Iqbal, Zahid, Muhammad Attique Khan, Muhammad Sharif, Jamal Hussain Shah, Muhammad Habib ur Rehman, and Kashif Javed. "An automated detection and classification of citrus plant diseases using image processing techniques: A review." *Computers and Electronics in Agriculture* 153 (2018): 12–32.
5. Kumar, Vinod, Sushma Lehri, Ashok Kumar Sharma, P. D. Meena, and Arvind Kumar. "Image based rapeseed-mustard disease expert system: An effective extension tool." *Indian Research Journal of Extension Education* 8, no. 3 (2016): 10–13.
6. "The state of food and agriculture," Food and Agriculture Organization of the United Nations, USA, 2009.
7. Balasubramaniam, P., and V. P. Ananthi. "Segmentation of nutrient deficiency in incomplete crop images using intuitionistic fuzzy C-means clustering algorithm." *Nonlinear Dynamics* 83 (2016): 849–866.
8. Owomugisha, Godliver, and Ernest Mwebaze. "Machine learning for plant disease incidence and severity measurements from leaf images." In *2016 15th IEEE international conference on machine learning and applications (ICMLA)*, pp. 158–163. IEEE, 2016.
9. Kaur, Sukhvir, Shreelekha Pandey, and Shivani Goel. "Plants disease identification and classification through leaf images: A survey." *Archives of Computational Methods in Engineering* 26 (2019): 507–530.
10. Gui, Jiangsheng, Li Hao, Qing Zhang, and Xiaoan Bao. "A new method for soybean leaf disease detection based on modified salient regions." *International Journal of Multimedia and Ubiquitous Engineering* 10, no. 6 (2015): 45–52.
11. Phadikar, Santanu, Jaya Sil, and Asit Kumar Das. "Rice diseases classification using feature selection and rule generation techniques." *Computers and Electronics in Agriculture* 90 (2013): 76–85.
12. Johannes, Alexander, Artzai Picon, Aitor Alvarez-Gila, Jone Echazarra, Sergio Rodriguez-Vaamonde, Ana Díez Navajas, and Amaia Ortiz-Barredo. "Automatic plant disease diagnosis using mobile capture devices, applied on a wheat use case." *Computers and Electronics in Agriculture* 138 (2017): 200–209.
13. Luo, Jing, Shuze Geng, Chunbo Xiu, Dan Song, and Tingting Dong. "A curvelet-SC recognition method for maize disease." *Journal of Electrical and Computer Engineering* 2015, no. 1 (2015): 164547.
14. Aasha Nandhini, S., Radha Hemalatha, S. Radha, and K. Indumathi. "Web enabled plant disease detection system for agricultural applications using WMSN." *Wireless Personal Communications* 102 (2018): 725–740.
15. Hassanien, Aboul Ella, Tarek Gaber, Usama Mokhtar, and Hesham Hefny. "An improved moth flame optimization algorithm based on rough sets for tomato diseases detection." *Computers and Electronics in Agriculture* 136 (2017): 86–96.
16. Aji, Alham F., Qorib Munajat, Ardhi P. Pratama, Hafizh Kalamullah, Jodi Setiyawan, and Aniati M. Arymurthy. "Detection of palm oil leaf disease with image processing and neural network classification on mobile device." *International Journal of Computer Theory and Engineering* 5, no. 3 (2013): 528.

17. Majumdar, Diptesh, Dipak Kumar Kole, Aruna Chakraborty, and Dwijesh Dutta Majumder. "An integrated digital image analysis system for detection, recognition and diagnosis of disease in wheat leaves." In *Proceedings of the Third International Symposium on Women in Computing and Informatics*, pp. 400–405. 2015.
18. Islam, Monzurul, Anh Dinh, Khan Wahid, and Pankaj Bhowmik. "Detection of potato diseases using image segmentation and multiclass support vector machine." In *2017 IEEE 30th Canadian Conference on Electrical and Computer Engineering (CCECE)*, pp. 1–4. IEEE, 2017.
19. Sharma, Anita, Kamlesh Lakhwani, and Harmeet Singh Janeja. "Plant disease identification using deep learning: a systematic review." In *2021 2nd International Conference on Intelligent Engineering and Management (ICIEM)*, pp. 222–227. IEEE, 2021.
20. Samanta, Debabrata, and Arya Ghosh. "Histogram approach for detection of maize leaf damage." *International Journal of Computer Science and Telecommunications* 3, no. 2 (2012): 26–28.
21. Zhang, S. W., Y. J. Shang, and L. Wang. "Plant disease recognition based on plant leaf image." *The Journal of Animal & Plant Sciences* 25, No. 3 Suppl. 1 (2015): 42–45.
22. Sanyal, Pritimoy, Ujjwal Bhattacharya, Swapan K. Parui, Samir Kumar Bandyopadhyay, and S. Patel. "Color texture analysis of rice leaves diagnosing deficiency in the balance of mineral levels towards improvement of crop productivity." In *10th International Conference on Information Technology (ICIT 2007)*, pp. 85–90. IEEE, 2007.
23. V. K. Shah, H. B. Prajapati, and J. P. Shah, "A Survey on Detection and Classification of Rice Plant Diseases," in *2016 IEEE International Conference on Current Trends in Advanced Computing (ICCTAC)*, pp. 1–8. IEEE, 2016.
24. Rangel, Baldomero Manuel Sanchez, Marco Antonio Aceves Fernández, Jesus Cardoso Murillo, Jesús Carlos Pedraza Ortega, and Juan Manuel Ramos Arreguín. "KNN-based image segmentation for grapevine potassium deficiency diagnosis." In *2016 International Conference on Electronics, Communications and Computers (CONIELECOMP)*, pp. 48–53. IEEE, 2016.
25. Sharma, Pushkara, Pankaj Hans, and Subhash Chand Gupta. "Classification of plant leaf diseases using machine learning and image preprocessing techniques." In *2020 10th International Conference on Cloud Computing, Data Science & Engineering (Confluence)*, pp. 480–484. IEEE, 2020.
26. Bandi, Sudheer Reddy, A. Varadharajan, and A. Chinnasamy. "Performance evaluation of various statistical classifiers in detecting the diseased citrus leaves." *International Journal of Engineering Science and Technology* 5, no. 2 (2013): 298–307.
27. Nandhini, M., S. V. Pream, and M. S. Vijaya. "Identification and classification of leaf diseases in turmeric plants." *Journal of Engineering Research and Applications* 6, no. 2 (2016): 48–54.
28. Kaur, Sukhvir, Shreelekha Pandey, and Shivani Goel. "Semi-automatic leaf disease detection and classification system for soybean culture," *IET Image Processing*, 12, no. 6 (2018): 1038–1048.
29. Sarkar, Chittabarni, Deepak Gupta, Umesh Gupta, and Barenya Bikash Hazarika. "Leaf disease detection using machine learning and deep learning: Review and challenges." *Applied Soft Computing* 145 (2023): 110534.
30. Meena, P. D., R. P. Awasthi, C. Chattopadhyay, S. J. Kolte, and Arvind Kumar. "Alternaria blight: a chronic disease in rapeseed-mustard." *Journal of Oilseed Brassica* 1, no. 1 (2016): 1–11.
31. Yun, Shi, Wang Xianfeng, Zhang Shanwen, and Zhang Chuanlei. "PNN based crop disease recognition with leaf image features and meteorological data." *International Journal of Agricultural and Biological Engineering* 8, no. 4 (2015): 60–68.

32. Biswas, Sandika, Bhushan Jagyasi, Bir Pal Singh, and Mehi Lal. "Severity identification of potato late blight disease from crop images captured under uncontrolled environment." In *2014 IEEE Canada International Humanitarian Technology Conference (IHTC)*, pp. 1–5. IEEE, 2014.
33. Urva, Prajna. "Detection and classification of grain crops and legumes disease: a survey." *Sparklinglight Transactions on Artificial Intelligence and Quantum Computing (STAIQC)* 1, no. 1 (2021): 41–55.
34. Resti, Yulia, Chandra Irsan, Mega Tiara Putri, Irsyadi Yani, Ansyori Ansyori, and Bambang Suprihatin. "Identification of corn plant diseases and pests based on digital images using multinomial naïve bayes and k-nearest neighbor." *Science and Technology Indonesia* 7, no. 1 (2022): 29–35.
35. Dhingra, Gittaly, Vinay Kumar, and Hem Dutt Joshi. "Study of digital image processing techniques for leaf disease detection and classification." *Multimedia Tools and Applications* 77 (2018): 19951–20000.
36. Asefpour Vakilian, Keyvan, and Jafar Massah. "An artificial neural network approach to identify fungal diseases of cucumber (Cucumis sativus L.) plants using digital image processing." *Archives of Phytopathology and Plant Protection* 46, no. 13 (2013): 1580–1588.
37. Sabrol, H., and S. Kumar. "Intensity based feature extraction for tomato plant disease recognition by classification using decision tree." *International Journal of Computer Science and Information Security* 14, no. 9 (2016): 622.
38. Sabrol, Hiteshwari, and Satish Kumar. "Fuzzy and neural network based tomato plant disease classification using natural outdoor images." *Indian Journal of Science and Technology* (2016).
39. Sannakki, Sanjeev S., Vijay S. Rajpurohit, V. B. Nargund, and Pallavi Kulkarni. "Diagnosis and classification of grape leaf diseases using neural networks." In *2013 Fourth International Conference on Computing, Communications and Networking Technologies (ICCCNT)*, pp. 1–5. IEEE, 2013.
40. Sullca, Cecilia, Carlos Molina, Carlos Rodríguez, and Thais Fernández. "Diseases detection in blueberry leaves using computer vision and machine learning techniques." *International Journal of Machine Learning and Computing* 9, no. 5 (2019): 656–661
41. Wang, Guan, Yu Sun, and Jianxin Wang. "Automatic image-based plant disease severity estimation using deep learning." *Computational Intelligence and Neuroscience* 2017, no. 1 (2017): 2917536
42. Mohanty, Sharada P., David P. Hughes, and Marcel Salathé. "Using deep learning for image-based plant disease detection." *Frontiers in Plant Science* 7 (2016): 215232.
43. Wang, Guan, Yu Sun, and Jianxin Wang. "Automatic image-based plant disease severity estimation using deep learning." *Computational Intelligence and Neuroscience* 2017, no. 1 (2017): 2917536.
44. Brahimi, Mohammed, Marko Arsenovic, Sohaib Laraba, Srdjan Sladojevic, Kamel Boukhalfa, and Abdelouhab Moussaoui. "Deep learning for plant diseases: detection and saliency map visualisation." *Human and Machine Learning: Visible, Explainable, Trustworthy and Transparent* (2018): 93–117.
45. Amara, Jihen, Bassem Bouaziz, and Alsayed Algergawy. "A deep learning-based approach for banana leaf diseases classification." In *Datenbanksysteme für Business, Technologie und Web (BTW 2017)-Workshopband*, pp. 79–88. Gesellschaft für Informatik eV, 2017.
46. Barbedo, Jayme Garcia Arnal. "Plant disease identification from individual lesions and spots using deep learning." *Biosystems Engineering* 180 (2019): 96–107.
47. Mohanty, Sharada P., David P. Hughes, and Marcel Salathé. "Using deep learning for image-based plant disease detection." *Frontiers in Plant Science* 7 (2016): 215232.

14 Digital Camouflage Generation by AI-Based Methods

A Survey of Recent Techniques

Vaishali Ganganwar, Parshant Rajput, Bhavana Nimmagadda, Swapnil Sinha, and Aashna Vaid

14.1 INTRODUCTION

The purpose of camouflage has been to conceal oneself, weapons, and other valuable installations from the enemy's sight. Camouflage has been used by humankind since time immemorial. Camouflage found its application in nature even before it was adopted by humans. The most dominant application of camouflage to humankind, however, has been in the military. The ability to remain hidden from the enemy is extremely valuable. Two British zoologists first proposed this tactic, in addition to animal or bird mimicry and means of hiding. Clothing with camouflage patterns is used in hunting as well to not stand out in the surroundings and not alert the prey.

Similarly, tigers and lions use camouflage in the wild. Camouflage is the cornerstone of the art of military deception. These patterns are used to conceal soldiers in terrain, preventing military bases from detection. Armed vehicles, tanks, fighter jets, and all such equipment are camouflaged to reduce their chances of detection with either blinds eye or cameras. The challenge now has been the creation of optimal camouflage patterns to keep pace with the detection technologies that now exist. The camouflage generation process is still widely manual and by no means ideal. And as such left much to be researched and developed, with the requirements only increasing.

It is time-consuming and is based on human hunches and judgment. However, these fail to be very useful with slight changes in scenery and weather. The other way to generate camouflage patterns is to use algorithms that have been created for this purpose [1–3]. The common technique here is to take an image of the terrain, analyze it, then extract colors and use them to generate patterns that would be appropriate for the terrain. There are several nuances here too for there are several

 DOI: 10.1201/9781032724508-16

different algorithms that can be used and several generation methods that can be developed. Considering the novelty of the use of technology in this domain, there is a need for assessment, innovation, and greater research. This chapter focuses on the study of different camouflage pattern generation techniques and the algorithms used to generate patterns.

Our aim is to identify different techniques across different areas of technology that have been in use in the past. It aims to give the reader a comprehensive understanding of each technique, methodology used, and their challenges if any.

This chapter is organized as follows: Section 14.2 describes different approaches to Camouflage Pattern Generation Algorithm, which further includes traditional approaches and digital approaches. Section 14.3 talks about evaluation metrics used for Generated Camouflage Pattern. Section 14.4 gives us various challenges in Camouflage Generation. Finally, Section 14.5 includes the conclusion and future scope in the field of Camouflage Pattern Generation.

14.2 DIFFERENT APPROACHES FOR CAMOUFLAGE PATTERN GENERATION

Nature used camouflage long before humans ventured into this field. This section on Camouflage Pattern Generation Approaches talks about early adaptions of camouflage generation in the sub-section of the Traditional Approach and some recent developments in this field in the sub-section of the digital/non-traditional approach.

14.2.1 Traditional Approach

S. J. Kim et al. [4] state that while digital and computer-generated camouflage patterns have become more prevalent in recent years, traditional hand-drawn methods still have a place in the design process due to their unique aesthetic qualities. The study involved creating several hand-drawn camouflage patterns by the authors, which were then analyzed for their visual characteristics and design elements. Table 14.1 summarizes different traditional approaches used for camouflage pattern generation.

The authors used both quantitative and qualitative methods to evaluate the patterns, including measurements of color distribution and spatial frequency, as well as surveys of perceived effectiveness and aesthetic appeal. While M. H. Lee et al. [5] begin by discussing the importance of effective camouflage in various military and civilian applications and note that traditional methods of generating camouflage patterns, such as hand-painting, can be time-consuming and difficult to reproduce. They propose a system based on stencils and spraying as a more efficient and precise method of generating patterns. It was noted that traditional printing methods have limitations in terms of color accuracy and resolution, which can result in poor-quality camouflage patterns.

Therefore, J. Kim et al. [6] proposes a new printing technology that can generate high-quality patterns using a combination of dithering and halftoning techniques for

TABLE 14.1
List of Traditional Approaches for Camouflage Generation

Study	Research Outcomes	Methodology Used	Applications
A Study on the Design Characteristics of Camouflage Pattern Using Hand-Drawn Method (2017)	Identified key design characteristics of camouflage patterns created with hand-drawn methods; enhanced pattern effectiveness	Hand-drawn design, visual analysis	Fashion, military, textiles
Development of the Camouflage Pattern Generation System Using Stencils and Spraying (2015)	Developed a new system for generating camouflage patterns using stencils and spraying; improved pattern precision	Stencil and spraying techniques	Military, automotive, outdoor equipment
A Study on the Camouflage Printing Technology for Fabric (2017)	Improved fabric camouflage printing technology; enhanced durability and pattern quality	Advanced printing techniques, fabric testing	Military, fashion, home textiles

which the proposed printing technology uses a set of predefined color palettes and a dither matrix to achieve accurate color representation. The dither matrix is generated by considering the characteristics of the human visual system, which helps to ensure that the printed pattern is perceived as having smooth color transitions.

14.2.2 Digital Approaches

Various algorithms and methodologies have come up in recent years in the field of Camouflage Generation. Some of the major ones that are discussed here are generative adversarial networks (GAN), adversarial autoencoder networks (ANNs), Fuzzy c-means, and Biased Random Walk algorithm. Table 14.2 describes the different approaches used for digital camouflage generation.

14.2.2.1 Generative Adversarial Networks

Talas et al. [7] use a GAN-based approach to evolve and optimize camouflage patterns that are effective at concealing a target in each environment. The CamoGAN method involves training a GAN on a dataset of natural images and then fine-tuning it on a small set of target images and their corresponding backgrounds. The GAN generates candidate camouflage patterns, which are then evaluated using an objective function that measures their effectiveness at concealing the target in the given environment.

TABLE 14.2
Digital Techniques of Camouflage Pattern Generation

Study	Research Outcomes	Methodology Used	Applications
Flexible Physical Camouflage Generation Based on a Differential Approach (2023) [20]	Developed an adversarial texture generation method using a differential mesh renderer and diffusion models.	Differentiable mesh rendering, diffusion-based models	Military, wildlife photography
Design and Evaluation of Digital Camouflage Patterns by Spot Combination (2023) [21]	Created more effective digital camouflage patterns by combining spot designs.	Spot combination algorithm	Military, tactical gear
Background Self-adapted Digital Camouflage Pattern Generation (2023) [22]	Developed an adaptive algorithm to match camouflage to various backgrounds.	K-means clustering, color feature extraction	Military, environment-specific concealment
CamoVeil: Digital Camouflage Pattern Generation System [23]	The method provides flexibility, productivity, and the opportunity for various pattern options, improving the efficiency of camouflage in a range of settings.	segmentation, identifying objects, extracting dominant colors, and creating patterns with the K-means++ algorithm.	Military, security, anti-surveillance
Camouflage Design, Assessment and Breaking Techniques (2023) [24]	Reviewed and classified existing camouflage design and assessment methods, proposed new evaluation metrics.	Literature review, comparative analysis	Military, security, anti-surveillance

The authors use an evolutionary algorithm to optimize the GAN's generator to produce more effective camouflage patterns over time. The Camouflage GAN [8] is trained on pairs of images, one of which is a "cover image" and the other is a "hidden image." The GAN is trained to generate an output image that looks like the cover image but also contains the hidden image. This is achieved by training the GAN with an objective function that minimizes the perceptual difference between the output image and the cover image while also maximizing the similarity between the output image and the hidden image. The authors evaluate the Camouflage GAN on a data set of cover and hidden images and show that it can effectively hide the hidden images within the cover images without significantly affecting their visual quality.

A very interesting application was observed in the paper [9] that proposes a method for generating dual adversarial camouflage patterns that can fool both human observers and object detection systems. The proposed method involves two stages. In the first stage, a differential evolution (DE) algorithm is used to optimize a set of parameters for the camouflage pattern. The objective function used by the DE algorithm includes both a measure of the camouflage's effectiveness in hiding a target object and a measure of its adversarial robustness to object detection algorithms.

In the second stage, a dual adversarial attack is performed using the optimized camouflage pattern. The attack is "dual" because it aims to fool both human observers and object detection systems. The camouflage pattern is designed to blend in with the background and hide the target object from human observers while also containing adversarial perturbations that can fool object detection systems. Another application observed was where the GAN is used to extract the deformation camouflage spot from the input image and reproduce it in a new image without the target object.

Considering how present-day reconnaissance deals with computer programs, it is a very specific use case that, as per us, has massive applicability. Stronger on the generation side, the method [10] for extracting and reproducing deformation camouflage spots GAN model displays one way of generation. The proposed method involves two stages. In the first stage, a GAN model is trained on a data set of deformation camouflage spots. The GAN is used to learn the distribution of the deformation camouflage spots and generate new spots that are visually like the training set.

In the second stage, the trained GAN model is used to extract and reproduce deformation camouflage spots. The input to the GAN is an image containing a target object and a deformation camouflage spot. The GAN is used to extract the deformation camouflage spot from the input image and reproduce it in a new image without the target object.

14.2.2.2 Adversarial Autoencoder Networks

Yang et al. [11] introduce a method for generating digital camouflage patterns using an adversarial autoencoder (AAN) network. The AAN network is trained on a dataset of existing camouflage patterns and then used to generate new patterns that are visually like the training set but not identical. The proposed method involves two stages. In the first stage, the AAN network is trained using a combination of a reconstruction loss and an adversarial loss. The reconstruction loss ensures that the generated patterns are like the training set while the adversarial loss encourages the generated patterns to be diverse and different from the training set.

In the second stage, the trained AAN network is used to generate new camouflage patterns. This is done by sampling random noise vectors and passing them through the AAN network to generate a new pattern. The generated pattern is then evaluated using a metric called the structural similarity index (SSIM) [12] to ensure that it is visually like the training set.

14.2.2.3 Fuzzy C-Means

Zhao et al. [13] presents a similar approach to [14] to automatic camouflage pattern generation using fuzzy cmeans (FCM) clustering. The algorithm starts by selecting an image of the environment in which the camouflage pattern will be used. The FCM algorithm is then applied to cluster the pixels in the image into different color groups. The centroids of these color groups are used as the initial color values for the camouflage pattern.

Next, the algorithm generates a random pattern consisting of small patches of different colors. The colors of these patches are then modified using the FCM algorithm to adapt to the colors of the surrounding environment. This process is repeated multiple times, with each iteration resulting in a new and improved version of the camouflage pattern. To evaluate the effectiveness of the proposed algorithm, the experiments were conducted on a dataset of natural images. The results showed that the generated patterns were able to blend into the environment more effectively than other state-of-the-art techniques.

One limitation of the proposed algorithm is that it requires a pre-selected image of the environment in which the camouflage pattern will be used. This limits the applicability of the algorithm in scenarios where the environment is constantly changing or where no prior information about the environment is available. Overall, the paper presents a novel approach to automatic camouflage pattern generation using FCM clustering.

The proposed algorithm has the potential to improve the efficiency and effectiveness of camouflage pattern design and could have applications in a variety of fields, including military, fashion, and art.

14.2.2.4 Biased Random Walk Algorithm

Zou et al. [15] introduces a new approach to automatic camouflage pattern generation that is based on a biased random walk algorithm. The algorithm is designed to generate camouflage patterns that can blend into different environments effectively. The process starts by placing seed points on a canvas that serves as starting points for the algorithm.

The biased random walk algorithm generates interconnected lines based on random variables that are biased toward the surrounding colors. The lines are then colored using a color mapping process that assigns colors based on the surrounding environment. The effectiveness of the algorithm was evaluated on a data set of natural images, which showed that the generated patterns were more effective than other techniques.

The chapter discusses the algorithm's potential applications in the military, fashion, and art but notes that it can be computationally expensive and has limited flexibility in terms of color values. Overall, the proposed algorithm provides a novel and promising approach to automatic camouflage pattern generation.

14.3 EVALUATION METRICS

In the process of camouflage pattern generation, it is important to evaluate the quality of the respective patterns to grade a generation method. Evaluation metrics for camouflage patterns depend on the specific use case and goals of the pattern. If the pattern is to be used on displaceable objects like clothes or vehicles, one will want to test the area of effectiveness of one pattern. While, if it is a static object, that will not be required to be tested.

- Visual Contrast: Visual contrast is the difference in brightness or color between the object and its background. Lower visual contrast can help a pattern blend in with the environment and make it less visible to the observer.
- Visual effectiveness: This is the degree to which the pattern reduces the detectability of the wearer or object in a specific environment. This can be measured by conducting visual surveys or experiments with human observers.
- Spatial frequency analysis: Camouflage patterns should have similar spatial frequencies to the background environment. Spatial frequency analysis can help evaluate whether the pattern has the correct amount of contrast and detail to blend into the background.
- Spectral characteristics: Camouflage patterns should blend with the spectral characteristics of the environment in which they are used. Spectral analysis can be used to evaluate whether the pattern matches the colors and textures of the environment.
- Performance in different environments: Camouflage patterns should be effective in a variety of environments, including different lighting conditions, seasons, and terrains. Testing the pattern in different environments can help evaluate its performance.
- Camouflage pattern recognition: The pattern should not be recognizable as a pattern or object by observers. Recognition tests can be used to determine whether the pattern is easily recognizable.
- Disruptive Coloration: Disruptive coloration is a pattern that uses high-contrast patches to break up the outline of an object and make it harder to detect. This can be evaluated using recognition tests or image analysis.
- Edge sharpness: The sharpness of the edges of the pattern can affect how well it blends into the background. A pattern with soft edges can be more effective in blending in with natural environments.
- Wear Resistance: The durability and wear resistance of a camouflage pattern are also important factors to consider. A pattern that is resistant to wear and tear will remain effective over its lifetime.

14.4 CHALLENGES

This chapter assimilates the challenges that were observed in the above surveyed articles.

- One of the major challenges for the algorithms is to finalize the shape and size of the camouflage pattern, as it strictly depends upon the terrain for which the pattern needs to be generated.
- The immense number of potential combinations of colors and textures can generate lot of different results that can sometimes make it strenuous for us to choose a suitable pattern and in turn makes testing difficult.
- The pattern needs to work well with color. If the terrain image is not preprocessed properly, the clustered colors could be off from reality to make the camouflage ultimately ineffective.
- The pattern needs to work well with depth perception. So as to look realistic when developed for larger areas. In such use cases, shadow and texture need to be added. These details are tricky to inculcate.
- The pattern needs to work for both sophisticated sensors and human viewers. To provide complete protection, a pattern needs to hide objects from the most common forms of detection, if not all forms. A human enemy as well as object detection needs to be deceived by the resultant camouflage pattern.
- Testing of the generated pattern is one of the biggest challenges, as camouflage detection is still in its early phase. Some solutions are detection system, such as the Faster-RCNN Inception V2 and Faster-R CNN ResNet101 recognition systems [16, 17] and validation by using humans as visual predators. Despite the isolation of evaluation metrics in this field, testing is challenging since all the metrics do not have a corresponding system in code form. Such systems should be complete and trustworthy enough to be considered as a valid criterion. For that, deep knowledge of computer vision is required.
- With further improvements in technology, AI will play a major role in shaping the functioning of many industries, including the defense sector. Data, as we know, is ever-growing and artificially intelligent systems will be key to working with big data [18].
- The quality of the input image is also an important factor, a high-resolution image will produce a better pattern than a low-resolution input image. For example, a vector output instead of an image could be very useful by virtue of its scalability.
- After the generation of the required patterns, it would be essential for us to understand the requirements as and when they occur and use smart and effective manufacturing techniques to fulfill the same [19].

14.5 FUTURE SCOPE

The field of digital camouflage pattern generation has seen significant advancements through various computational techniques such as GANs, AANs, fuzzy clustering, and biased random walk algorithms. Despite these advancements, several areas for future research and development are yet to be explored, presenting numerous opportunities for innovation and improvement in the field of camouflage pattern generation.

Current methodologies often rely on static environments or pre-defined scenarios, limiting their effectiveness in dynamic or rapidly changing conditions. One promising direction for future research could focus on developing adaptive camouflage systems capable of dynamically adjusting patterns in response to changing environmental factors such as lighting conditions, terrain variations, and seasonal changes. Integrating real-time data processing capabilities and sensor feedback could enable camouflage patterns to evolve continuously, ensuring optimal concealment in diverse operational environments.

Another critical area for future exploration lies in the integration of artificial intelligence (AI) and machine learning techniques with camouflage design. Deep learning models have shown promise in generating realistic and effective camouflage patterns by learning from large datasets of natural environments. Further advancements in AI algorithms could lead to the development of more sophisticated pattern generation models capable of simulating complex natural textures and patterns with enhanced realism and effectiveness.

Furthermore, there is a growing need to standardize evaluation metrics for camouflage patterns. Current evaluation criteria primarily focus on visual effectiveness, spectral characteristics, and adaptability across different environments. Future research could aim to develop standardized metrics that encompass a broader range of factors, including human visual perception, object detection system resilience, and durability under varying environmental conditions. Such standardized metrics would facilitate objective comparisons between different camouflage generation techniques and enable more reliable assessment of pattern effectiveness.

By addressing current limitations and exploring new research directions, we can contribute to the development of next-generation camouflage technologies that offer enhanced concealment capabilities across diverse operational contexts.

14.6 CONCLUSION

In this chapter, we discuss the literature around various camouflage pattern generation techniques. The older techniques use traditional pattern generation methods such as hand-drawn methods, Stencils or Sprays and Camouflage Printing Technology on fabrics. Then we focused on the digital ways of generating a pattern.

In that, specifically methods that were AI aided like image processing and machine learning techniques. We had a closer look at deep learning methods that are being explored to solve this problem. Out of all the methodologies that we have discussed, we find that GAN are well-suited to this task because they are able to learn the underlying patterns and textures of the natural environment and incorporate them into the generated camouflage. Additionally, GANs can learn to generate diverse and realistic patterns, making them useful for creating a range of camouflage designs. By training on large datasets of natural scenes, GANs can generate camouflage patterns that blend seamlessly into the surrounding environment, making them difficult to detect. Due to the flexibility of GAN, features SUCH AS color, texture and style can be isolated, which can later be reused or fine-tuned as per requirement.

With the increase in usage of technology in various domains, the area of Camouflage Pattern Generation is relatively new. For the same reason, we have highlighted some important elements of any pattern generation process through evaluation metrics and challenges in this field.

REFERENCES

1. Albert, Francisco, José María Gomis, Margarita Valor, and José Miguel Valiente. "Methodology for graphic redesign applied to textile and tile pattern design." In *Innovations in Applied Artificial Intelligence: 17th International Conference on Industrial and Engineering Applications of Artificial Intelligence and Expert Systems, IEA/AIE 2004*, Ottawa, Canada, May 17–20, 2004. Proceedings 17, pp. 876–885. Springer Berlin Heidelberg, 2004.
2. Yang, Hengfu, and Jianping Yin. "A digital camouflage generation algorithm using color similarity." *International Journal of Multimedia and Ubiquitous Engineering* 10, no. 6 (2015): 159–164.
3. Ostromoukhov, Victor. "Mathematical tools for computer-generated ornamental patterns." In International Conference on Raster Imaging and Digital Typography, pp. 193–223. Berlin, Heidelberg: Springer Berlin Heidelberg, 1998.
4. Kim, S. J., J. H. Kim, and J. H. Kim. "A Study on the Design Characteristics of Camouflage Pattern Using Hand-Drawn Method." *Journal of the Korea Society of Computer and Information* 22, no. 12 (2017): 97–104.
5. Kim, Y. H., M. H. Kim, and M. H. Lee. "Development of the Camouflage Pattern Generation System Using Stencils and Spraying." *Journal of the Korea Society of Computer and Information* 20, no. 9 (2015): 109–116.
6. Kim, J., Y. S. Kim, and K. H. Kwon. "A Study on the Camouflage Printing Technology for Fabric." *Journal of the Korea Society of Computer and Information* 22, no. 3 (2017): 159–166.
7. Talas, Laszlo, John Fennell, Karin Kjernsmo, Innes Cuthill, Nick Scott-Samuel, and Roland Baddeley. "CamoGAN: evolving optimum camouflage with Generative Adversarial Networks." *Methods in Ecology and Evolution* 11, no. 2 (2020): 240–247.
8. Liu, Xiyao, Ziping Ma, Xingbei Guo, Jialu Hou, Gerald Schaefer, Lei Wang, Victoria Wang, and Hui Fang. "Camouflage generative adversarial network: Coverless full-image-to-image hiding." In *2020 IEEE International Conference on Systems, Man, and Cybernetics (SMC)*, pp. 166–172. IEEE, 2020.
9. Sun, Jialiang, Wen Yao, Tingsong Jiang, Donghua Wang, and Xiaoqian Chen. "Differential evolution based dual adversarial camouflage: Fooling human eyes and object detectors." *Neural Networks* 163 (2023): 256–271.
10. Yang, Xin, Wei-dong Xu, Qi Jia, Ling Li, Wan-nian Zhu, Ji-yao Tian, and Hao Xu. "Research on extraction and reproduction of deformation camouflage spot based on generative adversarial network model." *Defence Technology* 16, no. 3 (2020): 555–563.

11. Yang, Xin, Wei-Dong Xu, Qi Jia, and Ling Li. "Research on digital camouflage pattern generation algorithm based on adversarial autoencoder network." *International Journal of Pattern Recognition and Artificial Intelligence* 34, no. 06 (2020): 2050017.
12. Wang, Zhou, Alan C. Bovik, Hamid R. Sheikh, and Eero P. Simoncelli. "Image quality assessment: from error visibility to structural similarity." *IEEE Transactions on Image Processing* 13, no. 4 (2004): 600–612.
13. Xiao, Weijie, Yan Zhao, Xiaohui Gao, Congwei Liao, Shengxiang Huang, and Lianwen Deng. "Implementation of fuzzy C-means (FCM) clustering based camouflage image generation algorithm." *IEEE Access* 9 (2021): 120203–120209.
14. Bian, Peng, Yi Jin, and Nai-ren Zhang. "Fuzzy c-means clustering based digital camouflage pattern design and its evaluation." In *IEEE 10th International Conference on Signal Processing Proceedings*, pp. 1017–1020. IEEE, 2010.
15. Gan, Yuanying, Chuntong Liu, Zhenxin He, Hongcai Li, and Zhongye Liu. "Digital Camouflage Pattern Design Based on the Biased Random Walk." *Modelling and Simulation in Engineering* 2022, no. 1 (2022): 2986346.
16. Ren, Shaoqing, Kaiming He, Ross Girshick, and Jian Sun. "Faster R-CNN: Towards real-time object detection with region proposal networks." *IEEE Transactions on Pattern Analysis and Machine Intelligence* 39, no. 6 (2016): 1137–1149.
17. Li, Zeming, Chao Peng, Gang Yu, Xiangyu Zhang, Yangdong Deng, and Jian Sun. "Light-head r-cnn: In defense of two-stage object detector." *arXiv preprint arXiv:1711.07264* (2017).
18. Rajendra, P., Mina Kumari, Sangeeta Rani, Namrata Dogra, Rahul Boadh, Ajay Kumar, and Mamta Dahiya. "Impact of artificial intelligence on civilization: Future perspectives." *Materials Today: Proceedings* 56 (2022): 252–256.
19. Rani, Sangeeta, Khushboo Tripathi, Yojna Arora, and Ajay Kumar. "Analysis of Anomaly detection of Malware using KNN." In *2022 2nd International Conference on Innovative Practices in Technology and Management (ICIPTM)*, Vol. 2, pp. 774–779. IEEE, 2022.
20. Li, Yang, Wenyi Tan, Chenxing Zhao, Shuangju Zhou, Xinkai Liang, and Quan Pan. "Flexible Physical Camouflage Generation Based on a Differential Approach." *arXiv preprint arXiv:2402.13575* (2024).
21. Jia, Qi, Wei Dong Xu, Jiang Hua Hu, Jun Liu, Xin Yang, and Li Yan Zhu. "Design and evaluation of digital camouflage pattern by spot combination." *Multimedia Tools and Applications* 79, no. 29 (2020): 22047–22064.
22. Lei, Qin, Duan Ze-wei, Hu Jiang-hua, Xu Wei-dong, and Zhang Xiang. "Background self-adapted digital camouflage pattern generation algorithm for fixed target." In *Image and Graphics: 9th International Conference, ICIG 2017*, Shanghai, China, September 13-15, 2017, Revised Selected Papers, Part II 9, pp. 545–551. Springer International Publishing, 2017.
23. Ganganwar, Vaishali, Parshant Rajput, N. Bhavana, Swapnil Sinha, and Aashna Vaid. "CamoVeil: Digital Camouflage Pattern Generation System." In *International Conference on Advances in Data-driven Computing and Intelligent Systems*, pp. 143–156. Singapore: Springer Nature Singapore, 2023.
24. Mondal, Ajoy. "Camouflage design, assessment and breaking techniques: a survey." *Multimedia Systems* 28, no. 1 (2022): 141–160.

15 Plant Disease Detection Using Deep Learning Techniques

A Comprehensive Review and Comparative Analysis

Shilpa Juneja, Parneeta Dhaliwal, Atul Srivastava, and Ajay Kumar

15.1 INTRODUCTION

Agriculture is the foundation of human civilization, providing sustenance, economic stability, and cultural identity on a global scale. The intersection of agriculture and technology has become a pivotal focal point as the global population continues to grow, placing increasing demands on our agricultural systems. The agricultural sector has seen a substantial change in recent years toward the use of new technologies to address critical issues such as increasing crop yields and food security for a growing global population. However, these systems face a multitude of challenges, with plant diseases being a significant and persistent threat. Agricultural production is negatively impacted by the occurrence of plant diseases. Diseases caused by pathogens such as viruses, fungi, and bacteria can disrupt the food supply chain, destroy crops, and reduce yields. Plant diseases not only reduce crop yields and compromise food quality but also lead to economic losses for farmers and destabilize local and global food supplies. Due to several plant diseases, farmers are still unable to enhance their income or production. For the production of healthy products, early diagnosis of plant illness is crucial. Traditional ways to plant disease detection frequently include human experts visually checking crops for disease indicators. This approach is time-consuming and prone to human error, which can lead to delayed or incorrect diagnosis. Furthermore, the introduction of novel and complex diseases needs increasingly advanced detection technologies capable of handling a wide range of infections and symptoms. Because of the small features, farmers can struggle to precisely diagnose the signs of plant illness. Additionally, a lot of farmers lack the expertise to diagnose the illness; thus, artificial intelligence (AI) can assist them in doing so. Applications for machine learning and AI have grown significantly in this

DOI: 10.1201/9781032724508-17

modern era [1]. However, the introduction of deep learning and AI approaches has resulted in a paradigm shift in the area of plant disease detection. This has sparked the creation of novel methodologies and models, which have given rise to a new class called deep learning [2].

Because of the automatic extraction of features and learning by self, deep learning has gained considerable interest these days [3, 52–54]. Deep learning technologies use multiple layers to describe the abstraction of data in computer models [4]. Deep learning techniques have the capability to detect diseases in plant leaves, which results in the advancement of new technologies and thus removes constraints on the selection of disease spot features, which was done artificially earlier, and thus increases the objectivity of plant disease feature extraction [5, 55]. These algorithms learn feature hierarchies from raw data, allowing them to automatically detect tiny differences that indicate disease presence. Researchers and practitioners have gone on a path to revolutionize disease detection in plants with the usage of deep learning. convolutional neural networks (CNNs), one of the remarkable deep learning models, has demonstrated amazing ability in automating the identification and classification of plant diseases from images, resulting in an efficient and accurate solution to this essential problem. CNN has three layers: convolution layers, pooling layers, fully connected layer [6]. The stacking of these layers forms the architecture of CNN. CNNs are basically utilized in pattern detection based on images. This is why CNNs are being employed to identify and diagnose leaf diseases using leaf images, whether healthy or diseased. Plants are susceptible to several diseases. Table 15.1 depicts the various diseases which harm various plant leaves.

To overcome the shortcomings of previous publications on disease detection, this study reviews modern and historical studies on disease detection in plants using different techniques, and also our comprehensive review compares their performance taking into account measures such as mAP, accuracy, and F1 score. In this study, we take a thorough look at the landscape of deep learning techniques to identify plant disease. Our major goal is to provide a thorough examination of the state-of-the-art models with their classification accuracy. We look at the intricacies of dataset curation, model architecture choices, and training methodologies that have resulted in achievements in this field. We hope to provide not only an up-to-date overview but also a critical analysis of the performance and application of various deep learning algorithms by compiling research results from multiple studies. We investigate improvements in computer vision, with a particular emphasis on the usage of deep learning algorithms to efficiently and effectively identify diseases in plants. Our review's framework includes a thorough categorization of the various deep learning architectures used to detect plant diseases. We give an in-depth examination of the most well-known CNN models, including AlexNet, VGGNet, ResNet, InceptionNet, and DenseNet, and analyze their advantages in consideration of plant disease classification. Our review seeks to be useful resource for scholars, practitioners interested in the interface of deep learning and agriculture. Dataset generation and curation are critical steps in efficiently training deep learning models. As a result, we examine the existing plant disease datasets such as PlantVillage dataset and self-acquired

TABLE 15.1
Plants with Diseases

Sr No.	Leaf/ Crop	Diseases
1.	Apple	Rust, Mosaic, Alternaria Leaf Spot, Brown Spot Black Star Disease, Grey Spot [7]
2.	Tomato	Leaf Mold, Bacterial Spot, Target Spot, Two Spotted Spider Mite, Mosaic Virus, Early Blight, Yellow Leaf Curl Virus, Late Blight, Septoria Leaf Spot [8]
3.	Potato	Late Blight, Alternaria, Early Blight [1]
4.	Cherry	Powdery Mildew [9]
5.	Cucumber	Anthracnose, Zucchini Yellow Mosaic Virus, Powdery Mildew, Melon Yellow Spot Virus, Downy Mildew, Target Leaf Spot [10]
6.	Grape	Mites [13], Black Rot, Isariopsis Leaf Spot [11], Leaf Blight, Esca Black Measles [12]
7.	Olive	Leaf Scorch [14]
8.	Corn	Common Rust, Leaf Spot, Northern Leaf Blight [15]
9.	Strawberry	Leaf Spot, Leaf Scorch [16], Powdery Mildew, Anthracnose [17]
10.	Citrus	Blackspot, Canker, Scab, Greening, Melanose [18]
11.	Wheat	Black Chaff, Leaf Blotch, Powdery Mildew, Stripe Rust, Leaf Rust, Smut [19]
12.	Rice	Rice Blast, Bacterial Blight, and Sheath Blight [20]
13.	Cotton	Cercospora, Alternaria Alternata, Bacterial Blight, White flies [21]
14.	Tulsi	Shoot Blight, Powdery Mildew, Fusarium Wilt [22]
15.	Peanut	Black Spot, Brow Spot, Anthracnose, Rust, Net Blotch [23]

dataset. We undertake a comparison analysis using common benchmark datasets to give a full evaluation of the various deep learning algorithms.

The next section elaborates related work. Section 15.3 defines metrics such as accuracy, precision, mean average precision used for comparative analysis. In Section 15.4, we have made comparison among different studies, and the same has been shown in a tabular form. Section 15.5 demonstrates the results of different studies which have been done in this area. Section 15.6 illustrates the conclusion and findings of our study.

15.2 RELATED WORK

A branch of machine learning called deep learning contains many artificial neural network layers. Deep learning approach applies non-linear transformations and high-level abstractions in big databases [24]. Accurate image classification is improved greatly by deep neural networks. For the identification of pattern, CNNs: AlexNet, ResNet, VGGNet, DenseNet, GoogLeNet, Inception V4, MobileNet [6] are only a few of the modern network structures used by existing plant diseases categorization networks. A lot of researchers have tried many things to prevent plant diseases because they pose a serious danger to productivity and quality. Kawasaki et al. [25] proposed a process that used CNN to identify diseases in cucumber leaves: Zucchini Yellow Mosaic Virus and Melon Yellow Spot Virus. CNN could automatically acquire the necessary features for classification using only training 800 images of cucumber leaves. An average accuracy of 94.9% was achieved by the system to classify leaves into two classes: diseased and non-diseased classes.

Mohanty et al. [26] used CNN architectures, namely GoogLeNet and AlexNet, which were applied on three versions of datasets (color, gray-scale, and segmented). Results showed that the model recognized 14 different crops and 26 different diseases. When comparing the performance of the AlexNet and GoogLeNet architectures, GoogLeNet regularly outperformed AlexNet and transfer learning was always more effective. The achieved accuracy was 99.35%. Plant disease classification research now has a new direction because of deep learning technologies in current years. But transparency and interpretability are lacking in DL classifiers. Classifying plant diseases requires high accuracy as well knowledge of how to detect them and also what symptoms are present in the plant. Because of this, many researchers have focused on the study of visualization approaches in recent years, including the development of salient and visual heat maps, so that diagnosis of plant diseases could be easily understood [27]. Brahimi et al. [28] used saliency maps to visualize the symptoms of plant diseases. The model produced the best results when colored version of the dataset was used.

By employing the VGG-FCN-VD16 and VGG-FCN-S models, Lu et al. [19] discovered a process for detecting wheat disease. The study used in-field dataset called WDD2017 (Wheat Disease Database 2017) and five-fold cross-validation was implemented on it [19]. The framework remarkably outperformed VGG-CNN-VD16 in recognition tasks even though it is built on a shallow model: VGG-FCN-S and also the best option for DMIL-WDDS (Deep Multiple Instance Learning-Wheat Disease Dataset) to increase recognition accuracy is SoftMax Aggregation.

Cruz et al. [14] developed a system that was vision-based to detect disease called leaf scorch on leaves of olive. This research demonstrated the possibility for rapid, automated, and cost-effective plant disease identification with 98.60% accuracy. Liu et al. [29] suggested a CNN—AlexNet—to identify four apple leaf diseases: Mosaic, Brown Spot, Rust, Alternaria Leaf Spot [29]. A dataset of 13,689 images of infected apple leaves was used in the study. The experimental findings demonstrated that this method achieved 97.62% accuracy. On comparing with standard AlexNet, there was a 10.83% improvement in this model's accuracy when used with pathological images. Ma et al. [10] used a CNN model for the identification of cucumber leaves diseases, i.e., Downy Mildew, Anthracnose, Target leaf spots and Powdery mildew. The detection accuracy achieved was 93.4%.

Zhang et al. [30] proposed a system to identify eight different maize diseases—Gray leaf spot, Rust, Round Spot, Curvularia leaf spot, Southern leaf blight, Dwarf Mosaic, Brown Spot, Northern leaf blight—in which improved GoogLeNet and Cifar 10 models based on deep learning were used. The Cifar 10 model had 98.8% average accuracy, while the improved GoogLeNet model had a 98.9% accuracy in identifying maize leaf diseases. Faster R-CNN two-stage network, SSD one-stage network, and YOLO one-stage network represent the network of detecting plant diseases. The two-stage network must first produce a candidate box before continuing the object detection process. On the other hand, the one-stage network makes use of the network's attributes extracted from it to immediately forecast the class and location of the lesions. Paired with VGG-Net and ResNet, Fuentes et al. [31] utilized Faster R-CNN at first in 2017 to spot tomato diseases and pests directly. The study considered a dataset having 5,000 tomato leaf diseases and nine classes of pests; the mAP value reached 85.98%. A quick method for detecting rice disease was published by Zhou et al. [20] and was based on the combination of Faster R-CNN and FCM-KM. The results revealed that the Bacterial blight, Rice blast, and Sheath blight detection accuracies are 97.53%, 96.71%, and 98.26%, respectively. Ozguven et al. [32] presented an R-CNN structure that is faster so that the identification of beetroot disease called Leaf Spot could be made; 155 pictures were tested and trained. The findings indicated that this method's total correct classification accuracy is 95.48%. Xie et al. [13] suggested a model called Faster DR-IACNN, which is a combination of Inception-V1 module, SE, Inception-V2 module, Faster R-CNN detection technique to detect diseases in grape leaves: Leaf Blight, Mites, Black Rot and Black Measles. Grape Leaf Disease Dataset was considered as a dataset for making this study successful. With a mAP of 81.1%, this study was able to extract more features with more efficiency. The two-stage detection network was being used for increasing detection speed. The single-stage detection network is separated into two varieties: SSD (Single Shot Detector) and YOLO (You Look Only Once). Both types used the entire image as their input and gave their output layer's bounding box's position and category in a direct manner. In order to detect maize leaf blight against a complicated background, Sun et al. [33] introduced an approach which was based on feature fusion with multi-scale to detect northern maize leaf blight. This method included disease detection, feature fusion, Data preprocessing and other processes were included in this method. In comparison to the old SSD model, this new model has a greater mAP (up from 71.80 to 91.83%).

YOLO has many benefits. It is capable of achieving global optimization, significantly enhancing detection speed, and achieving higher accuracy. Using images taken in tea gardens in improvised conditions. Prakruti et al. [34] described a procedure to identify tea leaf diseases and pests. To find pests and illnesses, YOLOv3 was employed; 86% mAP was accomplished while ensuring the availability of the system in real time. Chouhan et al. [35] suggested a CNN which was multi-layered to identify mango leaf disease such as anthracnose which is a fungal disease. The dataset used for this study was captured in real time. CNN was being trained and then tested using training images and testing images, respectively. Results were then validated and compared with other approaches. This work performed better achieving an accuracy of 97.13%. Davinder Singh et al. [36] created the PlantDoc dataset, which contains 2,598 data points from 13 plant species and 17 disease classes for detecting various plant diseases. Annotating images were obtained from the internet. Three models—MobileNet, VGG16, and Faster R-CNN with InceptionResNetV2 [36]—had been used for plant disease detection. Results show that Faster R-CNN with InceptionResNetv2 performed better and achieved mAP of 38.9. Zhang et al. [9] proposed a completely automatic CNN-based detection procedure to recognize cherry leaf disease powdery mildew. In this model GoogLeNet had been used as a CNN architecture. A comparison of this method against back propagation neural network, SVM, transfer learning techniques, K-means clustering had been made achieving 99.6% average accuracy. Agarwal et al. [8] suggested a model called ACNN with three convolution, three max pooling layers, and two fully connected layers to detect nine tomato leaf diseases: Early Blight, Septoria Leaf Spot, Two Spotted Spider Mite, Target Spot, Bacterial Spot, Mosaic, Late Blight, Leaf Mold and Yellow Leaf Curl. For testing purposes, a total of 500 samples were employed, and testing accuracy varies by class, ranging from 76% to 100%. The results revealed that 91.2% average accuracy has been achieved.

Gehlot et al. [37] recommended a system in which 14,529 plant village images and five different CNN models (GoogleNet, DenseNet-121, AlexNet, VGG16, ResNet-101) were utilized to classify ten tomato leaf diseases. Table 15.2 depicts the accuracy in percentage achieved by these five models.

Table 15.2 represents accuracy achieved when different five models are applied. Among all, ResNet-101 gives the greatest accuracy of 99.68% and 97.85%, and the lowest accuracy is achieved by AlexNet. Despite having a significant impact on crop disease image detection, CNNs still have several issues. The max-pooling preserves the maximum feature while discarding other data. The object might be given a different label as a result of the absence of rotational invariance. In order to solve this issue, Sabour et al. [38] suggested the capsule network in which replacement of the CNNs' scalar output with vector output was done. After that, max-pooling was replaced by dynamic routing because the spatial link between two portions of the item is ignored by the pooling layer. In this regard, Dong et al. [23] proposed a procedure to detect peanut leaf diseases: Rust, Anthracnose, Black Spot, Eyebrow Spot, and Pure Spot using two different types of capsule network. The first capsule network was obtained by changing some criterion for the images of peanut-diseased leaves and the second was obtained by adding three convolution layers to the original model. Three main layers of the capsule networks were the Primary Capsule Layer, the Convolutional Layer, and the Class Capsule Layer. Also, capsule networks used ReLu as an activation function. Results showed that capsule networks achieved

TABLE 15.2
Accuracy Achieved by Five CNN Models

Model Used	Accuracy (%)
ResNet-101	99.68
DenseNet-121	99.69
AlexNet	97.85
VGG-16	99.19
GoogleNet	98.52

82.17% accuracy, while adding convolution layers to capsule networks achieved 83.58%. Using CNNs only, diseases can be detected with 81.14% accuracy. Kwabena et al. [39] proposed Gabor-CapsNet (Gabor and Capsule Network) to identify tomato and citrus leaf diseases. This model outperformed AlexNet and GoogLeNet. The results achieved 98.13% test accuracy. The findings implied that Capsule Networks could perform better on challenging real-world datasets than other deep learning techniques. Additionally, they were able to identify diseased plants from a variety of angles and even under adverse weather and lighting circumstances. With its improved texture extraction capabilities, the Gabor-CapsNet architecture can recognize damaged or ill portions. Pham et al. [40] proposed a model named PhamModel using deep neural networks to detect multi-class mango leaf disease in which APGWO (Adaptive Particle-Grey Wolf) meta-heuristic feature selection approach had been used to extract 81 features out of the originally proposed 114 features [40]. These 81 features were then fed as input to multi-layer perceptron for the task of classification. PhamModel achieved higher accuracy, i.e., 89.41% on comparing with other popular CNN models like VGG16, AlexNet, ResNet50, whereas AlexNet achieved the lower accuracy, i.e., 78.64%.

Ji et al. [11] proposed an architecture based on CNN, UnitedModel, to identify three diseases in grape leaves: Isariopsis leaf spot, Black Rot, Esca. UnitedModel consisted of InceptionV3 and ResNet50 which enabled it to extract distinguished features. High-level feature fusion improved UnitedModel's representational capabilities, enabling it to perform at its peak level in the task of identifying grape diseases; 98.57% testing accuracy was achieved by this model, helping farmers to identify grape diseases. Ahmad et al. [41] emphasized on the recognition of tomato leaf diseases using four CNN techniques, namely Inception V3, ResNet, VGG-16, and VGG-19. The models were being tested on two datasets, a dataset based on laboratory and a dataset acquired by self from the field [41]. The results revealed that all architectures outperform field-based data on the laboratory-based dataset. A CNN-based model to detect apple leaf diseases was built by Xin Li et al. [42] using a dataset of healthy leaves and images of Cedar Rust, Black Star, and Apple Grey-Spot diseased leaves. SVM classifiers were used for picture segmentation, and VGG and ResNet CNN models were used for the classification of image. Because ResNet-18 has less layers than the other two models, the findings show that it performs predictions with an accuracy of 98.5. A CNN architecture—AlexNet—was used for the extraction of features and classification by Rao et al. [2] to identify Grape and Mango leaf diseases. PlantVillage dataset was used for training the model. MATLAB was used to develop the system with 99% accuracy and 89% accuracy for grape leaves and mango leaves, respectively.

Nigam et al. [43] created a CNN model to conduct identification of yellow rust disease in wheat crop with the help of images of infected wheat crop and healthy leaves. The CNN architecture-based model classified healthy leaves and leaves affected by yellow rust with 97.37% testing accuracy. There were 12,12,513 parameters in all that were trained for the model. The model is appropriate in image-based identification of diseases due to its significantly high success rate. Arshad et al. [44] suggested a model which was based on ResNet and Transfer Learning to detect 16 categories of potato, tomato, and corn diseases, which achieved accuracy of 98.7% with ResNet50. Srinidhi et al. [7] introduced a system in which EfficientNetB7 and DenseNet were used to detect apple leaf diseases, and categorized the apple leaf images into four categories: Scab, Rust, Multiple diseases, and Healthy. Data augmentation and image annotation methods, specifically Flipping, Blurring, and Canny Edge Detection, were utilized for the enhancement of the apple leaves dataset. Using an updated dataset, models utilizing EfficientNetB7 and DenseNet provided accuracy of 99.75% and 99.8%, respectively. Kirti et al. [12] proposed a method in which grape plant dataset from PlantVillage Database was taken to identify Esca Black measles disease in GrapeVines.1807 images (healthy and diseased) were used in the study. The findings were computed using Transfer Learning, Fine Tuning, and ResNet 50 architecture of deep neural network. The proposed system provided an accuracy of 100%.

Sachdeva et al. [6] suggested a CNN model that was incorporated with Bayesian learning to enhance the classification accuracy. The Bayesian approach helps in modeling uncertainty and obtaining better predictions. This study utilized a comprehensive dataset comprising of images of healthy leaves and diseased leaves. The 20,639 images from PlantVillage, which were divided into 15 different groups of healthy and diseased and pepper bell, tomato, potato leaf images, were utilized as the basis for this research [6]. The suggested method employed a Bayesian framework and largely used features created by CNN in various hierarchies to achieve 98.9% accuracy without overfitting. Bande et al. [45] employed a project in which ResNet-50, AlexNet, InceptionV3, DenseNet-169, and VGG-16 deep learning models were used to detect diseases from images of 38 different classes, including diseased and healthy leaves from the PlantVillage Dataset to classify them into two groups: infected and uninfected. The results revealed that among all ResNet-50 achieved 97.80% accuracy, which is the highest. PlantVillage dataset had been used for acquiring images of 38 classes of plants containing healthy and unhealthy leaves.

Wang et al. [15] proposed AT-AlexNet, an attention neural network based on down sampling attention, to identify six corn leaf diseases: Common rust, Own Spot, Bipolarize maydis, Sheath Blight, Northern Leaf Blight, Curvularia lunata (wakker) Boed Spot. The structure of AT-AlexNet had three parts: extraction module, fusion module, and fully connected classification module. Mish activation function was used in this model so that nonlinear expression could be improved. Results showed that this attention-based network model achieved 99.35% accuracy. Guo-feng et al. [16] suggested a system based on cloud system to detect strawberry leaf disease identification. To detect different types of strawberry diseases, the system was integrated with a multi-network fusion classification model which was self-supervised and comprised multiple networks, such as a feedback network, a classification network, and location network. This cloud-based system achieved 92.48% accuracy.

Wenchao et al. [17] suggested G-ResNet50, a model based on deep residual network with transfer learning to identify and classify strawberry diseases: Powdery mildew, Anthracnose, Leaf Spot. The convolutional layer and pooling layer of this CNN model inherited weight parameters that were pre-trained from ResNet50 model; throughout the training phase dataset used was PlantVillage. The model's results showed that G-ResNet50 model achieved 98.67% accuracy. Pandey et al. [18]

proposed a model named AResNet-50, which employed attention residual learning in the standard ResNet-50 CNN model so that citrus, guava, mango, and aubergine leaf diseases detection can be made possible successfully; 98.20% classification accuracy was achieved with this model. A light-weight deep learning approach based on Vision Transformer (ViT) was proposed by Borhani et al. [46] for real-time automated plant disease classification. And also, ViT with CNN had been combined to recognize plant disease. The concept was based on how people categorize images. Person concentrates on a certain portion of the image of his interest while looking at that image. This study focused on three different datasets: PlantVillage Dataset, Rice Leaf Disease Dataset, and Wheat Rust Classification Dataset [46]. This research used two main building blocks: CNN Block and Transformer Block. Eight different models were used in this study: Model 1, Model 2 and so on. Results showed that Model 4 consisting of only transformer blocks had achieved has highest precision and highest convergence score i.e., 0.95 was achieved by Model 4 as compared to other models.

Pandian et al. [47] proposed 14 layered DCNN (14-DCNN) having 5 convolution and 5 max-pooling layers [47] to identify plant diseased leaves. A new dataset was generated by combining several open datasets. The dataset contained 147,500 images representing diseased and healthy plant leaf classes. The 14-DCNN model outperformed well in contrast to the current transfer learning strategies; 99.9655% accuracy was achieved on the 8850 test images with this model. Shafi et al. [48] proposed a system in which four classes of wheat rust diseases had been detected. Four classes were Healthy, Resistant, Moderate, and Susceptible. The research aims to develop an embedded AI system capable of accurately classifying the types of wheat yellow rust infection directly on resource-constrained devices. This approach enabled timely detection and diagnosis of yellow rust in wheat crops, facilitating effective disease management and mitigation strategies. Tabbakh et al. [49] proposed a hybrid model of transfer learning and vision transformers named as TLMViT for plant disease classification. This model identified three diseased crops namely bell peppers, potatoes, tomatoes. For acquiring leaf images, 3 classes of Wheat Rust Dataset and 15 classes of PlantVillage Dataset were being used. TLMViT accurately classified plant diseases, achieving 98.81% accuracy for VGG19 and 99.86% for the ViT model on the PlantVillage and wheat datasets, respectively. Bi et al. [50] suggested improved MobileNetv3 model for the detection of corn diseases: Rust, Gray Spot, Dwarf Mosaic virus, Northern Leaf Blight and Mild Corn. Dilated convolutions had been introduced for increasing the receptive fields. Results showed that the model achieved 98.23% accuracy. S.Patil et al. [22] proposed a model based on CNNs to recognize diseases which harm the health and growth of Tulsi plant. Tulsi is used as an herbal therapy in the treatment of many diseases. This research was done on self-created database by collecting variety of images of Tulsi leaves and achieved 75% accuracy. Rajeena P.P. et al. [51] used a model called EfficientNet to identify common corn diseases: Fusarium Ear Rot, Southern Corn Leaf Blight, Gray Leaf Spot, Northern Corn Leaf Blight, and Goss's Wilt. Results showed that this model achieved 98.85% accuracy.

15.3 COMPARISON OF VARIOUS STUDIES

Table 15.3 depicts the work of various researchers in this field.

15.4 INTERPRETATION OF RESULTS

Different researchers have made great efforts and used different models and techniques in classifying and detecting diseases of some plants. Many researches have been done on Tomato leaf because it is cultivated worldwide and it is found in almost every kitchen. Moreover, India is the second largest producer of Tomato. CNN based approaches have remarkable results in finding Tomato leaf diseases. Table 15.4 shows the work of various researchers in identifying various diseases which harm tomato plant leaves with their results.

Besides Tomato, there are many plant leaves on which research based on Deep learning techniques is being done. Results and findings of these techniques vary in terms of mAP and accuracy. Figure 15.1 reveals the test accuracies obtained from different techniques implemented by different researchers in the field of identification and classification of plant diseases. Results show that system suggested by Kirti et al. [12] has achieved maximum accuracy of 100% till now. The study used color-modeled image segmentation techniques and deep neural network so that Black Measles disease of Grape leaf could be detected with 100% accuracy.

The above plot compares the results of numerous researches, and it reveals that S.Patil et al. [47] achieved the lowest accuracy of 75% in detecting Tulsi disease. To increase the accuracy rate, a large dataset is needed, and the model could be trained with different CNN architectures. Many researches have been done on PlantVillage dataset in which different plant species have been taken for the research. Accuracy achieved using PlantVillage dataset with the name of researcher is shown in Figure 15.2.

The above plot demonstrates that the highest and lowest accuracy achieved on the PlantVillage dataset is 100% and 89% when grape and mango leaves were taken for disease detection, respectively.

15.5 FUTURE SCOPE

Deep learning-based advances in plant disease identification may advance several important areas in the future. The main goal is to strengthen the robustness of the model against environmental changes that can affect the detection accuracy. In resource-limited agricultural environments, transfer learning offer the potential to use existing knowledge for creating flexible and scalable detection systems. Combining data from multiple sensor modalities with visual data can increase accurate and reliable disease detection. Interpretability and explicability are still important to build trust in deep learning models to explore ways to discover how models make decisions. Advances in edge computing and IoT technologies have increased the demand for lightweight models that can be used for real-time inference on edge devices. This enables distributed deployment and reduces dependency on cloud

TABLE 15.3
Comparison of CNN-based Techniques to Detect Various Leaf Diseases

Sr No.	Author & Year	Model/ Algorithm	Plant	Disease	Result (Accuracy/mAP)
1.	Kawaski et al. (2015)	4-fold Cross-Validation Strategy	Cucumber	Zucchini Yellow Mosaic Virus and Melon Yellow Spot Virus	94.9% accuracy
2.	Mohanty et al. (2016)	GoogleNet with transfer learning, AlexNet and training from scratch	14 Crop Species	26 Crop Diseases	99.35% accuracy
3.	Cruz et al. (2017)	Transfer Learning with Context Injection System	Olive	Xylella fastidiosa	98.60% accuracy
4.	Ma et al. (2018)	Deep CNN	Cucumber	Target Leaf Spot, Downy Mildew, Powdery Mildew, Anthracnose	93.4% accuracy
5.	Zhang et al. (2018)	GoogLeNet and Cifar 10	Maize	Curvularia leaf spot, southern leaf blight, dwarf mosaic, Rust, Gray leaf spot, Brown Spot Northern leaf blight, Round Spot	98.9% accuracy
6.	Rao. et al. (2018)	AlexNet with Transfer Learning	Grape and Mango	Grape Diseases- Blight, Black Measles, Black rot Mango Diseases-Bacterial Canker, Powdery Mildew, Scab	Mango leaves- 89% accuracy Grape Leaves-99%
7.	Chouhan et al. (2019)	Multi-layer CNN	Mango	Anthracnose	97.13% accuracy

(Continued)

TABLE 15.3 (CONTINUED)
Comparison of CNN-based Techniques to Detect Various Leaf Diseases

Sr No.	Author & Year	Model/ Algorithm	Plant	Disease	Result (Accuracy/mAP)
8.	Ji et al. (2019)	UnitedModel	Grape	Isariopsis leaf spot, Esca and Black Rot	98.57% accuracy
9.	Dong et al. (2019)	Capsule Network with Convolution layers	Peanut	Net Blotch, Rust, Brow Spot, Black Spot, and Anthracnose	83.58% accuracy
10.	Zhou et al. (2019)	FCM-KM and Faster R-CNN	Rice	Bacterial blight, Rice blast and sheath blight	Sheath Blight-98.26%,Bacterial Blight-97.53%, Rice Blast-96.71% accuracy
11.	Singh et al. (2019)	Faster R-CNN with MobileNet and InceptionResnetV2	13 plant species	17 classes of Diseases	mAP - 38.9 with Fater R-CNN with InceptionResnetV2
12.	Xie et al. (2020)	Faster DR-IACNN	Grape	Black measles, Mites, Black rot, Leaf blight	81.1% mAP
13.	Sun et al. (2020)	Feature Fusion Instance Multi-scale Detection Algorithm	Maize	Leaf Blight	91.83% mAP
14.	Prakruti et al. (2020)	YOLOv3	Tea	Tea Mosquito bugs, Red Spider Mites	86% mAP
15.	Pham et al. (2020)	Deep Neural Network with APGWO Feature Selection approach(Pham et al., 2020)	Mango	Powdery Mildew, Gall Midge, Anthracnose	89.41% accuracy
16.	Xin Li et al. (2020)	ResNet and VGG	Apple	Cedar Rust, Black Star, And Apple Grey-Spot Disease	98.5% accuracy by ResNet

(Continued)

TABLE 15.3 (CONTINUED)
Comparison of CNN-based Techniques to Detect Various Leaf Diseases

Sr No.	Author & Year	Model/ Algorithm	Plant	Disease	Result (Accuracy/mAP)
17.	Srinidhi et al. (2021)	EfficientNet and DenseNet	Apple	Scab, Rust and Multiple Diseases	EfficienNet-99.8% and DenseNet- 99.75%
18.	Nigam et al. (2021)	Deep CNN	Wheat	Yellow Rust	97.37%
19.	Kirti et al. (2021)	Transfer Learning, Fine Tuning, and ResNet 50	Grape	Esca Black measles disease	100% accuracy
20.	Shah et al. (2021)	Residual Teacher/ Student Architecture	14 plant species	Black spot	F1 score-0.991
21.	Pandian et al. (2022)	14-DCNN	16 plants	42 leaf diseases	99.9655% accuracy
22.	Borhani et al. (2022)	Vision Transformer (ViT) and CNN	Wheat, Rice, 14 crop Species,	26 different Infections	ViT based model has higher accuracy than CNN based Models
23.	Feng et al. (2022)	Multi-network Fusion classification model(YANG et al., 2022)	Strawberry	Leaf scorch	92.48%

(Continued)

TABLE 15.3 (CONTINUED)
Comparison of CNN-based Techniques to Detect Various Leaf Diseases

Sr No.	Author & Year	Model/ Algorithm	Plant	Disease	Result (Accuracy/mAP)
24.	Wenchao et al. (2022)	G-ResNet50	Strawberry	Powdery mildew, Strawberry Anthracnose, Leaf Spot	98.67%
25.	Pandey et al. (2022)	AResNet-50	Citrus, Guava, Mango, and Aubergine	15 different classes	98.20%
26.	Wang et. al (2022)	AT- AlexNet model	Corn	6 diseases	99.35%
27.	Bi et al. (2023)	Improved MobileNetv3(CD-MobileNetv3)	Corn	Northern Leaf Blight, Gray spot, Dwarf Mosaic virus, Mild Corn, Rust	98.23%
28.	S.Patil et al. (2023)	CNN with 2 Convolution and Max Pooling Layers	Tulsi	Fusarium Wilt, Powdery Mildew, Tulsi shoot blight	75%
29.	Rajeena P. P. et al. (2023)	EfficientNet	Corn	Leaf Blight, Goss's Wilt, Northern Corn Leaf, Southern Corn Fusarium Ear Rot, Gray Leaf Spot, Blight	98.85%

TABLE 15.4
Comparison of CNN-based Techniques to Detect Tomato Leaf Diseases

Sr No.	Author & Year	Models/ Algorithms	Plant	Disease	Accuracy (%)
1.	Brahimi et al. (2017)	AlexNet and GoogleNet	Tomato	Bacterial Spot, Leaf Mold, Early Blight, Spider Mite, Yellow Leaf Curl Virus, Late Blight, Septoria Spot, Tomato Mosaic Virus,Target Spot	99.18%
2.	Agarwal et al. (2019)	CNN with three Convolution and three Max Pooling Layers	Tomato	9 classes of Diseases	91.2 %
3.	Ahmad et al. (2020)	VGG-16, VGG-19, ResNet, and Inception V3	Tomato	Late blight, Septoria Leafspot, Bacterial speck, Bacterial spot, Early blight	Laboratory Dataset-99.60% and Field dataset-93.70% with Inception V3
4.	Kwabena et al. (2020)	Gabor-CapsNet	Tomato and Citrus	9 classes and 5 classes of Tomato and citrus diseased leaves resp.	98.13%
5.	Arshad et al. (2021)	ResNet50 with Transfer Learning	Potato, Tomato, Corn	16 classes of diseases	98.7%
6.	Sachdeva et al. (2021)	Bayesian Learning based deep CNN	Potato, Tomato, Pepper Bell	Leaf Spot, Rust Spores, Botrytis Blight, Powdery Mildew, Black spot	98.9%

(Continued)

TABLE 15.4 (CONTINUED)
Comparison of CNN-based Techniques to Detect Tomato Leaf Diseases

Sr No.	Author & Year	Models/ Algorithms	Plant	Disease	Accuracy (%)
7.	Bande et al. (2022)	DenseNet-169, AlexNet, VGG-16, ResNet-50, and InceptionV3	14different plants like Potato, Tomato, Apple, Cherry, Blueberry etc	20 different diseases	97.8% accuracy with ResNet-50
8.	Tabbakh et al. (2023)	Vision Transformer model with Transfer Learning (TLMViT)	Bell Peppers, Potatoes, Tomatoes	Septoria Leaf Spot, Late Blight, Spider Mites, Brown Rust, Bacterial Spot, Yellow leaf Curl, Yellow Rust, Mosaic, Early Blight, Target Spot, Leaf Mold	98.81% on PlantVillage Dataset and 99.86% on Wheat dataset.

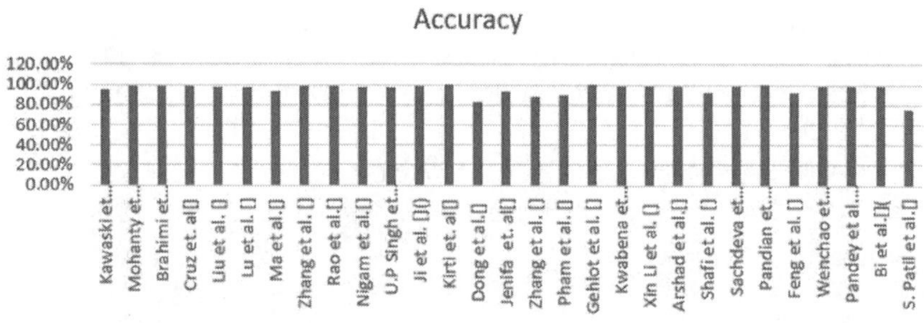

FIGURE 15.1 Performance comparison of different deep learning techniques implemented by different researchers in terms of accuracy

infrastructure. Advances in edge computing and IoT technologies have increased the demand for lightweight models that can be used for real-time inference on edge devices. This enables distributed deployment and reduces dependency on cloud infrastructure. Integrating precision agriculture approaches with common platforms

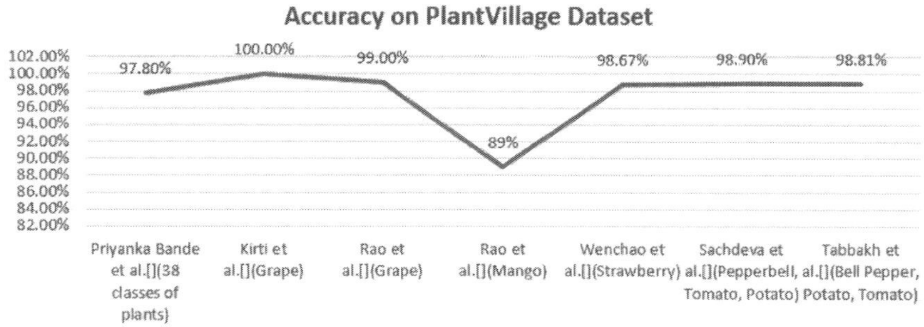

FIGURE 15.2 Performance comparison of the various deep learning techniques implemented on PlantVillage dataset in terms of accuracy

and shared datasets can facilitate community engagement and benchmarking efforts to optimize crop management strategies. Research on how automated disease detection systems impact communities, farmers, and food security and how to ensure equitable access and address issues of bias and privacy is needed.

15.6 CONCLUSION

This review chapter presented an overview of previous and current research on the disease detection of various plant leaves. If there is adequate training data, deep learning algorithms may be able to identify plant leaf diseases with high accuracy. We have found relevant publications and analyzed plant diseases, crop models, data sources, and overall performance based on different metrics. This review chapter examined researches based on deep learning models for recognizing and classifying various plant diseases. The review's findings reveal that deep learning techniques outperform other techniques that were used earlier for disease recognition and classification. The deficiency of benchmarking images in the agricultural area is a key impediment to advancing research in this domain. Most of these frameworks suggested in this review demonstrate effective identification of diseases on their datasets but not on other datasets. The diversity of the datasets used is restricted, despite the literature demonstrating very positive outcomes. Training a CNN model requires a large dataset containing hundreds and thousands of images. Unfortunately, researchers have not yet gathered such datasets. The performance of most of the DL models in the researches has been evaluated using PlantVillage Dataset. Most of the researches were carried out in China and other developed countries. This also explains the necessity for AI in crop disease identification in India, with a focus on grain crops. Thus, India has a poor degree of disease identification research, and lack of knowledge of applications in deep learning may be the main cause of India's low research output. To fully fulfill modern deep learning techniques in Indian agriculture, efforts are needed. The symptoms of various illnesses might resemble one another, making it difficult to determine which infection is present. Without a well-defined boundary condition, it will be a difficult task to differentiate between diseased and healthy regions. This chapter provides insight into the suitability of several deep learning models for plant disease identification and classification tasks for novice researchers. This review concludes that there is an urgent need to develop an agriculture system so that the quality of oil seed crops could be increased, as these are very important crops in India. Moreover, very less research has been done in the recognition of diseases of these crops. This requires the detection of diseases of such crops at the initial stage itself.

REFERENCES

1. Rajendra, P., Kumari, M., Rani, S., Dogra, N., Boadh, R., Kumar, A., and Dahiya, M. "Impact of artificial intelligence on civilization: Future perspectives." *Materials Today: Proceedings*, 56: 252–256 (2022). https://doi.org/10.1016/j.matpr.2022.01.113
2. Rani, S., Tripathi, K., Arora, Y., and Kumar, A. "A Machine Learning Approach to Analyze Cloud Computing Attacks." In *2022 5th International Conference on Contemporary Computing and Informatics (IC3I)*, pp. 22–26. IEEE, 2022. https://doi.org/10.1109/IC3I56241.2022.10073468
3. Shrivastava, V. K., Kumar, A., Shrivastava, A., Tiwari, A., Thiru, K., and Batra, R. "Study and trend prediction of Covid-19 cases in India using deep learning techniques." In *Journal of Physics: Conference Series* (Vol. 1950, no. 1, p. 012084). IOP Publishing, 2021. https://doi.org/10.1088/1742-6596/1950/1/012084
4. Singhal, A., Phogat, M., Kumar, D., Kumar, A., Dahiya, M., and Shrivastava, V. K. "Study of deep learning techniques for medical image analysis: A review." *Materials Today: Proceedings*, 56: 209–214 (2022). https://doi.org/10.1016/j.matpr.2022.01.071
5. Kumar, A., Kumar, P., Sharma, N., and Srivastava, A. K. (Eds). *3D Printing Technologies: Digital Manufacturing, Artificial Intelligence, Industry 4.0.* Walter de Gruyter GmbH & Co KG, 2024. https://doi.org/10.1515/9783111215112
6. Sachdeva, G., Singh, P., and Kaur, P. Plant leaf disease classification using deep Convolutional neural network with Bayesian learning. *Materials Today: Proceedings*, 45: 5584–5590 (2021). https://doi.org/10.1016/j.matpr.2021.02.312
7. Srinidhi, V. V, Sahay, A., and Deeba, K. Plant Pathology Disease Detection in Apple Leaves Using Deep Convolutional Neural Networks Apple Leaves Disease Detection using EfficientNet and DenseNet. In *Proceedings of the Fifth International Conference on Computing Methodologies and Communication (ICCMC 2021)*, p. 9, 2021. https://doi.org/10.1109/ICCMC51019.2021.9418268
8. Agarwal, M., Singh, A., Arjaria, S., Sinha, A., and Gupta, S. ToLeD: Tomato Leaf Disease Detection using Convolution Neural Network. *Procedia Computer Science*, 167: 293–301 (2020). https://doi.org/10.1016/j.procs.2020.03.225
9. Zhang, K., Zhang, L., and Wu, Q. Identification of cherry leaf disease infected by podosphaera pannosa via convolutional neural network. *International Journal of Agricultural and Environmental Information Systems*, 10(2): 98–110 (2019). https://doi.org/10.4018/IJAEIS.2019040105
10. Ma, J., Du, K., Zheng, F., Zhang, L., Gong, Z., and Sun, Z. A recognition method for cucumber diseases using leaf symptom images based on deep convolutional neural network. *Computers and Electronics in Agriculture*, 154(August): 18–24 (2018). https://doi.org/10.1016/j.compag.2018.08.048
11. Ji, M., Zhang, L., and Wu, Q. Automatic grape leaf diseases identification via UnitedModel based on multiple convolutional neural networks. *Information Processing in Agriculture*, 7(3): 418–426 (2020). https://doi.org/10.1016/j.inpa.2019.10.003

12. Kirti, K., Rajpal, N., and Yadav, J. Black measles disease identification in grape plant (vitis vinifera) using deep learning. In *Proceedings - IEEE 2021 International Conference on Computing, Communication, and Intelligent Systems, ICCCIS 2021* (pp. 97–101). 2021. https://doi.org/10.1109/ICCCIS51004.2021.9397205
13. Xie, X., Ma, Y., Liu, B., He, J., Li, S., and Wang, H. A Deep-Learning-Based Real-Time Detector for Grape Leaf Diseases Using Improved Convolutional Neural Networks. *Frontiers in Plant Science*, 11(June): 1–14 (2020). https://doi.org/10.3389/fpls.2020.00751
14. Cruz, A. C., Luvisi, A., De Bellis, L., and Ampatzidis, Y. Vision-based plant disease detection system using transfer and deep learning. In *2017 ASABE Annual International Meeting* (pp. 1–9). 2017. https://doi.org/10.13031/aim.201700241
15. Wang, Y., Tao, J., and Gao, H. Corn Disease Recognition Based on Attention Mechanism Network. *Axioms*, 11(9): (2022). https://doi.org/10.3390/axioms11090480
16. Yang, G. F., Yang, Y., He, Z. K., Zhang, X. Y., and He, Y. A rapid, low-cost deep learning system to classify strawberry disease based on cloud service. *Journal of Integrative Agriculture*, 21(2): 460–473 (2022). https://doi.org/10.1016/S2095-3119(21)63604-3
17. Wenchao, X., and Zhi, Y. Research on Strawberry Disease Diagnosis Based on Improved Residual Network Recognition Model. *Mathematical Problems in Engineering*, 2022 (2022). https://doi.org/10.1155/2022/6431942
18. Pandey, A., and Jain, K. Plant leaf disease classification using deep attention residual network optimized by opposition-based symbiotic organisms search algorithm. *Neural Computing and Applications*, 34(23): 21049–21066 (2022). https://doi.org/10.1007/s00521-022-07587-6
19. Lu, J., Hu, J., Zhao, G., Mei, F., and Zhang, C. An in-field automatic wheat disease diagnosis system. *Computers and Electronics in Agriculture*, 142(1): 369–379 (2017). https://doi.org/10.1016/j.compag.2017.09.012
20. Zhou, G., Zhang, W., Chen, A., He, M., and Ma, X. Rapid Detection of Rice Disease Based on FCM-KM and Faster R-CNN Fusion. *IEEE Access*, 7: 143190–143206 (2019). https://doi.org/10.1109/ACCESS.2019.2943454
21. Jenifa, A., Ramalakshmi, R., and Ramachandran, V. Classification of Cotton Leaf Disease Using Multi-Support Vector Machine. *IEEE International Conference on Intelligent Techniques in Control, Optimization and Signal Processing, INCOS 2019* (pp. 1–4). 2019. https://doi.org/10.1109/INCOS45849.2019.8951356
22. Patil, S. S., Patil, S. H., Bhall, A., Rajvaidya, A., Sehrawat, H., Pawar, A. M., and Agarwal, D. Tulsi Leaf Disease Detection using CNN. In *2022 IEEE Conference on Interdisciplinary Approaches in Technology and Management for Social Innovation, IATMSI 2022, December*, 2022. https://doi.org/10.1109/IATMSI56455.2022.10119405
23. Dong, M., Mu, S., Su, T., and Sun, W. Image recognition of peanut leaf diseases based on capsule networks. In *Communications in Computer and Information Science* (Vol. 1001). 2019. Springer Singapore. https://doi.org/10.1007/978-981-32-9298-7_4
24. Kota, V. M., Manoj Kumar, V., and Bharatiraja, C. Deep Learning - A Review. *IOP Conference Series: Materials Science and Engineering*, 912(3): 032068 (2020). https://doi.org/10.1088/1757-899X/912/3/032068
25. Kawasaki, Y., and Uga, H. Basic Study of Automated Diagnosis of Viral Plant Diseases Using Convolutional neural networks. In *Advances in Visual Computing: 11th International Symposium, ISVC 2015, Las Vegas, NV, USA, December 14–16, 2015, Proceedings, Part II 11* (pp. 638-645). Springer International Publishing, 2015. https://doi.org/10.1007/978-3-319-27863-6
26. Mohanty, S. P., Hughes, D. P., and Salathé, M. Using deep learning for image-based plant disease detection. *Frontiers in Plant Science*, 7(September): 215232 (2016). https://doi.org/10.3389/FPLS.2016.01419

27. Liu, J., and Wang, X. Plant diseases and pests detection based on deep learning: a review. *Plant Methods*, 17(1): 1–18 (2021). https://doi.org/10.1186/s13007-021-00722-9
28. Brahimi, M., Boukhalfa, K., and Moussaoui, A. Deep Learning for Tomato Diseases: Classification and Symptoms Visualization. *Applied Artificial Intelligence*, 31(4): 299–315 (2017). https://doi.org/10.1080/08839514.2017.1315516
29. Liu, B., Zhang, Y., He, D. J., and Li, Y. Identification of apple leaf diseases based on deep convolutional neural networks. *Symmetry*, 10(1): 11 (2018). https://doi.org/10.3390/sym10010011
30. Zhang, X., Qiao, Y., Meng, F., Fan, C., and Zhang, M. Identification of maize leaf diseases using improved deep convolutional neural networks. *IEEE Access*, 6: 30370–30377 (2018). https://doi.org/10.1109/ACCESS.2018.2844405
31. Fuentes, A., Yoon, S., Kim, S. C., and Park, D. S. A robust deep-learning-based detector for real-time tomato plant diseases and pests recognition. *Sensors (Switzerland)*, 17(9): 2022 (2017). https://doi.org/10.3390/s17092022
32. Ozguven, M. M., and Adem, K. Automatic detection and classification of leaf spot disease in sugar beet using deep learning algorithms. *Physica A: Statistical Mechanics and Its Applications*, 535: 122537 (2019). https://doi.org/10.1016/j.physa.2019.122537
33. Sun, J., Yang, Y., He, X., and Wu, X. Northern Maize Leaf Blight Detection under Complex Field Environment Based on Deep Learning. *IEEE Access*, 8: 33679–33688 (2020). https://doi.org/10.1109/ACCESS.2020.2973658
34. Bhatt, P., Sarangi, S., and Pappula, S. Detection of diseases and pests on images captured in uncontrolled conditions from tea plantations. In *Autonomous Air and Ground Sensing Systems for Agricultural Optimization and Phenotyping IV* (Vol. 11008, pp. 73–82). SPIE, 2019. https://doi.org/10.1117/12.2518868
35. Singh, U. P., Chouhan, S. S., Jain, S., and Jain, S. Multilayer Convolution Neural Network for the Classification of Mango Leaves Infected by Anthracnose Disease. *IEEE Access*, 7(March): 43721–43729 (2019). https://doi.org/10.1109/ACCESS.2019.2907383
36. Singh, D., Jain, N., Jain, P., Kayal, P., Kumawat, S., and Batra, N. PlantDoc: A dataset for visual plant disease detection. In *ACM International Conference Proceeding Series* (pp. 249–253). 2020. https://doi.org/10.1145/3371158.3371196
37. Gehlot, M., and Saini, M. L. Analysis of Different CNN Architectures for Tomato Leaf Disease Classification. In *2020 5th IEEE International Conference on Recent Advances and Innovations in Engineering, ICRAIE 2020 - Proceeding*. 2020. https://doi.org/10.1109/ICRAIE51050.2020.9358279
38. Sabour, S., Frosst, N., and Hinton, G. E. "Dynamic routing between capsules." *Advances in Neural Information Processing Systems*, 30 (2017). https://doi.org/10.48550/arXiv.1710.09829
39. Kwabena, P. M., Weyori, B. A., and Mighty, A. A. Gabor capsule network for plant disease detection. *International Journal of Advanced Computer Science and Applications*, 11(10): 388–395 (2020). https://doi.org/10.14569/IJACSA.2020.0111048
40. Pham, T. N., Tran, L. Van, and Dao, S. V. T. Early Disease Classification of Mango Leaves Using Feed-Forward Neural Network and Hybrid Metaheuristic Feature Selection. *IEEE Access*, 8: 189960–189973 (2020). https://doi.org/10.1109/ACCESS.2020.3031914
41. Ahmad, I., Hamid, M., Yousaf, S., Shah, S. T., and Ahmad, M. O. Optimizing pretrained convolutional neural networks for tomato leaf disease detection. *Complexity*, 2020(1): 8812019 (2020). https://doi.org/10.1155/2020/8812019

42. Li, X., and Rai, L. Apple Leaf Disease Identification and Classification using ResNet Models. In *ICEICT 2020 - IEEE 3rd International Conference on Electronic Information and Communication Technology* (pp. 738–742). 2020. https://doi.org/10.1109/ICEICT51264.2020.9334214
43. Nigam, S., Jain, R., Marwaha, S., Arora, A., Singh, V. K., Singh, A. K., Paul, R. K., and Kingsly, I. T. Automating yellow rust disease identification in wheat using artificial intelligence. *Indian Journal of Agricultural Sciences*, 91(9): 1391–1395 (2021). https://doi.org/10.56093/ijas.v91i9.116097
44. Arshad, M. S., Rehman, U. A., and Fraz, M. M. Plant Disease Identification Using Transfer Learning. In *2021 International Conference on Digital Futures and Transformative Technologies, ICoDT2 2021* (pp. 1–5). 2021. https://doi.org/10.1109/ICoDT252288.2021.9441512
45. Bande, P., and Dewangan, K. Plant Disease Detection using Deep Learning. In *2022 International Journal for Research in Applied Science & Engineering Technology (IJRASET)* (Vol. 10(VI), p. 10). 2022. https://doi.org/10.22214/ijraset.2022.43900
46. Borhani, Y., Khoramdel, J., and Najafi, E. A deep learning based approach for automated plant disease classification using vision transformer. *Scientific Reports*, 12(1): 1–10 (2022). https://doi.org/10.1038/s41598-022-15163-0
47. Pandian, J. A., Kumar, V. D., Geman, O., Hnatiuc, M., Arif, M., and Kanchanadevi, K. Plant Disease Detection Using Deep Convolutional Neural Network. *Applied Sciences (Switzerland)*, 12(14): 6982 (2022). https://doi.org/10.3390/app12146982
48. Shafi, U., Mumtaz, R., Qureshi, M. D. M., Mahmood, Z., Tanveer, S. K., Haq, I. U., and Zaidi, S. M. H. Embedded AI for Wheat Yellow Rust Infection Type Classification. *IEEE Access*, 11(March): 23726–23738 (2023). https://doi.org/10.1109/ACCESS.2023.3254430
49. Tabbakh, A., and Barpanda, S. S. A Deep Features Extraction Model Based on the Transfer Learning Model and Vision Transformer "TLMViT" for Plant Disease Classification. *IEEE Access*, 11(May): 45377–45392 (2023). https://doi.org/10.1109/ACCESS.2023.3273317
50. Bi, C., Xu, S., Hu, N., Zhang, S., Zhu, Z., and Yu, H. Identification Method of Corn Leaf Disease Based on Improved Mobilenetv3 Model. *Agronomy*, 13(2): 1–17 (2023). https://doi.org/10.3390/agronomy13020300
51. Rajeena P. P, F., Su, A., Moustafa, M. A., and Ali, M. A. S. Detecting Plant Disease in Corn Leaf Using EfficientNet Architecture—An Analytical Approach. *Electronics (Switzerland)*, 12(8): 1938 (2023). https://doi.org/10.3390/electronics12081938
52. Shrivastava, V. K., Shelke, C. J., Achary, R., David, M. P. P., Soliman, A. H., and Kumar, A. Optimised convolutional neural network model for multiclass acute lymphoblastic leukemia classification in in peripheral blood smear images. In *AIP Conference Proceedings* (Vol. 3217, No. 1). AIP Publishing, 2024.
53. Trivedi, N. K., Anand, A., Kumar, A., Lilhore, U. K., and Tiwari, R. G. Impacts on Health Frameworks of Big Data Analytics: A Review. In *Emerging Technologies in Data Mining and Information Security: Proceedings of IEMIS 2022* (Vol. 1, pp. 721–729). 2022.
54. Boadh, R., Grover, R., Dahiya, M., Kumar, A., Rathee, R., Rajoria, Y. K., Rawat, M., and Rani, S. Study of fuzzy expert system for the diagnosis of various types of cancer. Materials Today: Proceedings, 56: 298–307 (2022).
55. Kumar, A., Rajak, D.K., Kumar, P., and Kumar, A. (Eds.). *Advances in Sustainable Biomaterials: Bioprocessing 4.0, Characterizations, and Applications* (1st ed.). CRC Press, 2024. https://doi.org/10.1201/9781003434313

16 AI for Disaster prediction and Management

Mahesh K. Singh and Ankur Mittal

16.1 INTRODUCTION

For the past decade, over 2.6 billion people have been affected by catastrophic disasters, such as floods, tsunamis, cyclones, earthquakes, landslides, and many other pandemics. Disasters are the most common reason for some casualties in the last decade—i.e., a deadly disaster in the city of New Guinea was an earthquake that displaced 58,300 people, as stated by the displacement tracking matrix (DTM). Disasters are mainly due to environmental changes, while pandemics are mainly due to the superfast spread of any virus over a vast area.

There were many pandemics that spread across the world, including the American plague in the 16th century, yellow fever in Philadelphia in 1793, H1N1 swine flu in 2009–2010, Ebola pandemic in 2014—2016, and the most recent COVID-19. One of the deadly pandemics was Black Death, which spread in the 14th century all over Asia and Europe and caused plenty of deaths. In recent days, disasters and pandemics have been the hottest research area. Recently, there have been many research papers published on prevention related to COVID-19 and its management. Disasters generally are of two categories: natural disasters and man-made disasters. The modus operandi employed to forecast the expected outcome of a disaster plays a remarkable part in its management. The utilization of resources will be more effective and efficient if the prediction is accurate. Although we are in the 21st century, with a lot of technologies, plenty of people still suffer considerably because of less knowledge of a true pandemic and disaster management system.

Accurate forecasting of any type of natural disaster can be a helping hand in minimizing the effects and increasing disaster response. To reduce the effects of the outburst of a pandemic, a timely diagnosis is a requirement. We can use social media such as Facebook and Twitter to communicate valuable information regarding precautions and safety measures during the disaster, as well as manage hospital beds and equipment efficiently by artificial intelligence (AI).

16.2 AIM

- To predict natural disasters using AI and satellite images, such as landslides, volcano eruptions, floods, and move people as a precautionary measure from such areas.

DOI: 10.1201/9781032724508-18

- Apply AI to easily and quickly perform rescue and search operations to find and save the lives of victims of natural disasters and man-made disasters.
- Use AI to communicate information about precautions and safety measures of a pandemic through social media.
- Employ AI algorithms for the management of emergency services and for using services more efficiently during disasters.
- Utilize AI to detect fake messages that will reduce humors in case of disaster.

16.3 METHODS USED

We can predict natural disasters by using some AI approaches. Figure 16.1 shows that all the recent researchers used AI methods for disaster prediction and pre-disaster and post-disaster management of resources. The algorithms used are support vector machines (SVMs), random forest method (RFM), decision tree method, and convolution neural networks. AI gives an idea of promises to precisely predict natural calamities. Decision Tree comes under a supervised learning technique used for solving regression and classification problems. This method is a graphical way of getting every possible solution to a given problem.

The next algorithm is a random forest algorithm (RFA). The RFA is an advanced version of the decision tree algorithm. It is a classifier that carries more than one decision tree on numerous subsets of given datasets and calculates the average of the dataset to enhance predictive accuracy. Instead of dependency on a sole decision tree, random forest calculates the prediction of every single tree, and based on the maximum votes of prediction, it will predict the result. RFM can explore multiple data sources, which are satellite imagery, previous disaster data, and weather data, to forecast the possibility of a natural disaster occurring. As we increase the trees in

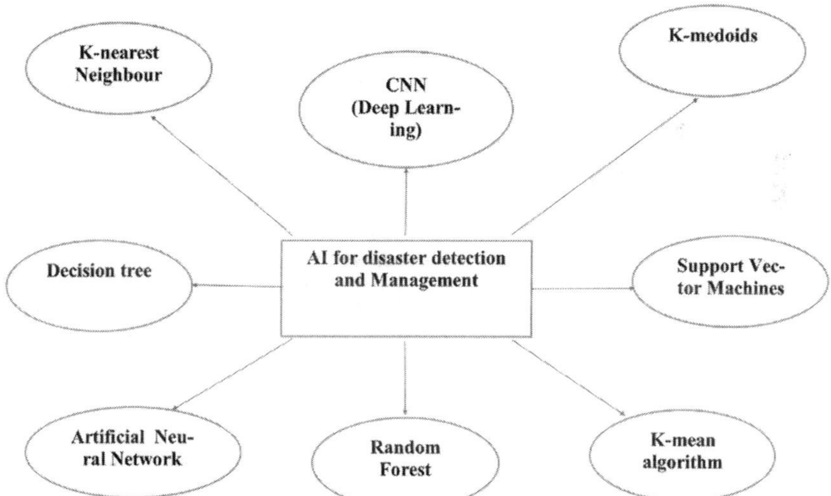

FIGURE 16.1 Various AI for disaster management and detection

the random forest, it will create more accuracy and decrease the chances of overfitting problems.

The third algorithm is the SVM. It is also associated with the group of supervised learning algorithms. The purpose of this method is to build the supreme line that will separate n-dimensional space into classes. The next algorithm is convolution neural networks in deep learning. This algorithm is effectively suited for analyzing visual data. The next algorithm is the k-mean algorithm. Its main goal is to partition a set of data points into distinct groups or clusters based on their similarities. The next algorithm is k-nearest neighbors (KNN). It assigns a label or value based on the most common label or the average of the neighbors. It is suitable for datasets with outliers or non-Euclidean distances. The last algorithm is an artificial neural network. Each neuron in an ANN receives input and applies a mathematical function. Table 16.1 illustrates the researcher's recent use of AI methods for disaster prediction and management with their applications in real scenarios.

It uses the concept from linear algebra, especially convolutions functioning to extract attributes and recognize patterns inside images. The paper "A UAV-Assisted Edge Framework for Real-Time Disaster Management" presents several practical implications that can significantly enhance disaster response efforts [16]. The framework allows for instantaneous data collection and choice selection abilities in disaster situations. By utilizing AI-powered UAVs, emergency responders can access hard-to-reach areas and obtain critical situational awareness, which is essential for

TABLE 16.1
AI Methods and Usage

Sr. No.	AI	Methods used for
1.	Artificial Neural Network [1–3]	Rainfall Prediction, forecast flood risk, Storm intensity, Detecting ISPA.
2.	Decision Tree [4]	Activating contraflows
3.	Random Forest [5–7]	Prediction of flood, classification of land cover, Detecting changes post disaster, detecting building damages post disaster.
4.	Support Vector Machines [8–11]	Flood areas detection, recognition of disaster, preventing crowds from disaster, determining the evacuation route, categorizing areas into favorable and unfavorable restoration, forecasting antigenic variants of H1N1 influenza, identifying ISPA.
5.	Convolutional Neural Networks [5, 12–14]	Prediction of floods, prediction of flood and landslides, classifying the crowd situation post disaster, future disaster risk evaluation.
6.	K-mean [8, 15]	Determine the evacuation route, classification of crowd situations post disaster, identifying regions impacted and affected by flooding.
7.	K-medoids	Determine the evacuation route post disaster.

planning and executing search and rescue missions effectively. The paper presents an innovative method to enhance the performance of locality preserving projections (LPPs) using heuristic methods, specifically focusing on the enhancement of the affinity matrix through generalized heat kernels [17]. The study demonstrates that using heuristic methods such as Harmony Search (HS) and Genetic Algorithm (GA) significantly improves classification accuracy in embedding spaces compared to traditional methods. The "CAMM: Cross-Attention Multimodal Classification of Disaster-Related Tweets" presents a novel approach to analyzing disaster-related data on social media platforms, particularly Twitter [18]. The CAMM model integrates both textual and visual data from tweets, allowing for a deeper insight into disaster scenarios. This can help humanitarian agencies respond more effectively by providing them with richer information about ongoing events.

The paper "Extracting Resource Needs and Availabilities from Microblogs for Aiding Post-Disaster Relief Operations" presents several practical implications for improving disaster response efforts through the effective use of microblogging data [19]. The paper "Facilitating mmWave Mesh Reliability in PPDR Scenarios Utilizing Artificial Intelligence" presents several practical implications for public protection and disaster relief (PPDR) scenarios, particularly in enhancing communication reliability through advanced technologies [20]. The use of mmWave technologies allows for high bandwidth communication, which is crucial in environments with obstacles such as smoke or water vapor. This technology can provide multiple communication paths, ensuring that even if one path is blocked, others remain available, thus maintaining connectivity during critical operations.

The chapter introduces an innovative fire detection system that has several practical implications for enhancing fire safety in indoor environments [21]. The suggested system aims to enable the early detection of fires, which is vital for reducing damage and ensuring safety. By integrating multiple sensor types, it can identify various components emitted from fires, allowing for quicker response times. One major challenge with current fire detection systems is the high rate of false alarms, often caused by relying on a single type of sensor, such as smoke detectors. The new system utilizes a similarity matching-based detection algorithm that effectively reduces false alarms by accurately classifying sensor signals as "fire" or "non-fire."

The chapter introduces an intelligent multisource remote sensing technique designed to enhance disaster detection in high-altitude mountain forest regions [22]. By utilizing a detection approach based on multisource images, the research addresses the challenges of detecting damaged objects in disaster areas. The method has shown a significant improvement in accuracy, with a reported 7% increase over existing single-source detection algorithms such as Yolov4/v5. This means that rescue teams can rely on more accurate data to make informed decisions.

16.3.1 Decision Tree

Decision tree is mostly preferred for the classification problems. For decision trees internal nodes work as attributes of datasets, branches work as decision rules, and leaf nodes work as outcomes. This algorithm asks a question, and depending on the response (Yes or No), this algorithm breaks the trees into sub-trees. Figure 16.2 explains the flow of the decision tree from the root node to the branch node and leaf node.

Decision nodes enable the creation of any finding and allow for a significant number of branches. A leaf node is at the end of a branch; it will not be further extended.

For calculating the category of the dataset, the decision tree begins at the root node of the tree. The decision tree method compares the values of the root element with the saved (real dataset) attribute value, and based on this comparison, it follows the branch and moves the subsequent node.

For the next node, this method once more matches the attribute value to the further sub-nodes and moves forward. It repeatedly does this process to reach the leaf node of every tree.

The study provides valuable insights into forecasting water accumulation in urban settings, offering practical applications for flood control and urban planning strategies [23]. It emphasizes the importance of accurate predictions of water accumulation, which can help urban water management authorities to respond effectively to flooding events. By utilizing the GBDT (gradient boosting decision tree) algorithm, the model can provide real-time predictions, allowing for timely interventions to reduce flood damage. The findings indicate that the GBDT model significantly improves prediction accuracy, with a mean relative error of 19.77% and a peak average relative error of 5.48% for maximum water accumulation depth. Such accuracy is essential for urban planners to make well-informed decisions about flood preparedness and response strategies.

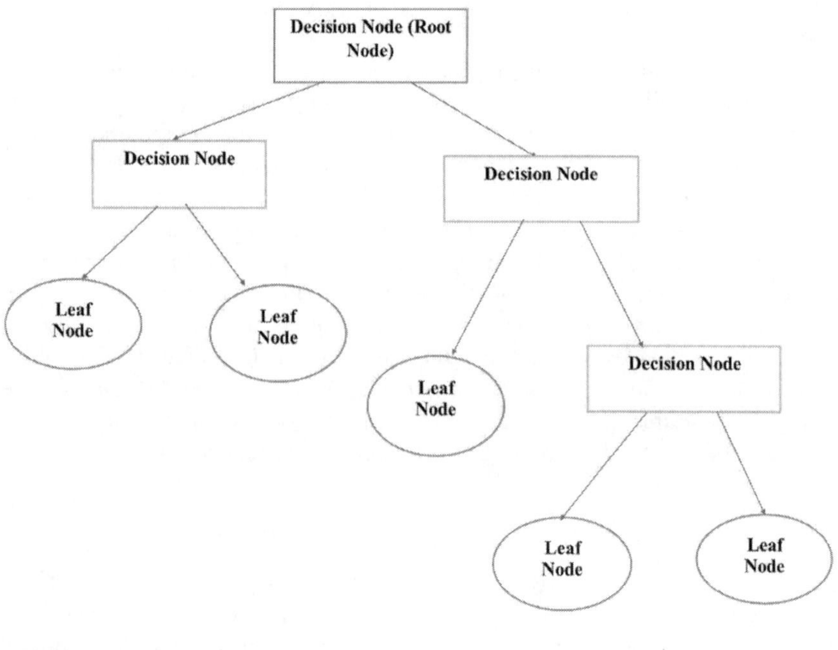

FIGURE 16.2 Decision tree

16.3.2 Random Forest Algorithm

Random forest contains more than one decision tree, each for a different subset based on a prescribed dataset, and calculates the means to enhance the precision of a given dataset. The increased number of trees utilized in the RFM enhances precision and mitigates the risk of overfitting. Figure 16.3 illustrates the workflow of the RFA. It has separate training and test data. Training data is distributed to *N* number of batches, and each batch creates a separate decision tree. The result of all decision trees is aggregated and a final prediction is achieved.

These are some points used in the RFM:

- RFA consumes a smaller amount of training time in comparison to other methods.
- RFA forecasts results with greater precision; even for a huge dataset it gives results efficiently.
- RFA can sustain precision for a large amount of data that is missing.

The research focused on Southwest China, utilizing the black marble day-level Nighttime Light (NTL) imagery from the VNP46A2 product to construct radiance multivariate features. This data was essential for analyzing forest fire occurrences in the region during 2021 and 2022 [24]. The effectiveness of the RFM was validated by comparing its outcomes with those from other machine learning (ML) models, such as eXtreme Gradient Boosting (XGBoost), Adaptive Boosting (AdaBoost), and K-Nearest Neighbor (K-NN).

The paper on RFM presents several practical implications for logistics companies and supply chain management [25]. The intelligent RFM enables quick detection and categorization of potential risks within the logistics supply chain network.

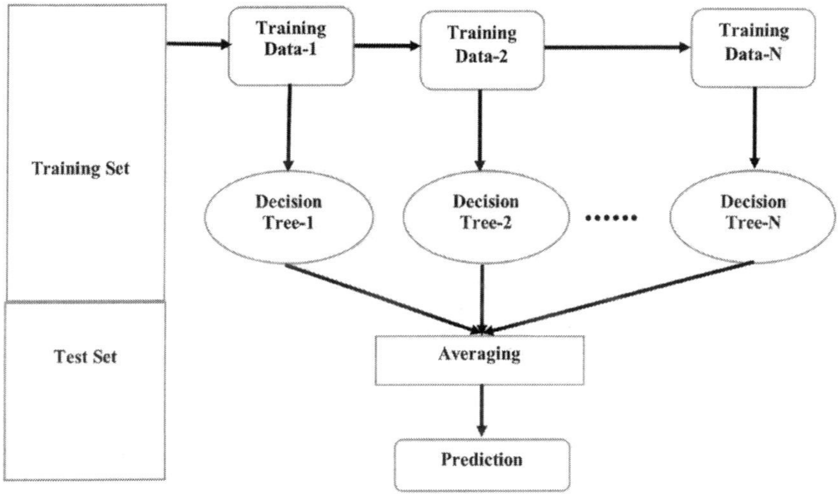

FIGURE 16.3 Random forest algorithm

This capability enables companies to proactively address risks before they escalate, thereby improving overall operational stability.

16.3.3 Support Vector Machine (SVM)

SVM belongs to supervised learning algorithms and is one of the popular methods. SVM is more famous for classification problems in AI. Figure 16.4 shows a diagram illustrating the characteristics and strengths of an SVM and highlights key features of SVMs. The objective of this algorithm is to create an ideal boundary that separates n-dimensional space into distinct classes. The SVM method manages to identify a hyper-plane that categorizes the data points. It focuses on finding a plane with the maximal range. For example, SVM algorithms can be used for determining whether a person is suffering from acute respiratory infections (ISPA).

The paper presents a hybrid model for classification in remote sensing data, combining the strengths of two methods: Least Squares Support Vector Machine (LSVM) and Support Vector Subspace Analysis (SVSA) [26]. The findings suggest that further research could explore the hybrid model's application across different datasets and its potential enhancements, particularly in optimizing classification accuracy without significant computational costs. The study demonstrates that the newly designed Selective Arithmetic Mean (SAM) approach significantly enhances the accuracy and reliability of air temperature (Ta) retrieval compared to traditional methods. The SAM model yielded lower mean absolute error (MAE) and root mean square error (RMSE) values, demonstrating improved performance in predicting *Ta* under different conditions [27]. The model effectively identifies and discards problematic predictions, which helps in minimizing the uneven error distribution that was a limitation in previous SVM models.

The paper presents a novel approach to assessing geo-hazards risk using the RFM combined with geographic information systems (GIS) [28]. The RFM demonstrated substantially improved accuracy compared to the SVM model, with an out-of-bag (OOB) error of just 3.6%. This indicates that the RFM has broad applicability for geo-hazards risk assessment, providing reliable results for disaster management strategies.

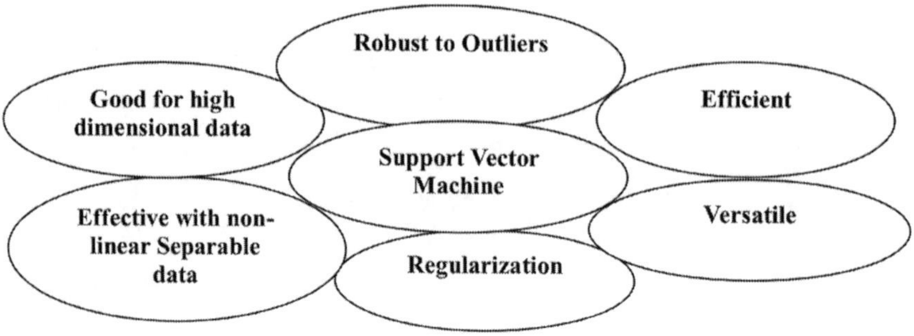

FIGURE 16.4 Support vector machine

The paper presents a ML that achieved an impressive accuracy of 99.8%, outperforming the other models [29]. It discusses various methodologies for landslide susceptibility mapping, highlighting the practical implications of these methods in real-world applications [30]. The study highlights the application of ML methods, specifically SVM and adaptive network-based fuzzy inference system (ANFIS), which have shown to outperform traditional methods such as decision trees (DT) and linear regression (LR) in terms of accuracy in landslide susceptibility mapping. This improvement can lead to better risk assessment and management in areas prone to landslides.

16.3.4 Convolutional Neural Network (CNN)

CNN is a kind of deep learning model, which belongs to a deep learning algorithm family, and it is suitable for exploring visual data. It is sometimes called convents (convolution networks) and uses the concept from linear algebra, especially convolution functioning, for extracting the attributes and identifying the patterns within images. CNNs are chiefly employed in applications like image classification and object detection but can also be adapted for other domains, such as audio signal processing. The structure of CNNs is designed to efficiently analyze visual data, enabling them to process entire images at once. This makes CNNs a powerful tool for computer vision applications, including self-driving vehicles, medical image analysis, and facial recognition. Unlike traditional neural networks, which often required processing visual data incrementally with lower-resolution inputs, CNNs excel by processing high-resolution data holistically. This capability allows them to outperform traditional neural networks in tasks related to images and, to a lesser extent, in areas like audio and speech processing. Figure 16.5 shows the architecture of a basic Convolutional Neural Network (CNN) used for image processing and classification tasks. It has three operations convolution, pooling, and fully connected layers repeatedly used.

The proposed deep multi-instance convolutional neural network (DMCNN) effectively addresses the challenges of conventional disaster categorization methods, particularly in high-definition remote-sensing images. It overcomes issues such as complex scene variations, diverse feature distributions, and missing sample class labels [31]. Despite the advancements, the DMCNN model still encounters occasional errors in classification issues within the categorized images. Future work will focus on developing a more robust network model that can better detect and classify disasters, particularly in challenging environments characterized by random distributions and small targets.

FIGURE 16.5 Convolutional neural network

16.3.5 K-Nearest Neighbors (KNN)

KNN is a position-determining method used for classification as well as used for regression. The KNN can be applied in synchronization with some other methods to find ISPA, and likewise, it has been applied to diagnose influenza in patients. The KNN method calculates the resemblance between the new case and existing cases and after that places the new case into the most similar category from the available categories. The paper presents a method for estimating the execution time of various machine learning classifiers using a metric called FPTC (Function of Parameters and Time Complexity) [32]. The research provides a systematic approach to evaluate the time efficiency of five widely used classifiers (KNN, LR, CART, RF, and SVM). This allows practitioners to filter algorithms that not only perform well but also have lower FPTC values, making it easier to choose efficient algorithms for real-time applications.

16.3.6 K-mean Algorithm

K-mean is a clustering type of algorithm belonging to unsupervised learning. It can group the unlabeled datasets into different clusters. It is based on the centroid approach and each cluster is concerning centroid. The main objective of the k-mean is to minimize the sum of distances between the corresponding cluster and the data points. The input is an unlabeled data set, and the process is to split the data set into n numbers of clusters and continue to repeat this process until the best cluster is achieved.

There are various mechanisms for electing manager nodes within these spatio-temporal components. These mechanisms are based on centrality measures such as degree centrality, eigenvector centrality, and the K-means paradigm. The goal is to select nodes that hold structural importance, thereby minimizing the spread of faulty information [33]. Extensive simulations were conducted to estimate the probabilities of managing network nodes effectively. The authors express intentions to explore alternative mobility models and develop an information-theoretic framework to enhance the probabilistic management scheme. The paper presents significant findings regarding the management of drone small cells (DSCs) using a machine learning-based framework [34]. The affinity propagation algorithm is specifically aimed at mitigating inter-DSC interference, while the K-means algorithm is employed to determine optimal positions for DSCs to enhance signal quality for ground users.

The paper "Joint Optimization of Real-Time Deployment and Resource Allocation for UAV-Aided Disaster Emergency Communications" presents several practical implications that can significantly enhance emergency response efforts during disasters [35]. The proposed framework utilizes UAVs as flying base stations to restore and maintain network connectivity in disaster-stricken areas. This is crucial for effective communication among emergency responders and affected individuals, ensuring timely information exchange and coordination.

16.3.7 Artificial Neural Network

Artificial neural networks have a basic unit called artificial neurons. These artificial neurons are arranged in a sequence of layers that make the whole arrangement called artificial neural networks. A layer can take a few artificial neurons or have millions of artificial neurons, and as the number of neurons is increasing the complexity is also increasing. The input layer has outside world data that a neural network needs to learn after that input layer information passes through hidden layers and these layers transform the data into valuable form for the output. The output layer will provide the output after processing input data. The concept of ANN is like the human neurons' serve system. ANN training is done by a training set. Once an artificial neural network is trained with enough data, after that it will predict the output with a new set of inputs. Figure 16.6 illustrates a fully connected artificial neural network (ANN). It showcases the flow of data through the network, beginning at the input layer, progressing through one or more hidden layers, and concluding at the output layer.

The paper presents several practical implications for enhancing emergency rescue operations in urban waterlogging scenarios [36]. The study introduces a wargame-based evaluation approach that allows emergency responders to simulate various scenarios. This helps in making informed decisions quickly, which is crucial during urban flooding situations where time is of the essence. By utilizing advanced analytical tools like neural networks, the research aims to minimize costs associated with emergency rescue operations. The paper presents significant findings regarding the use of deep learning systems for use in emergency response scenarios utilizing UAVs (Unmanned Aerial Vehicles) [37]. The research emphasizes the development of an effective deep learning system capable of real-time recognition and classification of disaster events from UAVs. This system is designed to operate effectively under the constraints typical of emergency scenarios. UAVs equipped with advanced deep learning systems for emergency response, showcasing significant improvements in

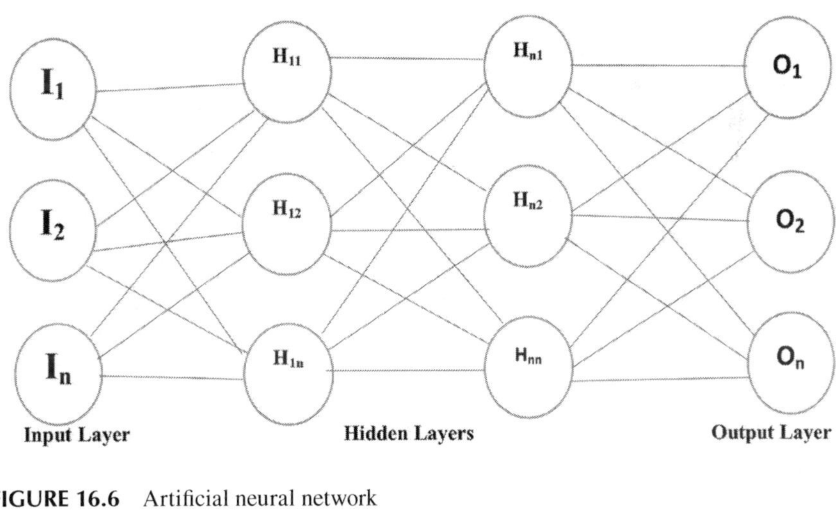

FIGURE 16.6 Artificial neural network

processing speed. The paper presents significant findings regarding the prediction of debris flow scales using a novel model [38]. The research introduces a hybrid model that combines the Cubic map Improved Harris Hawk Optimization (CIHHO) algorithm with a Back Propagation (BP) neural network. This model aims to enhance the precision of debris flow scale predictions, addressing the limitations of existing models.

The paper presents several important conclusions regarding the evaluation of government emergency management systems, especially in the context of the COVID-19 pandemic [39]. The study demonstrates that the BP neural network method is highly effective for establishing a smart assessment framework for governmental crisis management. It shows significant advantages over other evaluation methods, particularly in terms of adaptability and accuracy in real-time data processing.

16.3.8 K-Medoids Algorithm

This algorithm is also known as partitioning around medoid. This method is a clustering approach used as a basis for variation between data points. K-medoids run smoothly with outliers. A medoid is a point in the cluster such that its diversity with each of the supplementary points in the group is minimal.

An "Overview of the Integration of Social Media and AI" presents several important conclusions regarding the fusion of edge systems in disaster management [40]. The survey highlights the critical role of social media analytics and AI in emergency situations. Social media serves as a rich data source, while AI is crucial for handling the vast volumes of data produced by smart devices. This combination is vital for predicting, detecting, managing information, and enabling authorities to act efficiently to emergencies.

16.4 CONCLUSION

The integration of AI into disaster prediction and management marks a groundbreaking shift in how we approach mitigating the impacts of both natural and man-made disasters. AI's ability to process large volumes of real-time data, detect patterns, and generate accurate predictions is instrumental in providing early warnings, which can significantly reduce the loss of life. By leveraging technologies such as machine learning, satellite imagery, and sensor networks, AI can forecast disasters like earthquakes, floods, and hurricanes with greater precision, enabling timely responses and risk reduction strategies.

Beyond just predicting these catastrophic events, AI is essential in managing disasters. During crises, AI-powered systems are essential for enhancing decision-making under pressure, particularly in resource allocation, emergency response coordination, and logistics management. AI-driven drones, for example, can quickly assess disaster zones, providing critical information about damage and helping locate survivors. Additionally, AI tools like chatbots and predictive models ensure seamless communication and efficient supply chain operations, helping to distribute resources where they are most needed in real time.

Despite these promising capabilities, the effective use of AI in disaster scenarios requires collaboration between governments, organizations, and tech providers to ensure ethical deployment and equitable access. Building robust infrastructure, prioritizing inclusivity, and addressing potential biases are essential for maximizing the impact of AI in disaster situations. Moreover, ongoing investments in AI technologies will help create resilient systems that can not only predict and respond to disasters more efficiently but also facilitate long-term recovery efforts. By fostering the development and application of AI in disaster management, we can ensure a more prepared, responsive, and sustainable future, where communities around the world are more prepared to tackle the challenges presented by natural and human-made disasters.

REFERENCES

1. Wang, Qingqing, Hong Liu, Kaizhou Gao, and Le Zhang. "Improved multi-agent reinforcement learning for path planning-based crowd simulation." *IEEE Access* 7 (2019): 73841–73855.
2. Singh, Krishna Kant, M. J. Nigam, Kirat Pal, and Akansha Mehrotra. "A fuzzy Kohonen local information c-means clustering for remote sensing imagery." *IETE Technical Review* 31, no. 1 (2014): 75–81.
3. Rodriguez-Galiano, Victor Francisco, Bardan Ghimire, John Rogan, Mario Chica-Olmo, and Juan Pedro Rigol-Sanchez. "An assessment of the effectiveness of a random forest classifier for land-cover classification." *ISPRS Journal of Photogrammetry and Remote Sensing* 67 (2012): 93–104.
4. Anzengruber, Bernhard, Danilo Pianini, Jussi Nieminen, and Alois Ferscha. "Predicting social density in mass events to prevent crowd disasters." In *Social Informatics: 5th International Conference, SocInfo 2013*, Kyoto, Japan, November 25–27, 2013, Proceedings 5, pp. 206–215. Springer International Publishing, 2013.
5. Juhana, Tutun. "Mobile application design of disaster rescue team tracking." In *2014 8th International Conference on Telecommunication Systems Services and Applications (TSSA)*, pp. 1–3. IEEE, 2014.
6. Le, Van-Minh. "Machine learning methods for optimization in multi-agent decision support system: Application to sign placement for tsunami evacuation." PhD dissertation, Université Sorbonne Paris Cité, 2016.
7. Dey, Shopan, Murshed Nawaz Kamal, Soumya Dutta, Abhishek Tiwari, Santanu Ray, Md Jamil Moatasimbillah, Nabanit Saha et al. "Ad-hoc networked UAVs as aerial mesh network for disaster management application and remote sensing: An approach." In *2017 8th IEEE Annual Information Technology, Electronics and Mobile Communication Conference (IEMCON)*, pp. 301–304. IEEE, 2017.
8. Xu, Xiaolong, Lei Zhang, Stelios Sotiriadis, Eleana Asimakopoulou, Maozhen Li, and Nik Bessis. "CLOTHO: A large-scale Internet of things-based crowd evacuation planning system for disaster management." *IEEE Internet of Things Journal* 5, no. 5 (2018): 3559–3568.
9. Cho, Sung-Bae. "Exploiting machine learning techniques for location recognition and prediction with smartphone logs." *Neurocomputing* 176 (2016): 98–106.

10. Namoun, Abdallah, Aabid Mir, Ahmad B. Alkhodre, Ali Tufail, Ahmed Alrehaili, Mohammad Farquad, Majid Alwaqdani, Turki Alghamdi, and Mohamed Benaida. "A multi-agent architecture for evacuating pilgrims in panic and emergency situations: The Hajj scenario." *Journal of Theoretical and Applied Information Technology* 96, no. 20 (2018): 6665–6676.
11. Hassija, Vikas, Vikas Saxena, Vinay Chamola, and F. Richard Yu. "A parking slot allocation framework based on virtual voting and adaptive pricing algorithm." *IEEE Transactions on Vehicular Technology* 69, no. 6 (2020): 5945–5957.
12. Amit, Siti Nor Khuzaimah Binti, and Yoshimitsu Aoki. "Disaster detection from aerial imagery with convolutional neural network." In *2017 International Electronics Symposium on Knowledge Creation and Intelligent Computing (IES-KCIC)*, pp. 239–245. IEEE, 2017.
13. Yabe, Takahiro, and Satish V. Ukkusuri. "Integrating information from heterogeneous networks on social media to predict post-disaster returning behavior." *Journal of Computational Science* 32 (2019): 12–20.
14. Wei, Guanghui, and Zhou Sheng. "Image quality assessment for intelligent emergency application based on deep neural network." *Journal of Visual Communication and Image Representation* 63 (2019): 102581.
15. Shibata, Kyosuke, and Hiroshi Yamamoto. "People crowd density estimation system using deep learning for radio wave sensing of cellular communication." In *2019 International Conference on Artificial Intelligence in Information and Communication (ICAIIC)*, pp. 143–148. IEEE, 2019.
16. Ijaz, Haris, Rizwan Ahmad, Rehan Ahmed, Waqas Ahmed, Yan Kai, and Wu Jun. "A UAV-assisted edge framework for real-time disaster management." *IEEE Transactions on Geoscience and Remote Sensing* 61 (2023): 1–13.
17. Taşkın, Gülşen, and Oğuzhan Ceylan. "An adaptive affinity matrix optimization for locality preserving projection via heuristic methods for hyperspectral image analysis." *IEEE Journal of Selected Topics in Applied Earth Observations and Remote Sensing* 12, no. 12 (2019): 4690–4697.
18. Khattar, Anuradha, and S. M. K. Quadri. "Camm: cross-attention multimodal classification of disaster-related tweets." *IEEE Access* 10 (2022): 92889–92902.
19. Basu, Moumita, Anurag Shandilya, Prannay Khosla, Kripabandhu Ghosh, and Saptarshi Ghosh. "Extracting resource needs and availabilities from microblogs for aiding post-disaster relief operations." *IEEE Transactions on Computational Social Systems* 6, no. 3 (2019): 604–618.
20. Pirmagomedov, Rustam, Dmitri Moltchanov, Aleksandr Ometov, Khan Muhammad, Sergey Andreev, and Yevgeni Koucheryavy. "Facilitating mmWave mesh reliability in PPDR scenarios utilizing artificial intelligence." *IEEE Access* 7 (2019): 180700–180712.
21. Baek, Jaeseung, Taha J. Alhindi, Young-Seon Jeong, Myong K. Jeong, Seongho Seo, Jongseok Kang, and Yoseob Heo. "Intelligent multi-sensor detection system for monitoring indoor building fires." *IEEE Sensors Journal* 21, no. 24 (2021): 27982–27992.
22. Wang, Haifeng, Hualong Cao, Yan Kai, Haicheng Bai, Xuefeng Chen, Yang Yang, Lin Xing, and Chengjiang Zhou. "Multi-source remote sensing intelligent characterization technique-based disaster regions detection in high-altitude mountain forest areas." *IEEE Geoscience and Remote Sensing Letters* 19 (2022): 1–5.
23. Saengtabtim, Kumpol, Natt Leelawat, Jing Tang, Wanit Treeranurat, Narunporn Wisittiwong, Anawat Suppasri, Kwanchai Pakoksung, Fumihiko Imamura, Noriyuki Takahashi, and Ingrid Charvet. "Predictive analysis of the building damage from the 2011 Great East Japan tsunami using decision tree classification related algorithms." *IEEE Access* 9 (2021): 31065–31077.

24. Yu, Yuehan, Lili Liu, Zhijian Chang, Yuanqing Li, and Kaifang Shi. "Detecting Forest Fires in Southwest China From Remote Sensing Nighttime Lights Using the Random Forest Classification Model." *IEEE Journal of Selected Topics in Applied Earth Observations and Remote Sensing* (2024).
25. Xu, Zhiming, Weichuan Ni, Shaojiang Liu, Feng Wang, and Jing Li. "Logistics Supply Chain Network Risk Prediction Model Based on Intelligent Random Forest Model." *IEEE Transactions on Engineering Management* (2023).
26. Kaya, Gulsen Taskin. "A hybrid model for classification of remote sensing images with linear SVM and support vector selection and adaptation." *IEEE Journal of Selected Topics in Applied Earth Observations and Remote Sensing* 6, no. 4 (2013): 1988–1997.
27. Qin, Jiang-Lin, Xiu-Hao Yang, Ji-Tong Luo, He Fu, Xiu-Feng Lei, Jun Wei, Yuan-Rui Qin, and Zi-Xin Yang. "An improved novel nonlinear algorithm of area-wide near-surface air temperature retrieval." *IEEE Journal of Selected Topics in Applied Earth Observations and Remote Sensing* 11, no. 3 (2018): 830–844.
28. Huang, Pei, Li Peng, and Hongyi Pan. "Linking the random forests model and GIS to assess geo-hazards risk: a case study in Shifang County, China." *IEEE Access* 8 (2020): 28033–28042.
29. Kumar, M. Ankush, and A. Jaya Laxmi. "Machine learning based intentional islanding algorithm for ders in disaster management." *IEEE Access* 9 (2021): 85300–85309.
30. Jebur, Mustafa Neamah, Biswajeet Pradhan, and Mahyat Shafapour Tehrany. "Manifestation of LiDAR-derived parameters in the spatial prediction of landslides using novel ensemble evidential belief functions and support vector machine models in GIS." *IEEE Journal of Selected Topics in Applied Earth Observations and Remote Sensing* 8, no. 2 (2014): 674–690.
31. Li Chengfan, Zixuan Zhang, Lan Liu, Jung Yoon Kim, and Arun Kumar Sangaiah. "A Novel Deep Multi-Instance Convolutional Neural Network for Disaster Classification From High-Resolution Remote Sensing Images." *IEEE Journal of Selected Topics in Applied Earth Observations and Remote Sensing* (2023).
32. Zheng, Xiaorou, Jianxin Jia, Shanxin Guo, Jinsong Chen, Luyi Sun, Yingfei Xiong, and Wenna Xu. "Full parameter time complexity (FPTC): A method to evaluate the running time of machine learning classifiers for land use/land cover classification." *IEEE Journal of Selected Topics in Applied Earth Observations and Remote Sensing* 14 (2021): 2222–2235.
33. Badonnel, Remi, Radu State, and Olivier Festor. "A probabilistic approach for managing mobile ad-hoc networks." *IEEE Transactions on Network and Service Management* 4, no. 1 (2007): 39–50.
34. Wang, Li-Chun, Yung-Sheng Chao, Shao-Hung Cheng, and Zhu Han. "An integrated affinity propagation and machine learning approach for interference management in drone base stations." *IEEE Transactions on Cognitive Communications and Networking* 6, no. 1 (2019): 83–94.
35. Do-Duy, Tan, Long D. Nguyen, Trung Q. Duong, Saeed R. Khosravirad, and Holger Claussen. "Joint optimisation of real-time deployment and resource allocation for UAV-aided disaster emergency communications." *IEEE Journal on Selected Areas in Communications* 39, no. 11 (2021): 3411–3424.
36. Zhang, Han, Yanyan Huang, Ze Geng, and Tiande Chen. "Effectiveness evaluation of emergency rescuing plans oriented to urban waterlogging based on a neural network model." *IEEE Access* (2024).
37. Kyrkou, Christos, and Theocharis Theocharides. "EmergencyNet: Efficient aerial image classification for drone-based emergency monitoring using atrous convolutional feature fusion." *IEEE Journal of Selected Topics in Applied Earth Observations and Remote Sensing* 13 (2020): 1687–1699.

38. Qiang, Yue, Xinlong Xu, Li Li, Siyu Liang, Xi Wang, Cheng Chen, and Xinyi Tan. "High-precision Debris flow scale prediction model based on CIHHO algorithm combined with multilayer perceptron neural network." *IEEE Access* (2024).
39. Jia, Dian, and Zhaoyang Wu. "Intelligent evaluation system of government emergency management based on BP neural network." *IEEE Access* 8 (2020): 199646–199653.
40. Aboualola, Mohamed, Khalid Abualsaud, Tamer Khattab, Nizar Zorba, and Hossam S. Hassanein. "Edge technologies for disaster management: A survey of social media and artificial intelligence integration." *IEEE Access* (2023).

17 Investigation of Stress among University Students Using PSS and AI-Based Analysis

Dhruva Bisht, Kushagra Trivedi, Akshay Patwal, Ashutosh Farswan, and Vivek Tomar

17.1 INTRODUCTION

Stress is a pervasive experience that affects individuals across various contexts and stages of life. This includes fields such as medical management, husbandry, and conveyance and thus is not limited to sole education [1]. For undergraduate students, stress is a common occurrence and can be attributed to academic, social, and personal factors. While some levels of stress can be motivating and necessary for optimal performance, prolonged or excessive stress can have detrimental effects on an individual's mental health and academic outcomes. Given the significant impact of stress on undergraduate students, extensive research has been conducted to explore its causes, consequences, and potential interventions. Despite this research, stress continues to be a pervasive issue, and the mechanisms underlying its development and persistence remain unclear. Undergraduate students are under a lot of stress because of their rigorous academic schedules, intense competition, and long study sessions. Stress has a significant impact on mental health since it can cause unhelpful behaviors, emotional suffering, and defensive reactions [2]. This might offer elucidation as to the rationale behind the prevalence of health ailments among numerous university students. When stress spirals into an unrestrained state, it commences to exert a profound influence on human well-being, engendering a myriad of maladies such as disrupted slumber, cardiovascular afflictions, or even cerebral apoplexy [3 –5]. Recently, the COVID-19 pandemic has also caused an increase in student stress [5]. Using a commonly reported stress scale questionnaire, the aim of this study is to gather information on undergraduate students' reported stress levels. To find trends and patterns in the individual's stress levels, the data will be analyzed. The rationale of this investigation is to gain profound insights into the stress thresholds of the colleagues.

DOI: 10.1201/9781032724508-19

17.2 LITERATURE SURVEY

This research review investigates the effect of perceived stress on undergraduate student's mental health. According to the available research, undergraduate students frequently experience perceived stress, especially in times of major stresses such as the COVID-19 pandemic [6]. It was reported that 10.8% of students were in the high stress level range, while 32.7% were in the moderate range. This research was focused on the medical students with a heavy percentage of female students amounting to 60.8% of the entire data.

Most of the research and reviews analyzed found negative psychological outcomes such as post-traumatic stress disorder symptoms [7], bewilderment, and hostility. Sources of stress encompassed extended periods of quarantine, apprehensions regarding illness, exasperation, ennui, inadequate provisions, erroneous information, financial deprivation, and social stigma. Additionally, it has been hypothesized that women displayed greater stress levels as compared to men [8].

However, there is little data and study available on undergraduates taking technical courses. By gathering information from undergraduates, particularly those enrolled in technical courses, this project seeks to close this knowledge gap while also raising awareness. Positive emotion in the study of foreign languages has been studied thoroughly over the past decade in a variety of academic fields, including Foreign Language Enjoyment (FLE). When a person's psychological needs are satisfied, they experience FLE, which has been defined as a non-negative disposition in this research, and the observed degrees of Pressure (PS), Multilingualism (DM), impacts on Foreign Language Education (FLE) during second language acquisition and Subjective Well-Being (SWB) are thoroughly investigated. The data were amassed from the pupils who were studying either Italian or French (35 trilingual and 33 bilingual). Statistics revealed no relationship between SWB and PS. In the context of FLE, SWB and PS were found to be the most potent predictors, where PS had a noteworthy albeit slight influence. Although there was a mathematically significant variance observed for FLE among trilingual participants and their monolingual counterparts, such distinctions were not detected for SWB or PS.

Upon review of the outcomes within the scope of related research, it can be observed that achieving contentment in the process of learning a fresh language is heavily influenced by a favorable level of Subjective Wellbeing (SWB), Multilingualism (DM), and a low degree of Stress. This study's implications for vernacular teaching and the development of vernacular policies are examined [9]. Research titled "Psychometric appraisal among individuals with persistent ailment," which employed Korean adaptations of the PSS and its variations, was also performed, with the aim of evaluating the psychometric qualities of the Korean versions of PSS and its variations among those with prolonged diseases. The PSS and its variations were transformed into Korean through a combination of forward and backward translations [10]. A Psychometric Scrutiny of the PSS Among Salubrious University Students was undertaken, however no female participants were included because the study was conducted on male campuses. The goals and techniques of the study were communicated to all respondents. If a participant had a history of stress-related problems, they were omitted. Respondents were instructed to complete the

PSS version of English [10]. The model's performance may be assessed using three challenging benchmark datasets of varied quality in terms of RMSE, Pearson-R, MAE and R2, and a variety of machine learning approaches can be used. The behaviors of stressed-out students can also be observed using face-tracking devices [11]. Utilizing the dataset as the foundation, an array of machine learning (ML) methodologies was employed for regression and categorization analysis, facilitating the anticipation and classification of psychological distress. Deep neural networks aid in complex decision-making processes [12].

Python simulation regression by machine learning was executed, and performance metrics such as root mean squared error (RMSE), R2 score, mean absolute error (MAE), mean percentage absolute error (MAPE), as well as machine learning classifiers accuracy, F1 score, recall, and precision were scrutinized [6]. The analysis of a student's mental stress can thus greatly benefit from ML-based stress analysis. The multiple-layer artificial neural network gives better results as compared to single-layer artificial neural network [13]. An examination Finding signs of psychological stress among South African university students in this research, a group of university students was employed to assess the factorial composition, consistent measurement (based on sex and ethnicity), and dependability of the PSS10. Also looked at was the PSS-10 measurement legitimacy across different populations.

Five different models were tested for measuring PSS-10, including a single-factor model, a two-factor model with correlation, a bifactor model with two factors specific to domains, a bifactor model with a perceived self-efficacy factor, and a bifactor model with a distress factor. The confirmatory factor analysis and Rasch analysis were conducted to cross-validate the models. In addition to outperforming the competing latent structures, the two-factor model was reliable and consistent among groups with various ethnic and gender identities. This study confirms the PSS-10's applicability to a range of student demographics [14]. Andreou et al. scrutinized the psychodiagnostics qualities of the PSS; corroborative analyzing factors and measures of internal consistency were also utilized in the analysis [15]. Table 17.1 showcases the outcomes of the literature review conducted above.

TABLE 17.1
Outcomes of the Literature Review Table

Outcome	Details
High Stress Levels	10.8% of students
Moderate Stress Levels	32.7% of students
Focus Group	60.8% female medical students
Negative Outcomes	PTSD symptoms, bewilderment, hostility
Stress Sources	Quarantine, illness apprehensions, exasperation, ennui, inadequate provisions, misinformation, financial issues, social stigma
Gender Difference	Higher stress in women
Data Gap	Lack of data on technical course students
Foreign Language Enjoyment (FLE)	SWB, DM, and low stress as predictors; slight influence of PS
Participant Data	35 trilingual, 33 bilingual students

17.3 SCOPE

Data collection involved administering the PSS questionnaire to a sample of undergraduate students, after which the resulting data was analyzed to determine the distribution of stress levels among the students. To better understand this distribution, the labeled data was then used to create visualizations and graphs. Fuzzy The overarching objective of the research was to achieve a comprehensive understanding of the stress levels of collegiate students and provide a framework for future research into the topic. Fuzzy logic modeling forecast is the most used prediction technique [16]. As a part of this investigation, the study also assessed the credibility and dependability of the PSS questionnaire when used for Indian undergraduate students.

Cronbach's alpha was used in the study to evaluate reliability. The values of the intraclass correlation coefficient were satisfactory, and Cronbach's alpha PSS-10 was higher than the commonly identified threshold of 0.70, as represented in Table 17.2. Overall, this study contributes to a greater understanding of stress levels among undergraduate students, particularly in the Indian context, and provides insight into the credibility and validity of the PSS question set in this population. The study's results also lay a foundation for future research into stress management strategies and interventions for undergraduate students.

17.4 THE PROPOSED METHOD

Data for the study was gathered through Google Forms Survey. The Perceived Stress Scale (PSS), a popular 10-item questionnaire developed by Shelden Cohen [17] to measure stress levels, served as the basis for the first section of the study. Figure 17.1 depicts the flowchart of the methodology followed.

The PSS measures how stressed people view their current circumstances, including whether they feel overburdened or out of control. PSS is intended for people who have completed junior high school or above. Each item's response options are never, almost never, sometimes, pretty frequently, and very frequently. Adding the scores for each item results in a final score that ranges from 0 to 40. Scores between 0 and 13

TABLE 17.2
Cronbach Alpha Reliability Table

No.	Coefficient of Cronback's Alpha	Reliablity Level
1	More than 0.89	Excellent
2	0.79–0.88	Good
3	0.69–0.78	Acceptable
4	0.59–0.68	Questionable
5	0.49–0.58	Poor
6	Less than 0.48	Unacceptable

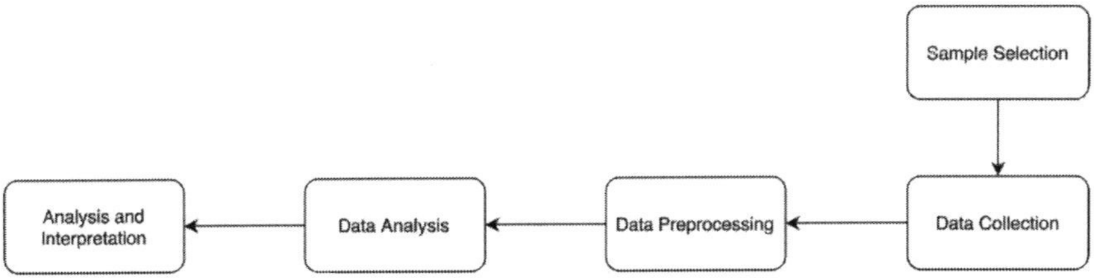

FIGURE 17.1 Flowchart of methodology followed

denote modest stress levels, whereas higher scores suggest higher perceived stress. Students between the ages of 14 and 26 are categorized as moderately stressed, while those between the ages of 27 and 40 are classified as very stressed.

The methodology, originated in 1983, continues to be consistently employed in elucidating the intricate interplay between diverse occurrences and their influence on our emotional states and stress perception. It is imperative to evaluate the cognitive and affective aspects spanning the previous month. Inquiries will be made regarding the frequency with which you have experienced or contemplated each sentiment or subject matter. Even though some of the questions are similar, you should still approach each one uniquely because there are differences between them. It's best to respond swiftly. Instead of attempting to tally how many times you experienced a certain feeling, suggest the one that sounds like an acceptable estimate.

The PSS 10 score is calculated by following these instructions: To begin, Flip your responses to questions 4, 5, 7, and 8 to start. Add up your marks for each question to reach a total. Individual scores on the scale, which ranges from 0 to 40, on the PSS reflect greater perceived stress. Scores falling within the range of 0 to 13 would signify diminished experiential stress, while scores within the range of 14 to 26 would signify a moderate degree of apprehended stress, and scores between 27 and 40 would denote an elevated level of perceived stress. The study was divided into the following portions:

- **Sample selection:** Random selection of undergraduate students from various faculties and programs.
- **Data Collection:** Standardized Perceived Stress Scale questionnaire to be administered to the participants for data collection.
- **Data Pre-processing:** Data cleaning and preprocessing will be performed to remove any missing or inconsistent data.
- **Data Analysis:** Basic statistical analysis will be performed on the collected data to obtain descriptive statistics, such as mean, median, standard deviation, and variance.
- **Analysis and Interpretation:** The final step will be to analyze and interpret the results obtained from the study, comparing the stress levels of undergraduate students from different years and genders.

17.5 STATISTICAL ANALYSIS AND RESULTS

The responses of students received through online Google Forms were transformed into an Excel sheet. Based on their PSS overall results, students were split into three groups: those who perceived low, moderate, and high stress. Prevalence of perceived stress as per categories was expressed in data analysis, we used the collected data and analyzed the results so produced. Around 72% of the students who participated in the survey belonged to the 4-year bachelors' program. In our study we have given the calculated scores of PSS the following ranges: A. 0-13 score means low stress. B. 14-26 score means medium stress. C. 27-40 score means high stress. The average value of stress as found via the PSS 10 scale was found to be 21.50, which indicated moderate stress levels among the undergraduate students. The Geometric Mean and the Median values stood at 20.79 and 20 respectively. The max value for the stress was 39 and the minimum value was 12 for the PSS 10 scale.

Online Google Forms were used to collect student responses, which were then converted into an Excel sheet. Based on their PSS total scores, the students were categorized into three groups: low, moderate, and high stress perceivers. Frequency and percentages were used to express the prevalence of perceived stress according to categories. Participants' average PSS scores were presented as mean and standard deviation. The study encompassed individuals spanning the age spectrum of 18 to 23 years, with a mean age of 20 years. Males made up 68.1% of the participants. The percentages of students in the first, second, pre-final, and final years were 9.57%, 12.17%, 36.52%, and 41.74%, respectively, as represented in Figure 17.2. The geometric mean score of perceived stress scale was 20.79. Based on the done classification, 7 (6.09%) had low experiential stress, 86 (74.78%) had moderate apprehended stress and 22 (19.13%) had elevated apprehended stress, as represented in Figure 17.3.

There is no correlation between stress levels and academic year, as denoted by the P-value surpassing the threshold of 0.05. The average level of stress was found to be 21.50; the least and greatest values were 12 and 39, respectively. Female students made up 32.45% of the student body, and their average stress score was determined to be 23.16, higher than the male students' average stress score of 20.74. Second-year undergraduate students experienced more stress than fourth-year students, whose average stress score is 21.25, at an average stress score of 24.82.

Cronbach's alpha for PSS 10 was 74.5%, higher than the necessary threshold of 70% as represented in Table 17.2. Further, expert comprehension can be incorporated in the superintended state [18].

17.6 CONCLUSION AND FUTURE SCOPE

Our study sought to determine the frequency of stress among graduate students as well as the contributing variables to stress. Students filled out online Google Forms to provide data, which we gathered and converted into an Excel file for study. Our study offers insightful information about the severity of stress among graduate students and the contributing causes. According to our findings, a sizable percentage of graduate students' report feeling moderately to severely stressed. Particularly,

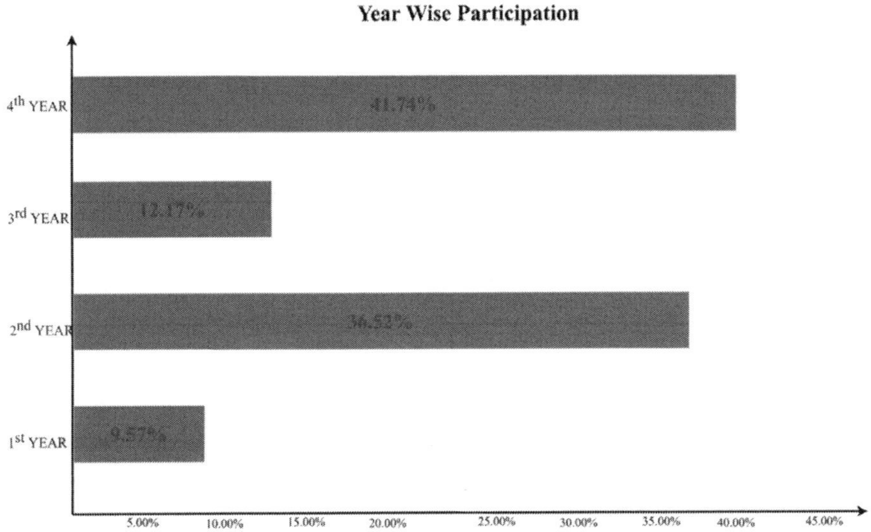

FIGURE 17.2 Percentage distribution of students of different years

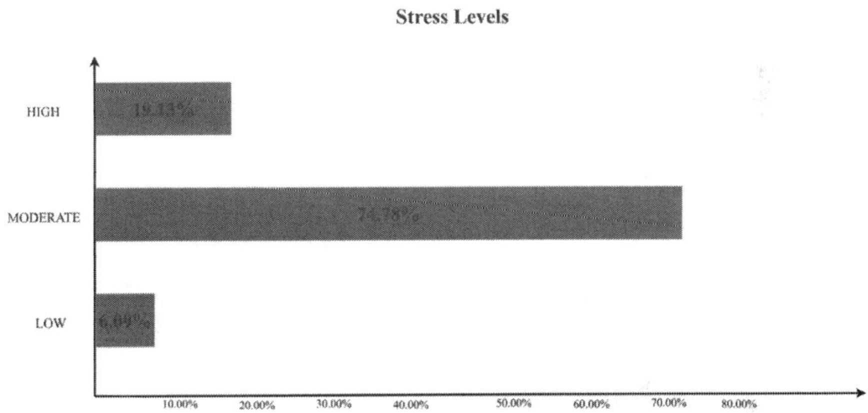

FIGURE 17.3 Percentage distribution of stress level of students

19.13% of pupils and 74.78% of students reported high and moderate levels of stress, respectively.

These results are alarming since stress can have detrimental effects on mental health, academic performance, and general well-being. In addition, we observed that female students are more stressed than male students. This finding highlights the need for gender-specific stress management strategies to cater to the unique needs of female students. Our research also discovered that the PSS 10 demonstrated high legitimacy, as evidenced by a Cronbach's alpha score of 74.5%. This suggests that

TABLE 17.3
Stress Levels and Year of Study

		Level of Stress							
		Low		Moderate		High			
Sno.	**Year of Study**	**No**	**Percentage**	**No.**	**Percentage**	**No.**	**Percentage**	**X^2 Value**	**P Value**
1.	First Year	1	14.29%	7	8.14%	3	13.65%	10.56	0.102
2.	Second Year	2	14.29%	31	36.05%	11	54.54%		
3.	Pre-final Year	2	28.57%	10	11.63%	2	13.66%		
4.	Final Year	3	42.86%	38	44.19%	8	18.15%		
	Total	7	100%	86	100%	22	100		

the scale can be used to measure stress levels in graduate students. Conclusively, the study provides an important comprehensive understanding of the prevalence of stress among graduate students and highlights the need for targeted interventions and support systems to address this issue. By providing students with effective stress management strategies and support, universities can improve their students' overall well-being and academic performance. Our findings can help inform the development of such interventions and support systems.

- Online counseling sessions may be scheduled to instruct students on behavior modification techniques, including regular exercise, eating a nutritious diet, getting enough sleep, and cultivating a happy outlook.
- This analysis can be used during admission counseling to help students find a suitable course based on their stress-handling capabilities, thus leading to less anxious behaviors of the students in the future.
- We can use this Indian college student's data to understand the psychology of Indian students, and based on the categorization of degree, we can analyze which course is leading to more pressure and stress.
- We can collect more data on several other degrees to establish a more concrete relationship of these parameters on Stress.
- The questionnaire can be further improved by scientific translation of the questions into other Indian languages.
- Artificial intelligence is readily available on mobile computing devices; it should hence be utilized to its fullest for real-time analysis of PSS in university students and to get relevant ways to address it [19].
- Facial biometrics has been gaining fame among researchers as it finds use cases in almost all domains [20]. Which can be used in the future to gain deeper insight into the rising stress levels.
- Cloud-based consumers, owing to increasing gateway toward indirect assets, can be used to the advantage for faster remote applications [18, 21, 22].
- The survey can be used by researchers to delve deeper into scientific developments in medical industries [23].

REFERENCES

1. Rani, S., Kumar, A., Bagchi, A., Yadav, S., & Kumar, S. RPL Based Routing Protocols for Load Balancing in IoT Network. In *Journal of Physics: Conference Series* (Vol. 1950). IOP Publishing Ltd, 2021. https://doi.org/10.1088/1742-6596/1950/1/012073.
2. Asmundson, G. J. G., & Taylor, S. How health anxiety influences responses to viral outbreaks like COVID-19: What all decision-makers, health authorities, and health care professionals need to know. *Journal of Anxiety Disorders*, 71, 102211 (2020). https://doi.org/10.1016/j.janxdis.2020.102211.
3. Levine, G. N. Psychological Stress and Heart Disease: Fact or Folklore? *American Journal of Medicine*, 135(6), 688–696 (2022). https://doi.org/10.1016/j.amjmed.2022.01.053.

4. Dimsdale, J. E. Psychological Stress and Cardiovascular Disease. *Journal of the American College of Cardiology*, 51(13), 1237–1246 (2008). https://doi.org/10.1016/j.jacc.2007.12.024.
5. Booth, J., Connelly, L., Lawrence, M., Chalmers, C., Joice, S., Becker, C., & Dougall, N. Evidence of perceived psychosocial stress as a risk factor for stroke in adults: A meta-analysis. *BMC Neurology*, 15(1), 1–12 (2015). https://doi.org/10.1186/s12883-015-0456-4
6. Rahman, A. A., Siraji, M. I., Khalid, L. I., Faisal, F., Nishat, M. M., Ahmed, A., & Mamun, M. A. A. Perceived Stress Analysis of Undergraduate Students during COVID-19: A Machine Learning Approach. In *MELECON 2022 - IEEE Mediterranean Electrotechnical Conference, Proceedings* (pp. 1129–1134). Institute of Electrical and Electronics Engineers Inc., 2022. https://doi.org/10.1109/MELECON53508.2022.9843081.
7. Kumar, A., Rajak, D. K., Kumar, P., & Kumar, A. (Eds.). *Advances in Sustainable Biomaterials: Bioprocessing 4.0, Characterizations, and Applications* (1st ed.). CRC Press, 2024. https://doi.org/10.1201/9781003434313.
8. Graves, B. S., Hall, M. E., Dias-Karch, C., Haischer, M. H., & Apter, C. Gender differences in perceived stress and coping among college students. *PLoS One*, 16(8 August), e0255634 (2021). https://doi.org/10.1371/journal.pone.0255634
9. Proietti Ergün, A. L., & Ersöz Demirdağ, H. The predictive effect of subjective well-being and stress on foreign language enjoyment: The mediating effect of positive language education. *Frontiers in Psychology*, 14, 1007534 (2023). https://doi.org/10.3389/fpsyg.2023.1007534.
10. Lee, E. H., Chung, B. Y., Suh, C. H., & Jung, J. Y. Korean versions of the Perceived Stress Scale (PSS-14, 10 and 4): Psychometric evaluation in patients with chronic disease. *Scandinavian Journal of Caring Sciences*, 29(1), 183–192 (2015). https://doi.org/10.1111/scs.12131.
11. Tomar, A., Kumar, S., Pant, B., et al. Dynamic Kernel CNN-LR model for people counting. *Apple Intelligence* 52, 55–70 (2022). https://doi.org/10.1007/s10489-021-02375-6.
12. Shrivastava, V. K., Kumar, A., Shrivastava, A., Tiwari, A., Thiru, K., & Batra, R. Study and Trend Prediction of Covid-19 cases in India using Deep Learning Techniques. *Journal of Physics: Conference Series*, 1950, 012084 (2021). https://doi.org/10.1088/1742-6596/1950/1/012084.
13. Jain, S., Rathee, S., Kumar, A., Sambasivam, A., Boadh, R., Choudhary, T., Kumar, P., & Singh, P. K. Prediction of temperature for various pressure levels using ANN and multiple linear regression techniques: A case study. *Materials Today: Proceedings*, 56, 194–199 (2022). https://doi.org/10.1016/j.matpr.2022.01.067.
14. Makhubela, M. Assessing psychological stress in South African university students: Measurement validity of the perceived stress scale (PSS-10) in diverse populations. *Current Psychology* 41, 2802–2809 (2022). https://doi.org/10.1007/s12144-020-00784-3.
15. Andreou, E., Alexopoulos, E. C., Lionis, C., Varvogli, L., Gnardellis, C., Chrousos, G. P., & Darviri, C. Perceived Stress Scale: Reliability and validity study in Greece. International Journal of Environmental Research and Public Health, 8(8), 3287–3298 (2011). https://doi.org/10.3390/ijerph8083287.
16. Boadh, R., Grover, R., Dahiya, M., Kumar, A., Rathee, R., Rajoria, Y.K., Rawat, M., & Rani, S. Study of fuzzy expert system for the diagnosis of various types of cancer. *Materials Today: Proceedings*, 56, 298–307 (2022). https://doi.org/10.1016/j.matpr.2022.01.161,
17. Cohen, S., Kamarck, T., & Mermelstein, R. A global measure of perceived stress. Journal of Health and Social Behavior, 24(4), 385–396 (1983). https://doi.org/10.2307/2136404.

18. Saran, M., Yadav, R. K., & Tripathi, U. N. Machine Learning based Security for Cloud Computing: A Survey. *International Journal of Applied Engineering Research*, 17(4), 338 (2022). https://doi.org/10.37622/ijaer/17.4.2022.338-344.
19. Rajendra, P., Kumari, M., Rani, S., Dogra, N., Boadh, R., Kumar, A., & Dahiya, M. Impact of artificial intelligence on civilization: Future perspectives. *Materials Today: Proceedings*, 56, 252–256 (2022). https://doi.org/10.1016/j.matpr.2022.01.113.
20. Tomar, V., Kumar, N., & Srivastava, A. R. Single sample face recognition using deep learning: a survey. *Artificial Intelligence Review*, 56(Suppl 1), 1063–1111 (2023). https://doi.org/10.1007/s10462-023-10551-y.
21. Phogat, M., Kumar, A., Nandal, D., & Shokhanda, J. A Novel Automating Irrigation Techniques based on Artificial Neural Network and Fuzzy Logic. In *Journal of Physics: Conference Series* (Vol. 1950). IOP Publishing Ltd, 2021. https://doi.org/10.1088/1742-6596/1950/1/012088.
22. Jain, R., & Sharma, N. A Deadline-Constrained Time-Cost-Effective Salp Swarm Algorithm for Resource Optimization in Cloud Computing. *International Journal of Applied Metaheuristic Computing*, 13(1), 1–21 (2022). https://doi.org/10.4018/ijamc.292509.
23. Rani, S., Tripathi, K., & Kumar, A. Machine learning aided malware detection for secure and smart manufacturing: a comprehensive analysis of the state of the art. *International Journal on Interactive Design and Manufacturing (IJIDeM)*, 1–28 (2023). https://doi.org/10.1007/s12008-023-01578-0.

18 An Experimental Overview of Assessment of Authenticity of Face Recognition by AI Techniques in Smartphones

Ritika Tehlan, Bhoopesh Kumar Sharma, Megha Walia, Deepika Dhillon, and Deepak Kumar Saini

18.1 INTRODUCTION

Facial recognition technology (FRT) is a method employed in various smart devices to ascertain and authenticate the identification of an individual through biometric software. FRT is currently employed in various smart devices to authenticate the identification of an individual for accessing these devices. It compares patterns from the digital images or video frames with stored facial data by employing intricate algorithms and deep learning techniques. The technology is utilized in a wide range of industries, such as consumer electronics, healthcare, law enforcement, and security, to provide personalized user experiences, streamlined authentication processes, and enhanced security measures [1]. FRT is a subset of computer visualization that makes use of optical input to analyze images with specific attention to faces that may be present. It compares patterns from digital images or video frames to stored facial data by employing intricate algorithms and deep learning techniques [2].

FRT many times raises substantial ethical and privacy concerns, despite its advantages. Robust regulatory frameworks and ethical guidelines are necessary to ensure responsible deployment, given concerns such as potential biases, data security vulnerabilities, and the implications for civil liberties. In 2017, the Federal Bureau of Investigation (FBI) revealed that the facial-recognition software had just an 85% chance of appropriately identifying a person within its 50 best choices from a bigger database [3].

DOI: 10.1201/9781032724508-20

A biometric identification system has the capability to utilize various physiological parameters, such as fingerprints, iris patterns, and facial features, as well as certain behavioral factors, including handwriting and voice for the purpose of identifying individuals [4, 38]. The technique mainly works using image capture, face detection, feature extraction, and matching of these features with the known enrolled features. Undoubtedly, the technique is more advanced with convenience, enhanced security, and automation processes [5]. Among the commercial implementations of this technology are access control systems that utilize real-time matching with still images or video image layouts. This technology is also employed in other commercial applications for the static matching of photographs on credit cards, ATM cards, passports, driver's licenses, and other photo IDs. Each program has its own distinct set of processing restrictions. The two primary components of each facial recognition algorithm are as follows [6]:

A. Detection and Normalization of Facial Features
B. Feature Identification.

The utilization of FRT is already well-known globally. Nevertheless, it is crucial to recognize that this is just one scenario of the technology's application. Facial recognition systems typically identify and authenticate a single user as the sole proprietor of the smart device, thereby restricting access to others, in contrast to extensive image databases used for individual identification [7].

Face recognition technology surpasses the mundane act of unlocking phones by comparing descriptions of individuals with the faces of those traveling through cameras. The watch lists may contain images of any individual, including those who are not suspected of committing a crime, and descriptions from any other source, including their social media accounts, may incite them [8]. Although there may be exceptions, FRT typically functions as shown in Figure 18.1. The following are the steps in any FRT process:

Step 1: The initial step involves the recognition and localization of a facial image captured by the camera installed in the smart device. The image potentially exhibits the individual assuming a traditional or side-view pose.

Step 2: The individual's face is captured and thereafter subjected to a comprehensive examination using an inbuilt FRT. The predominant utilization of 2D photographs in FRT is due to their enhanced compatibility with web images or database images, as opposed to 3D photographs [9]. Several essential criteria for defining facial features include the interocular distance, the depth of eye sockets, the distance from the forehead to the maxilla, the contour of the cheekbones, and the shape of the lips, ears, and maxilla [9]. The aim of this study was also to identify the unique facial characteristics that constitute a person's face and are recorded by the FRT.

Step 3: The third step involves the conversion of the image into binary data. The face capture approach utilizes distinctive facial landmarks to convert analog facial features into a digital dataset. A mathematical formula is

FIGURE 18.1 Typical Functioning of Facial Recognition Technology (FRT) in Any Smart Device

commonly employed for the purpose of facial analysis. The numerical representation is commonly referred to as a facial imprint [10].

Step 4: The facial imprint is compared to an existing repository of recognized facial profiles. To provide an example, it is worth noting that the Federal Bureau of Investigation (FBI) possesses the capability to retrieve and analyze a substantial collection of approximately 650 million photographs sourced from various government databases [11]. Images that are labeled with an individual's name on social media platforms such as Facebook, Instagram, and Snapchat are subsequently included in databases, which possesses the capability of doing facial recognition. According to earlier studies, facial recognition technology is utilized by more than 50% of the global population [11].

FRT is widely used in several other mobile applications such as Amazon, Facebook, Google, Snapchat, Apple, Cigna (healthcare insurance provider), security applications, etc. [12]. A few of these applications are listed below.

18.1.1 Face Recognition in Mobile Phones

FRT has been integrated into Apple's iOS and Android OS smartphones in recent years. In addition to this, the fingerprint and touchpad identification methods have also been established for a considerable period. The FRT is used in various applications in a smartphone for not only unlocking them but also for banking services, e-commerce applications, and social media applications [13].

18.1.2 Facial Recognition versus Touchpad ID

Both fingerprint recognition and facial recognition have their advantages and disadvantages. Theoretically, both are quicker and safer than just entering a passcode. Fingerprint identification technology utilizes the conductivity variations resulting from the intricate ridges on an individual's finger, then compares this data with a pre-existing image of the fingerprint recorded previously in a memory database. On the other side, Touchpad identification has been a prevalent feature in most of the smartphones globally. Notably, the iPhone 5s, which was unveiled in 2013, became the pioneering device on a prominent U.S. carrier to provide this functionality. The mechanism of security in question is well acknowledged and regarded as confidential in such devices. Despite several limitations, Touch ID was found to be a potential and highly efficient technology. The probability of another individual's fingerprint successfully unlocking a phone is approximately 1 in 50,000, which can be attributed to factors such as dirt, oil, gloves, or damage affecting the user's finger [14].

Mobile facial recognition offers several advantages [15, 16], including no requirements of buttons to unlock the phone, providing quick and suitable access to the user, and comprehensive analysis of several facial attributes simultaneously [17]. These attributes are derived into a single code, which serves as a distinctive identifier for an individual.

The functionality can be utilized in conjunction with the integrated camera of one's mobile device. The likelihood of an unfamiliar individual successfully unlocking one's phone is estimated to be one in a million or even less.

18.1.3 Safety Issues with Facial Recognition Technology

FRT is not 100% secure against hackers. Even identical twins share similar facial features and can unlock a device with each other's faces [17]. Also, as per the previous research, one in a million may have similar facial features. Additionally, there are certain privacy concerns as well. For example, how accessible is the facial data to third parties, and where is a particular mobile application going to store the data? Apple usually protects this privacy by encrypting it on their smartphones [18, 19].

Apple's Face ID is widely regarded as one of the most robust facial recognition systems available currently. It employs advanced technology, including an infrared camera, a depth sensor, and a dot projector, to meticulously map about 30,000 distinct facial locations. The software application is utilized for the purpose of producing a simulated 3D scan image that possesses a sufficient level of security to enable the unlocking of one's mobile device and the verification of online transactions conducted through Apple Pay or any such mobile application as and when required [20, 21].

Similar to the prior technique, the first type of infrared-based facial recognition includes obtaining a two-dimensional (2D) image of the face in the infrared range. The main advantage is that infrared cameras can work in low-light conditions as well and don't require a face to be properly lit. Infrared cameras provide a picture using thermal energy; they are also significantly more resistant to disturbance attempts [21].

Although 2D Infrared Facial Recognition Technology (IRFRT) has made a significant advancement as compared to conventional camera-based systems, there remains ongoing potential for further enhancements in the technology. For instance, Apple's Face ID employs a diverse array of sensors to capture a three-dimensional (3D) representation of an individual's facial features. This is achieved through the utilization of a flood illuminator and a dot projector, which display numerous little dots onto the facial surface, rendering them indistinguishable. The positioning of the dots is subsequently assessed by an infrared sensor, resulting in the generation of a depth map representing the contours and dimensions of facial structure [22].

On the other hand, 3D systems have two major advantages: they can operate in the dark and they are very difficult to trick [23]. Based on the data provided by the International Data Corporation (IDC), it can be observed that Android commanded a global market share of 23.7% for mobile operating systems in the first quarter of 2022, while iOS, developed by Apple, secured the second position with a market share of 18% approximately [24, 25].

Despite its advantages, FRT raises significant ethical and privacy concerns. To guarantee responsible deployment, it is necessary to establish comprehensive regulatory frameworks and ethical guidelines, which address concerns such as potential biases, data security vulnerabilities, and the implications for civil liberties [26].

Although facial recognition technology offers a multitude of security and authentication advantages, it is also susceptible to a variety of forms of deception and abuse. Fraudsters exploit vulnerabilities in these systems to circumvent security measures, resulting in substantial privacy and security concerns. The challenges posed by the various methods of fraud that involve facial recognition techniques are the focus of this abstract [27].

Presentation attacks, which involve the use of photographs, videos, or 3D masks to deceive facial recognition systems, are prevalent forms of fraud. The objective of these attacks is to deceive the technology into recognizing a false identity. Deepfakes are an additional approach that employs sophisticated machine learning techniques to generate fraudulent images or videos of individuals that are exceedingly realistic. Deepfakes can be used to impersonate an individual, thereby obtaining unauthorized access to systems or disseminating misinformation [28, 29, 36].

18.2 LITERATURE REVIEW

In a study conducted by Kostka, G. et.al. in 2023 on the cross-country attitudes toward facial recognition technology, the authors have demonstrated that respondents' sentiments are most influenced by privacy concerns, particularly those related to FRT violating one's privacy. The findings of the study have significantly contributed in regard to the FRT acceptability and usage and indicate that legislators must act quickly to fill the present legislative void in respect to the usage of FRT [17].

According to another study conducted by Cao and Ma et al. in the year 2022, it was found that the predominant focus of existing face forgery detection methods lies on specific counterfeit patterns, such as local texturing, noise characteristics, or frequency statistics. The presence of unknown patterns poses a challenge in the detection of forgeries, as trained representations tend to become specialized in recognizing known forging patterns that are included in the training set [18].

Dang et al. [23] have emphasized the growing importance of identifying manipulated facial images and videos in the field of digital media forensics. The development of innovative artificial face representations has been facilitated by the emergence of sophisticated face synthesis and modification techniques. Significant apprehension has been generated regarding their implementation on social media platforms as a result [22].

In 2020, Jiang and Li et al. conducted research on the development of a comprehensive benchmark for detecting facial forgery. In their publication, the authors have addressed the growing concern among private and governmental authorities about the widespread presence of deepfake media. According to the authors, creating defenses against fake accounts on social media is very crucial. Multi-face forgery detection and separation were discussed in the study as a preventative measure with thorough examination [25].

According to Luo et al. [28], it was found that contemporary face forgery detection algorithms exhibit high levels of accuracy when both training and test forgeries are generated using the same technique inside a given database. Researchers have proposed the utilization of high-frequency sound waves for the purpose of detecting facial features. The suggestion is based on the observation that image noises can expose disparities between authentic and manipulated areas, hence eliminating color textures [27].

In 2013, Soliman et al. [31] explored various face recognition methods specifically designed for mobile devices. The researchers conducted experiments on various face detection approaches, including color segmentation and template matching, as well as face recognition algorithms such as Eigen and Fisher. The methods were first analyzed in MATLAB before being put into practice on the DROID smartphone. They said that during the development of the facial recognition system on a mobile phone with constrained hardware, one has to compromise between accuracy and computational challenge when implementing the algorithms. The researchers drew the conclusion that the face detection system possesses numerous limitations. The color segmentation technique exhibits limitations in its ability to accurately identify instances where the background color of a picture closely resembles the color of human skin. The findings of this study indicated that the effectiveness of the blueprint-matching algorithm is dependent on various factors, such as the specific template chosen (including orientation), lighting conditions, and other relevant parameters. Furthermore, the system encountered difficulties in accurately recognizing and distinguishing faces associated with specific ethnicities [30].

In the study conducted in 2021, Zhao and Zhou et al. developed a multi-attentional face forgery detection system. This system effectively combines the texture data and high-level semantic information from various local sections to accurately classify samples as either real or fake [32].

In another study carried out by Qian Yin et al. [35], the authors presented a model that utilizes frequency information to effectively distinguish between genuine and counterfeit faces. The researchers concluded that their methodologies mostly depend on the acquired forgery patterns seen in the training samples. Subsequently, they saw a noticeable decline in performance when confronted with unfamiliar forging patterns [34].

18.3 THE EXPERIMENT

18.3.1 ENROLMENT

The face enrolment process was followed as per the instruction from the selected smartphones.

As seen in Figure 18.2, first, in the settings of the smartphone, the Biometrics and Security tab was opened, and the face recognition was accessed to enroll the face, and the instructions were followed. In case if the smartphone does not have a screenlock set-up, the individual will be asked to set it up first. Once the screenlock was set up, the researcher asked the subject if he wearing glasses or not. Accordingly, the face was enrolled. The smartphone began to scan the face in various dimensions to record and enroll it. All the on-screen instructions were followed as directed.

18.3.2 VALIDATION WITH ENROLLED PARTICIPANT

The first participant (sister) was enrolled for facial recognition on her Android smartphone and iPhone with the above-mentioned steps as discussed in the enrolment process. When she was asked to unlock the smartphones, she was able to unlock the phones through the Facial Recognition Technique (FRT) as usual.

18.3.3 VALIDATION WITH NON-ENROLLED PARTICIPANTS

In the next step, the second participant (brother) who has not enrolled for facial recognition on both the selected smartphones was also asked to unlock the smartphones in the similar manner, and as an astonishing result, he was also able to unlock the phones through the facial recognition feature, even though the gender was different, the age was different, and they were not identical twins. However, as physically

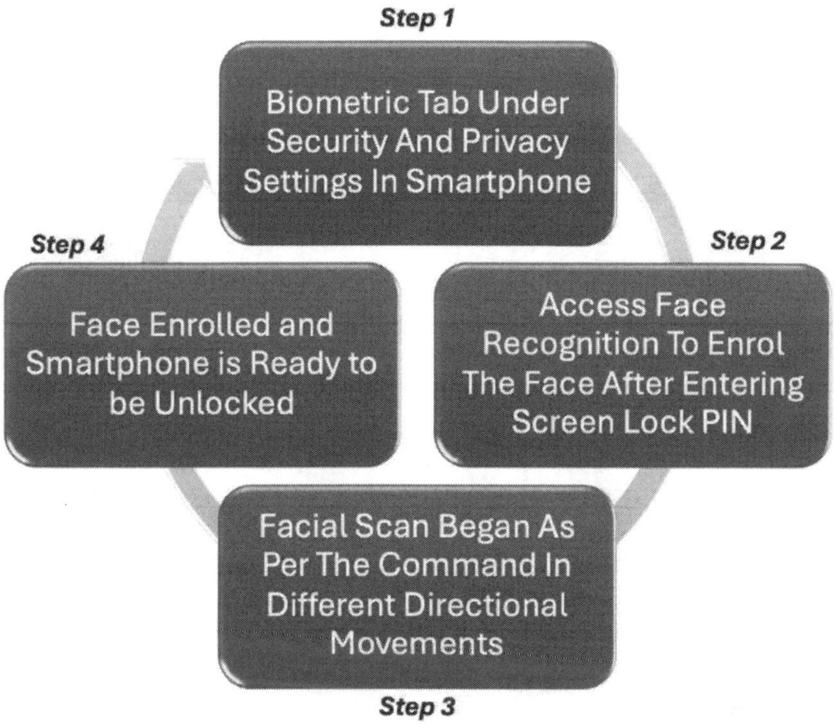

FIGURE 18.2 Enrolment Process of the Face in a Smartphone Using Facial Recognition Technology

observed, they both possess some facial similarities in respect to their eyebrows, eye dimensions, chin, and cheek similarities, along with quite similar color complexions.

18.4 RESULTS AND DISCUSSION

The primary objective of the present study was to evaluate the efficacy of facial recognition technology in the context of siblings with similar facial features, utilizing two distinct smartphone sets—i.e., an Android and an iPhone. The disclosure of smartphone identities is withheld in the study due to ethical concerns. As a matter of fact, the identical twins usually have facial similarities in terms of their physical appearance and other facial dimensional features [33, 37].

The siblings in the present study were non-identical; they were not twins and were having an age difference of five years approximately. The male participant (brother) was younger to the female (sister) participant. Findings of the study revealed instances where the biometric facial recognition system exhibited inaccuracies in correctly identifying the enrolled face (Figures 18.3 and 18.4) while successfully verifying the non-registered face and unlocking the smartphone. Initially, the occurrence of unlocking the smartphones was regarded as a mere happenstance;

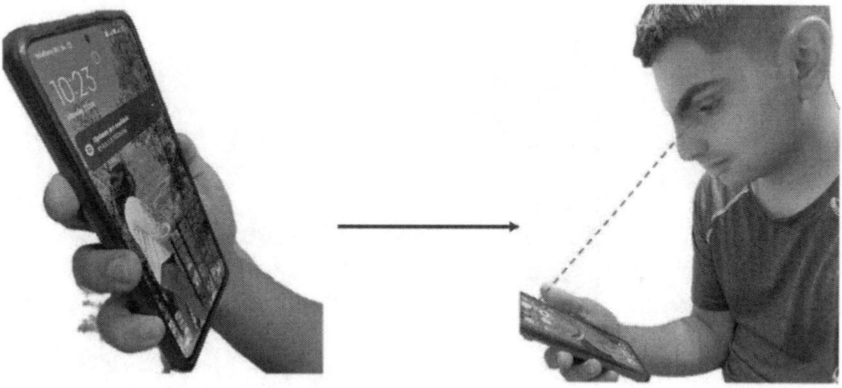

FIGURE 18.3 (a) Showing the locked home screen of an Android Phone with sister's enrolled face Figure 18.3 (b) Un-enrolled face unlocking the smart phone through facial recognition

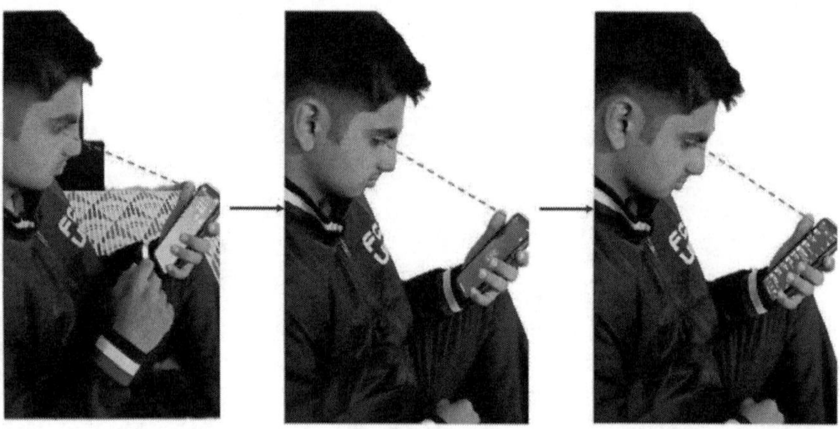

FIGURE 18.4 (a) Showing the locked home screen of an iPhone with sister's enrolled face. (b) Unlocking process of iPhone through facial recognition, and (c) Showing the unlocked iPhone

nevertheless, subsequent experiments revealed the consistent malfunctioning of facial recognition systems in these smartphones. As per one of the explanations given by Apple support on "Community Discussion" was that "When any of the sibling uses the same smartphone and attempts to open the smartphone first time, and if the enrolled sibling opens up the smartphone for the other who is not enrolled in the Face ID system, observes the phone, and subsequently enters the passcode, the Face ID system will acquire knowledge of the facial features of that specific individual in addition to the utilization of the passcode" [34, 36]. In the community forum it was further discussed that when Face ID is enabled, it is advisable to refrain from permitting someone to access the phone using the passcode while observing the screen. Though this explanation does not solve the problem with the Face ID system, however, it may be taken as a precautionary step for security of privacy. As per the Apple Community Forum, if such a thing happens, one must reset Face ID and then activate it again with the enrolled face [35].

18.5 CONCLUSION

Facial Recognition Technology possesses the capability to discern human faces inside photographs or videos, ascertain the similarity between faces in two distinct images to establish if they correspond to the same individual, or conduct a search for a specific face among an extensive repository of pre-existing images. Despite its benefits, FRT raises significant ethical and privacy concerns. Issues such as potential biases, data security vulnerabilities, and the implications for civil liberties necessitate robust regulatory frameworks and ethical guidelines to ensure responsible deployment. As it was observed in the study that in certain cases the biometric FR system failed to precisely recognize the enrolled face, whereas it verified the non-enrolled face to unlock the phone. Firstly, it was considered a coincidence, later, after numerous experiments showed the failure of the facial recognition system in these smartphones of high repute, they suggested a crucial quality check and update in FR technology.

Future Scope: This study discusses the potential risks associated with fraud in facial recognition systems and underscores the need for ongoing research and development to enhance security measures and protect users from malicious activities.

REFERENCES

1. Rajendra, P., Kumari, M., Rani, S., Dogra, N., Boadh, R., Kumar, A., & Dahiya, M. Impact of artificial intelligence on civilization: Future perspectives. *Materials Today: Proceedings*, 56, 252–256 (2022).
2. Rani, S., Kumar, A., Bagchi, A., Yadav, S., & Kumar, S. RPL based routing protocols for load balancing in IoT network. In *Journal of Physics: Conference Series*, Vol. 1950 (1), p. 012073). IOP Publishing (2021).
3. Gonzalez-Sosa, E., Fierrez, J., Vera-Rodriguez, R., & Alonso-Fernandez, F. Facial soft biometrics for recognition in the wild: Recent works, annotation, and COTS evaluation. *IEEE Transactions on Information Forensics and Security*, 13(8), 2001–2014 (2018).
4. Chen, B., Shen, J., & Sun, H. A fast face recognition system on mobile phone. In *2012 International Conference on Systems and Informatics (ICSAI2012)*, pp. 1783–1786, IEEE (2012).
5. Jain, R., & Sharma, N. A quantum inspired hybrid SSA–GWO algorithm for SLA based task scheduling to improve QoS parameter in cloud computing. *Cluster Computing*, 1–24 (2022).
6. Li, L., Bao, J., Yang, H., Chen, D., & Wen, F. Advancing high fidelity identity swapping for forgery detection. In *Proceedings of the IEEE/CVF Conference on Computer Vision and Pattern Recognition*, pp. 5074–5083 (2020).

7. Phogat, M., Kumar, A., Nandal, D., & Shokhanda, J. A novel automating irrigation technique based on artificial neural network and fuzzy logic. In *Journal of Physics: Conference Series*, Vol. 1950 (1), p. 012088. IOP Publishing (2021).
8. Da Silva, S., & Agbinya, J. I. Applications of face recognition with mobile phones. In *Proc. South African Telecommunications Networks and Applications Conference (SATNAC)*, Vol. 2, pp. 373–374 (2003).
9. Han, S. Y., Park, H. A., Cho, D. H., Park, K. R., & Lee, S. Face recognition based on near-infrared light using mobile phone. In *Adaptive and Natural Computing Algorithms: 8th International Conference, ICANNGA 2007*, Warsaw, Poland, April 11–14, 2007, Proceedings, Part II 8, pp. 440–448. Springer Berlin Heidelberg (2007).
10. Findling, R. D., & Mayrhofer, R. Towards face unlock on the difficulty of reliably detecting faces on mobile phones. In *Proceedings of the 10th International Conference on Advances in Mobile Computing & Multimedia*, pp. 275–280 (2012).
11. Rani, S., Tripathi, K., Arora, Y., & Kumar, A. Analysis of Anomaly detection of Malware using KNN. In *2022 2nd International Conference on Innovative Practices in Technology and Management (ICIPTM)*, Vol. 2, pp. 774–779, IEEE (2022).
12. Rani, S., Tripathi, K., Arora, Y., & Kumar, A. A Machine Learning Approach to Analyze Cloud Computing Attacks. In *2022 5th International Conference on Contemporary Computing and Informatics (IC3I)*, pp. 22–26. IEEE (2022).
13. Choi, K., Toh, K. A., & Byun, H. Realtime training on mobile devices for face recognition applications. *Pattern Recognition*, 44(2), 386–400 (2011).
14. Jain, R., & Sharma, N. A deadline-constrained time-cost-effective salp swarm algorithm for resource optimization in cloud computing. *International Journal of Applied Metaheuristic Computing (IJAMC)*, 13(1), pp. 1–21 (2022).
15. Hassan, G., & Elgazzar, K. The case of face recognition on mobile devices. In *2016 IEEE Wireless Communications and Networking Conference*, pp. 1–6. IEEE (2016).
16. Dave, G., Chao, X., & Sriadibhatla, K. *Face Recognition in Mobile Phones*. Department of Electrical Engineering Stanford University, USA, pp. 7–23 (2016).
17. Wang, C., & Deng, W. Representative forgery mining for fake face detection. In *Proceedings of the IEEE/CVF Conference on Computer Vision and Pattern Recognition*, pp. 14923–14932 (2021).
18. Kostka, G., Steinacker, L., & Meckel, M. Under big brother's watchful eye: Cross-country attitudes toward facial recognition technology. *Government Information Quarterly*, 40(1), 101761 (2023).
19. Cao, J., Ma, C., Yao, T., Chen, S., Ding, S., & Yang. X End-to-end reconstruction-classification learning for face forgery detection. In *Proceedings of the IEEE/CVF Conference on Computer Vision and Pattern Recognition*, pp. 4113–4122 (2022).
20. Singhal, A., Phogat, M., Kumar, D., Kumar, A., Dahiya, M., & Shrivastava, V. K. Study of deep learning techniques for medical image analysis: A review. *Materials Today: Proceedings*, 56, 209–214 (2022).
21. Tariq, S., Lee, S., Kim, H., Shin, Y., & Woo, S. S. Detecting both machine and human created fake face images in the wild. In *Proceedings of the 2nd International Workshop on Multimedia Privacy and Security*, pp. 81–87 (2018).
22. Chen, L., Zhang, Y., Song, Y., Liu, L., & Wang, J. Self-supervised learning of adversarial example: Towards good generalizations for deepfake detection. In *Proceedings of the IEEE/CVF Conference on Computer Vision and Pattern Recognition*, pp. 18710–18719 (2022).
23. Dang, H., Liu, F., Stehouwer, J., Liu, X., & Jain, A. K. On the detection of digital face manipulation. In *Proceedings of the IEEE/CVF Conference on Computer Vision and Pattern recognition*, pp. 5781–5790 (2020).

24. Adjabi, I., Ouahabi, A., Benzaoui, A., & Taleb-Ahmed, A. Past, present, and future of face recognition: A review. *Electronics*, 9(8), 1188 (2020).
25. Hassan, G., & Elgazzar, K. The case of face recognition on mobile devices. In *2016 IEEE Wireless Communications and Networking Conference*, pp. 1–6 (2016).
26. Jiang, L., Li, R., Wu, W., Qian, C., & Loy, C. C. Deeperforensics-1.0: A large-scale dataset for real-world face forgery detection. In *Proceedings of the IEEE/CVF Conference on Computer Vision and Pattern Recognition*, pp. 2889–2898 (2020).
27. Le, T. N., Nguyen, H. H., Yamagishi, J., & Echizen, I. Openforensics: Large-scale challenging dataset for multi-face forgery detection and segmentation in-the-wild. In *Proceedings of the IEEE/CVF International Conference on Computer Vision*, pp. 10117–10127 (2021).
28. Luo, Y., Zhang, Y., Yan, J., & Liu, W. Generalizing face forgery detection with high-frequency features. In *Proceedings of the IEEE/CVF Conference on Computer Vision and Pattern Recognition*, pp. 16317–16326 (2021).
29. Aralikatti, A., Appalla, J., Kushal, S., Naveen, G. S., Lokesh, S., & Jayasri, B. S. Real-time object detection and face recognition system to assist the visually impaired. In *Journal of Physics: Conference Series*, Vol. 1706(1), p. 012149. IOP Publishing (2020).
30. Jassim, S., & Sellahewa, H. A wavelet-based approach to face verification/recognition. In *Unmanned/Unattended Sensors and Sensor Networks II*, Vol. 5986, pp. 77–86. SPIE (2005).
31. Soliman, Hassan, Ahmed Saleh & Eman, Fathi. Face recognition in mobile devices. *International Journal of Computer Applications*, 73(2), (2013).
32. Cao, J., Ma, C., Yao, T., Chen, S., Ding, S., & Yang, X. End-to-end reconstruction-classification learning for face forgery detection. In *Proceedings of the IEEE/CVF Conference on Computer Vision and Pattern Recognition*, pp. 4113–4122 (2022).
33. Zhou, T., Wang, W., Liang, Z., & Shen, J. Face forensics in the wild. In *Proceedings of the IEEE/CVF Conference on Computer Vision and Pattern Recognition*, pp. 5778–5788 (2021).
34. Zhao, H., Zhou, W., Chen, D., Wei, T., Zhang, W., & Yu, N. Multi-attentional deepfake detection. In *Proceedings of the IEEE/CVF Conference on Computer Vision and Pattern Recognition*, pp. 2185–2194 (2021).
35. Qian, Y., Yin, G., Sheng, L., Chen, Z., & Shao, J. Thinking in frequency: Face forgery detection by mining frequency-aware clues. In *European Conference on Computer Vision*, pp. 86–103. Cham: Springer International Publishing (2020).
36. Allywilsoon. Available at: https://discussions.apple.com/thread/8251423 (accessed 27 July 2023).
37. Sehrawat, S., Kumar, A., Grover, S., Dogra, N., Nindra, J., Rathee, S., ... & Kumar, A. Study of 3D scanning technologies and scanners in orthodontics. *Materials Today: Proceedings*, 56, pp. 186–193 (2022).
38. Kumar, A., Kumar, P., Rathee, S., & Kumar, B. (Eds.). *Mechatronics: Concepts, Tools, Applications, and New Trends* (1st ed.). CRC Press (2025). https://doi.org/10.1201/9781003494478.

Index

Note: Page numbers in **bold** reference tables and Page numbers in *italics* reference figures